The Association for Science Education

Teaching Secondary Chemistry
Third Edition

Series Editor: Chris Harrison
Editors: Vanessa Kind, Katherine Aston

The publishers would like to thank specialist subject advisers Helen Harden and Ann Childs for their contribution to this title.

Titles in this series
Teaching Secondary Biology 978 1 5104 6256 4
Teaching Secondary Chemistry 978 1 5104 6257 1
Teaching Secondary Physics 978 1 5104 6258 8

The Publishers would like to thank the following for permission to reproduce copyright material.

Acknowledgements
Every effort has been made to trace all copyright holders, but if any have been inadvertently overlooked, the Publishers will be pleased to make the necessary arrangements at the first opportunity.

Although every effort has been made to ensure that website addresses are correct at time of going to press, Hodder Education cannot be held responsible for the content of any website mentioned in this book. It is sometimes possible to find a relocated web page by typing in the address of the home page for a website in the URL window of your browser.

Hachette UK's policy is to use papers that are natural, renewable and recyclable products and made from wood grown in well-managed forests and other controlled sources. The logging and manufacturing processes are expected to conform to the environmental regulations of the country of origin.

Orders: please contact Hachette UK Distribution, Hely Hutchinson Centre, Milton Road, Didcot, Oxfordshire, OX11 7HH. Telephone: +44 (0)1235 827827. Email education@hachette.co.uk Lines are open from 9 a.m. to 5 p.m., Monday to Friday. You can also order through our website: www.hoddereducation.co.uk

The authorised representative in the EEA is Hachette Ireland, 8 Castlecourt Centre, Dublin 15, D15 XTP3, Ireland (email: info@hbgi.ie)

ISBN: 978 1 5104 6257 1

© Association for Science Education 2022
First published in 2000.
Second edition published in 2012.

This edition published in 2022 by
Hodder Education,
An Hachette UK Company
Carmelite House
50 Victoria Embankment
London EC4Y 0DZ

www.hoddereducation.co.uk

Impression number   10 9 8 7 6 5 4 3

Year   2025 2024

All rights reserved. Apart from any use permitted under UK copyright law, no part of this publication may be reproduced or transmitted in any form or by any means, electronic or mechanical, including photocopying and recording, or held within any information storage and retrieval system, without permission in writing from the publisher or under licence from the Copyright Licensing Agency Limited. Further details of such licences (for reprographic reproduction) may be obtained from the Copyright Licensing Agency Limited, www.cla.co.uk

Cover photo ©berkay08 - stock.adobe.com
Illustrations by Integra Software Services Pvt., Ltd.
Typeset in India by Integra Software Services Pvt., Ltd.
Printed and bound by CPI Group (UK) Ltd, Croydon CR0 4YY

A catalogue record for this title is available from the British Library.

Contents

	Contributors	iv
	Photo credits	vii
	Health and safety	viii
1	The principles behind secondary chemistry teaching Vanessa Kind	1
2	Chemical change Vanessa Kind	26
3	Particle theory Maurice M.W. Cheng and Vanessa Kind	60
4	The periodic table Vicky Wong	89
5	Chemical bonding Hannah Sevian and Edenia Amaral	123
6	Energetics, rate and extent of chemical change Ann Childs and Neil Dixon	152
7	Acids and alkalis Jane Essex and Despoina O'Flynn	196
8	Redox reactions and electrolysis Nicklas Lindström and Katherine Aston	225
9	Chemical analysis Sheila Curtis	270
10	Organic chemistry David Paterson	313
11	Earth science Stuart Jones and Christopher Saville	347
	Index	387

Contributors

Vanessa Kind is Professor of Education and Head of the School of Education at the University of Leeds. Vanessa studied biochemistry and worked in molecular biology before becoming a chemistry teacher. She taught in schools in London prior to a doctorate study at York University investigating post-16 students' learning of chemistry concepts. After her first academic post at UCL Institute of Education, Vanessa became principal of an international school in Norway. She joined Durham University in 2005, continuing in teacher education followed by leading Science Learning Centre North East as Director. Vanessa was a Deputy Head of Faculty/Executive Dean for five years prior to joining Leeds University in January 2022.

Edenia Amaral is Professor of Chemistry and Science Education at the Post Graduate Programs in Science and Mathematics Education, Rural Federal University of Pernambuco, Brazil. Edenia Maria Ribeiro do Amaral's background is in chemical engineering. She holds a PhD in Science Education and develops research in three science education postgraduate programmes. Edenia's research themes include heterogeneity of thinking and language from which individuals can express different ways of thinking and speaking as they engage in science teaching and learning activities. Through her work on conceptual profiles she has expanded investigations on interrelations among cognition, language and culture, leading to implications for contextualised teaching design and activity systems for transformative science education. She has undertaken studies on discursive interactions in the science classroom and on the integration of knowledge types in the meaning-making process in scientific education.

Katherine Aston is a Lecturer in Chemistry Education at King's College, London. Katherine leads the PGCE Chemistry course at King's College London, teaches and tutors across all PGCE Science courses and supervises doctoral students in chemistry education. Previously, she was a secondary chemistry teacher, teacher coach and educational researcher. Katherine has worked extensively in science education research and development including curriculum development for schools, professional development programmes, and national and international assessment programmes. Her current research focuses on chemistry education at secondary and undergraduate levels, the nature of chemistry and teacher learning.

Maurice M.W. Cheng is Associate Professor at Te Hononga School of Curriculum and Pedagogy, University of Waikato, New Zealand. Maurice trained originally as a pharmacist before becoming a chemistry teacher. His main research interest is students' learning of science, particularly their visual learning of chemical ideas, and ways that verbal and visual representations facilitate meaning making. Maurice is also interested in student learning of socio-scientific issues. He has 20 years of experience in national and international assessment in science, and is is the National Research Coordinator (Science) for the TIMSS Hong Kong study 2019. Maurice also serves in the Science Extended Expert Group for OECD's PISA 2024.

Ann Childs completed her PhD in chemistry at Birmingham University in 1982 and then trained to be a science teacher at Oxford University. She taught for 11 years in Oxfordshire and in Sierra Leone for Voluntary Services Overseas (VSO). During her work as a teacher in Oxfordshire she mentored beginning science teachers on the

Contributors

Oxford Internship Scheme. She took up her current post as an Associate Professor in Science Education in 1997 where she now teaches on the PGCE and is director of the Masters in Teacher Education. Her key research interests are in the professional development of science teachers.

Sheila Curtis worked as a Science educator in a variety of settings, including teaching secondary science in north east London; as a continuing professional development provider through the UK's network of Science Learning Centres; and as a PGCE tutor at UCL Institute of Education. Sheila retired in 2020.

Neil Dixon completed his BA in Natural Sciences at Cambridge University in 1999 and then his PGCE at Oxford University in 2003. Since then, he has held a number of roles at South Bromsgrove High School. Neil has written textbooks and revision guides for secondary science and he works as a science education consultant for the BBC. He is also a Subject Expert for Ofqual, the UK exams regulator.

Jane Essex is a Senior Lecturer in the Department of Education, University of Strathclyde. Jane taught science, specialising in chemistry, in four state schools in England, before moving to an Initial Teacher Education role at the University of Keele. She now works on the PGDE chemistry course at Strathclyde. Jane developed teacher subject knowledge enhancement courses in chemistry at Keele and is committed to helping teachers develop subject expertise. Her major research area and professional practice specialism is science education for learners with special/additional support needs, specifically intellectual disabilities. Jane was awarded the Royal Society of Chemistry's Inclusion and Diversity Award in 2019 in recognition of her work enhancing diversity, equality and inclusion in all aspects of chemistry education. She is founder and co-chair of the Association for Science Education Special Educational Needs/Additional Support Needs interest group which is evaluating microscale chemistry as a technique enhancing practical chemistry experiences for diverse learners.

Stuart Jones is an Associate Professor of Sedimentology in the Department of Earth Sciences at Durham University. Stuart undertook his first degree in geology and closely followed by a PhD in sedimentology working on ancient fluvial sediments and their evolution in the Spanish Pyrenees. After a few years working in the hydrocarbon industry as a geoscientist he moved back into academia first at the University of Southampton and then to his current position. Stuart is passionate about Earth Sciences and has taught the subject to undergraduate and postgraduate students and has led many field trips and training courses for industry. He regularly undertakes Earth Science outreach into schools and has won many awards for his innovative outreach and teaching.

Nicklas Lindström is a Senior Lecturer in secondary science education at the University of Roehampton, London. Nicklas currently teaches on the secondary PGCE programme at the University of Roehampton. Prior to this he worked at King's College London on their PGCE programme and has been a teacher of science and chemistry for over 20 years in schools across the midlands. Nicklas is currently studying for a PhD at the University of Birmingham based on epistemic framing of prior knowledge of secondary school chemistry learners. He has previously published on the Structure of Observed Learning Outcome (SOLO) taxonomy and worked on a major research project investigating assessment of practical work in GCSE science.

Contributors

Despoina O'Flynn is a Science Teacher at Ellen Wilkinson School for Girls, London. Despoina obtained a degree in medicinal chemistry followed by a PhD in chemistry from University College London. She qualified as a Chemistry teacher in 2015 and works as a secondary school teacher in west London. Despoina is interested in issues of inclusion and diversity, and is experienced in working with children with autism and emotional and behavioural difficulties. Associated with this, she has a professional focus on raising the attainment of students with a history of low academic achievement.

David Paterson is Head of Digital Learning and Chemistry Teacher at Aldenham School, Elstree and Chemistry Adviser at CLEAPSS. He holds degrees in Chemistry from the University of Exeter, Biochemistry from Oxford University and in Education from Cambridge University. His teaching career includes running two successful state school science departments in Hertfordshire. Following two years at a UK examination board, David moved to teaching at Aldenham while also providing guidance to teachers and technicians through CLEAPSS. David's long-standing interest is improving practical work in schools. His classroom-based research, published in the *Journal of Chemical Education* contributed to his winning the 2020 Royal Society of Chemistry (RSC) Schools Education Award. David contributes to teacher education and professional development through the Association of Science Education, and as an RSC Scholar Mentor. He authors and edits for *Education in Chemistry*, and published *Understanding Chemistry through Microscale Practical Work* with Bob Worley.

Christopher Saville is an Associate Professor in the Earth Science department at Durham University. While studying natural sciences as an undergraduate at the University of Cambridge, Chris discovered the subject 'Geology' that he had never studied before. It slowly took over and he ended up specialising in it by the end of his degree. He then went on to work as a secondary science teacher, while also undertaking a part-time MA in education at Durham University. After that he returned to full-time study to do a PhD in Tectonic geomorphology at Durham. Chris now works as a teaching-focused academic at Durham University's Earth Sciences department. He has taught on courses covering; basic Earth Science systems; sedimentary environments; field mapping; taking geophysical measurements; public engagement in the sciences; and mathematical application in the sciences.

Hannah Sevian is Associate Provost and Professor of Chemistry at the College of Science and Mathematics, University of Massachusetts. Hannah joined the University of Massachusetts Boston in the US in 2001. Her research focuses on proactively inclusive chemistry learning that, recognizes and builds on students' cultural wealth and knowledge, and creates opportunities for learning so students can actively connect school to their lived worlds. Hannah studies how students develop chemical thinking from secondary to tertiary chemistry; green chemistry influences students' learning of chemistry; and how scientists and teachers develop responsive classroom assessment practices that promote students' chemistry meaning making. Hannah's career includes seven years as a public high school chemistry and physics teacher in a Spanish bilingual program in the Boston area and two years serving as a program officer at the US National Science Foundation. Her doctoral and postdoctoral training were in theoretical chemical physics.

Vicky Wong is Senior Lecturer in science education at the University of Exeter. After studying for a degree in chemistry, Victoria trained as a teacher and taught science and chemistry in England, Spain and New Zealand. In 2020 Victoria returned to the classroom to gain up-to-date experience, teaching science for two years in a state school in England. She has worked as a teacher educator at King's College London and the University of Oxford. Victoria has also held roles as an independent science education consultant for the Royal Society of Chemistry and the Nuffield Foundation. Her research interests include how science education policy is made and enacted and students' use of mathematics within science.

Photo credits

Figure 3.5 Philip Johnson, Figure 3.7 © nikkytok/stock.adobe.com, Figure 4.2 © Science History Images / Alamy Stock Photo, Figures 7.2 and 7.3 Pamela Tait, University of Strathclyde, Figures 8.13 and 8.14 Nicklas Lindström, Figure 9.15 © Phil Degginger / Alamy Stock Photo, Figure 11.2a © vvoe/stock.adobe.com, Figure 11.2b © Givaga/stock.adobe.com, Figure 11.3a © geoz / Alamy Stock Photo, Figire 11.3b © aleks-p/stock.adobe.com, Figure 11.4a © Björn Wylezich/stock.adobe.com, Figure 11.4b © Tyler Boyes/Shutterstock.com, Figure 11.13 © aleks-p/stock.adobe.com, Figure 11.14 © JOHN/stock.adobe.com

Figures 3.1, 3.3, 11.7, 11.9, 11.11, 11.13, 11.15, 11.16, 11.17 and photos in Tables 9.9 and 9.10 are the authors' own images.

Thanks to Wynne Harlen and The Association for Science Education for permission to reproduce extracts from Harlen, W. (ed.) (2010) Principles and Big Ideas of Science Education. Hatfield: The Association for Science Education and from Harlen, W. (ed.) (2015) Working with Big Ideas of Science Education. Trieste: InterAcademy Partnership on pages 5, 6 and 8 of this book.

Thanks to Vanessa Kind and The Association for Science Education for permission to reproduce extracts from Kind, V. and Kind, P.M. (2008) Teaching secondary science: How science works. London: Hodder Education on pages 68 and 69 of this book.

Health and safety

For all practical procedures described in this publication, we have attempted to ensure that:

- all recognised hazards have been identified;
- appropriate precautions are suggested;
- where possible procedures are in accordance with commonly adopted risk assessments;
- if a special risk assessment is likely to be necessary this is highlighted.

However, errors and omissions can be made, and employers may have adopted different standards. Therefore, before any practical activity, teachers should always check their employer's assessment. Any local rules issued by their employer must be obeyed, whatever is recommended in this publication.

Unless the context dictates otherwise it is assumed that:

- practical work is conducted in a properly equipped laboratory;
- any mains-operated and other equipment is properly maintained;
- any fume cupboard operates at least to the standard of CLEAPSS Guide G9;
- care is taken with normal laboratory operations such as heating substances or handling heavy objects;
- good laboratory practice is observed when chemicals or living organisms are handled;
- eye protection is worn whenever there is any recognised risk to the eyes;
- fieldwork takes account of any guidelines issued by the employer;
- pupils are taught safe techniques for such activities as heating chemicals or smelling them, and for handling microorganisms.

Readers requiring further guidance are referred to:

Hazcards (CLEAPSS, 2016 and updates)

Topics in Safety, 3rd edn (ASE, 2011); updates available at www.ase.org.uk/resources/topics-in-safety

Safeguards in the School Laboratory, 12th edn (ASE, 2020)

Preparing Risk Assessments for Chemistry Project Work in Schools & Colleges (SSERC, 2020)

SSERC hazardous chemicals database (www.sserc.org.uk/health-safety/chemistry-health-safety/hazchem_database-2/)

Be Safe! Health and Safety in School Science and Technology for Teachers of 3- to 12-Year-olds, 4th edn (ASE, 2011)

The principles behind secondary chemistry teaching

Vanessa Kind

Introduction

This book is written to help chemistry teachers enable their secondary school students, particularly those aged 11–16, to learn and understand the subject, connect ideas from disparate topics and enhance their interest in chemistry. The book contains eleven chapters including this one. Each of the remaining ten discusses how to teach a chemistry topic, such as particle theory, chemical change and acids and alkalis.

The book is one of a series of three Association for Science Education (ASE) handbooks; the others being parallel volumes in biology and physics. The first edition was published in 1999, followed by the second edition in 2012. This third edition retains the structure of previous editions but features new authors and substantially revised and updated chapters.

Throughout the book, the authors have kept in mind the needs of a secondary chemistry teacher confronted with the task of teaching a specific topic and the preparation required to deliver lessons that are satisfying educational experiences for teacher and students. Meeting the need for effective lessons that engage learners and enhance and sustain their curiosity lies at the heart of the book. Authors are aware that teachers approach lesson preparation with varied subject knowledge backgrounds: some have a deep understanding of chemistry, while others have little knowledge beyond their school experiences. Teachers with chemistry degrees may be specialists in sub-disciplines, with stronger subject knowledge of some topics than others. Our hope is that all teachers of secondary school chemistry, including those with extensive teaching experience, will find much of value in here.

This chapter examines the discipline of chemistry and discusses approaches to teaching that enable students to engage with the discipline, build their identity as chemists and learn conceptual ideas.

 ## What is chemistry?

Chemistry is the scientific study of matter. Matter is anything that occupies space. Chemistry involves studying and understanding the behaviour of matter at micro- and nanoscale levels. This means understanding that all matter consists of a finite set of materials, the chemical elements, which themselves are constructed from tiny atoms and molecules. Many types of matter occur naturally. Others are developed in chemical reactions and processes devised through innovation, experiment and technology. Understanding and gaining control of factors such as temperature, pressure, rate of change and the amount of matter present is part of chemistry. Chemistry explains how matter is constructed, formed, changed and used; provides information about structures that matter adopts; informs measurement and control of the manufacture of new types of matter and discusses consequences arising from chemical processes. Part of the fascination of the subject is realising that chemistry is involved in almost every aspect of life on Earth.

A history of chemistry

Origins of the chemical elements

Chemistry begins with stars. The chemical elements, from which all matter is made, originate in stars. The simplest element, hydrogen, has been present in the Universe from the Big Bang onwards. Intense heat and pressure within a star cause atoms of hydrogen to collide with enormous force and fuse. Fusion creates atoms of many elements, including oxygen, nitrogen, carbon and iron. The heaviest atoms, including uranium, gold and platinum, form under extreme conditions in supernovae, which appear as bright, short-lived stars visible to the naked eye. In a supernova, the outer layers of an ageing star collapse inwards, creating a huge 'star quake' which expels matter into space at high velocities. In these conditions, heavy atoms collide and fuse, creating atoms that are heavier still.

The Earth formed about 4.5 billion years ago, most likely from a supernova, as a molten planet. On cooling, the Earth's surface solidified to a hard crust, retaining liquid surface water, a thin gaseous atmosphere and a molten core. Life began to evolve about 4 billion years ago and complex life about 500 million years ago. About 1.5 million years ago, Homo erectus was the first human ancestor to control fire, a man-made chemical combustion reaction. The ability to control fire led to chemical processes to make tools, weapons, plates, coins and jewellery from metals including copper, gold, iron and bronze. By the fifteenth century CE, very high-quality steel called *tamahagane* was being made for Samurai swords in Japan; intricate figures and jewellery were being made from gold in South America; and in China, beautifully glazed ceramic jars and ornaments were manufactured. Wealth became associated with many of these items, particularly those made from precious elements such as gold and silver. The desire for wealth led to humans seeking methods for changing low-value base metals such as lead into rare 'noble' metals such as gold. Treating diseases that caused ill-health and death led humans to make medicines from plant and animal sources. An elixir was sought that would prolong life into eternity. These quests gave rise to alchemy, a proto-science practised from around 300 BCE onwards in Europe, Asia, the Middle East and on the Indian subcontinent. Alchemists devised techniques for measurement, heating and distilling. Their discoveries and practices laid the foundations for chemistry as a science based on observation and experiment.

Making discoveries in chemistry: products and techniques

In Europe, chemistry developed from this primitive state alongside huge changes wrought by the first industrial revolution that spanned the late-eighteenth to mid-nineteenth centuries. The invention and application of steam engines transformed small-scale local production of handmade items into large-scale mechanised processes, with factories manufacturing iron, fabrics, concrete, glass, paper and ceramics in huge quantities. These and other items were traded globally, contributing to enormous economic growth in many countries. The mid-nineteenth century brought a second industrial revolution in Europe, characterised by production of cheap steel of consistent quality made via the Bessemer process. The availability of steel transformed building and engineering, enabling construction of high-rise towers, longer and larger bridges and stable, long-distance railways. Discoveries of fossil fuels in multiple locations worldwide generated petroleum-based industries, leading to the development of the internal combustion engine and new materials based on crude oil. Huge expansion of chemical industries followed, as methods emerged for producing many new substances including fabric dyes, paints and pigments, fertilisers, explosives, soap, cosmetics, synthetic rubber and fabrics, pharmaceuticals, and foodstuffs including flavourings and colourants, factory-made sweets, chocolate and bread.

Throughout this era, chemists made fundamental discoveries that advanced understanding of the subject itself. Over half of the 92 naturally occurring chemical elements were isolated and identified in the nineteenth century as chemists mastered heating, reacting, electrolysing, liquefying and measurement, and applied emerging analytical techniques such as spectroscopy, chromatography and X-ray diffraction. These prompted improvements in observations of phenomena and understanding of molecular structures. Theories were proposed and tested, and rigorous experimental practices that significantly improved understanding of chemical reactions were developed. Physical chemistry grew from work on the energetics involved in industrial processes. Inorganic chemistry arose from organisation and classification of the chemical elements via the periodic table and industrial processes developed for the production of acids, fertilisers, paints and other substances. Organic chemistry was founded on the petroleum industry and increasingly detailed knowledge gained from the study of living organisms. This led to chemists understanding the properties of carbon-based compounds, how they were formed, and how they could be adapted to make new consumer products.

Connecting to physics and mathematics

In the twentieth century, chemistry became increasingly sophisticated, incorporating discoveries and developments from physics and mathematics. Identification of sub-atomic particles and the application of quantum physics to electron arrangements led to new understandings of atomic and molecular structures. Chemists realised that molecules, atoms and ions combine via different types of chemical bond, so understood with greater precision how chemical and physical properties of substances are different and are related to the position of an element in the periodic table. Applications of mathematics enabled understanding of chemical equilibria, electrolysis, acid–base chemistry and molecular structures. Chemists applied newly developed techniques to chemical reactions in living organisms and biochemistry came to be recognised as the biological branch of chemistry.

New materials, new approaches

Within the last fifty years, chemists have applied understanding of the behaviour of matter to develop new substances including polymers, pharmaceuticals, fabrics, food additives and novel materials. Amongst the most significant developments are the computer chip, which uses the semiconductor properties of silicon, and revolutionised computing by enabling production of small-scale and hand-held machines; confirmation in the 1990s of the existence of a ball-shaped structure of 60 carbon atoms named buckminsterfullerene; and, in 2004, discovery of graphene, a one-atom thick sheet of carbon atoms arranged in an infinite network of hexagons. In pharmacology, new drugs – such as statins, anti-retroviral and cancer treatments – and new vaccines have contributed to lengthening the lifespan of people in many countries. In chemical engineering, solar and photovoltaic cells, hydrogen fuel cells and battery-driven cars have been developed and are becoming commonplace as dependency on fossil fuels wanes in the light of evidence of climate change and as crude oil sources become depleted. Chemical processes have become greener, adapting to minimise waste, pollution and environmental impact. Analytical techniques and some reactions have been transformed by new technologies, including the scanning tunnelling microscope which allows chemists to 'see' atoms of chemical elements and perform reactions atom by atom.

Knowledge of chemistry is recognised as a foundation for understanding many subjects including medicine, materials science, geology, pharmacology, biochemistry and biology, forensics and chemical engineering. Chemical products contribute significantly to the economies of many countries: a recent report (Oxford Economics, 2019) suggests chemistry contributes around $5.7 trillion dollars to global gross domestic products and supports 120 million jobs. Chemistry is a global enterprise. Studying and working in chemistry is productive and rewarding. Teaching chemistry means setting students on a path to understanding principles that underpin this industry and field of research, as well as how chemistry applies to their everyday lives.

Big ideas in chemistry

Two influential reports connect so-called 'big ideas' in science with the school science curriculum (Harlen et al., 2010, 2015). Big ideas are aspects of science with one or more of these four qualities:

→ Explanatory power; that is, understanding a big idea leads students to develop knowledge of a range of phenomena, objects and events.
→ Provision of a basis for understanding societal issues.
→ A sense of satisfaction, as big ideas prompt answers to questions people ask about the world.
→ Cultural significance; that is, a big idea describes or reflects human achievement in science and how this impacts the environment.

In England, Northern Ireland and Wales, it is hoped that professional organisations for biology, chemistry and physics will work with curriculum developers to apply the reports in writing future versions of the science National Curriculum. For science teachers, these reports may facilitate departmental curriculum planning, ensuring coherence in students' experiences across age ranges, as the example in Table 1.1 illustrates. The progression shown here extends to post-16 and indicates that effective and constructive 11–16 chemistry teaching builds students' understanding of concepts in a way which prepares them well for studying post-16 chemistry.

Often big ideas in science proposed by the 2010 report (Harlen et al., 2010), two relate directly to chemistry:

→ All matter in the Universe is made of very small particles.
→ Chemical reactions involve rearrangements of the particles of reactants to form new substances.

Each is discussed briefly below. A third big idea, that the periodic table shows all known chemical elements in atomic-number order, is also introduced.

All matter in the Universe is made of very small particles

Understanding that all matter is made of discrete particles too small to be visible to the naked eye is the starting point for chemistry. This notion originated in ancient Greece when the philosopher Democritus (460–370 BCE) taught that matter was made of atoms ('atom' meaning 'indivisible') which were infinite in number, everlasting and uncreated. The Greeks thought matter could be divided down to atoms and no further. Democritus believed an object's properties derive from its atoms. Chemists now accept some of Democritus's principles as correct. Chemical elements sub-divide to atoms. This leads to a significant big idea in chemistry: that one atom of a chemical element has the chemical properties of that element. This means one atom of sodium has the chemical properties of sodium, one atom of platinum has the chemical properties of platinum, and so on for all chemical elements. This lays the foundation for understanding events occurring in chemical reactions.

Democritus's thinking remained current in the Middle East but was unfashionable in Europe until the seventeenth century, when scientists Isaac Newton and Robert Boyle stated their support for the concept. Formalisation of atomism in Western European science developed from 1808, when John Dalton (1766–1844) proposed his atomic theory. He emphasised that atoms could not be created or destroyed and that atoms combine to form chemical compounds with constant composition. Dalton reasoned that every water particle contains the same atoms combined in a fixed ratio. Dalton's initial principles were not entirely correct: at the end of the nineteenth century experiments showed that atoms had sub-components, namely protons and neutrons in a nucleus surrounded by electrons. In the twentieth century, advances in understanding of the properties of electrons and mathematics led to the notion that electrons are arranged in 'shells' or 'orbitals'. Sub-atomic particles do not have the properties of individual chemical elements. Protons, electrons and neutrons are held in atoms by electrostatic and nuclear forces of attraction. Electrostatic attraction between particles of opposite charge is responsible for the chemical bonding that holds matter together and contributes to its physical and chemical properties.

Understanding that all matter, living and non-living, is formed from tiny discrete particles invisible to the human eye requires a mental leap from the observation of matter as continuous. Chemistry teachers must make atoms 'visible' to students. Chapter 3 discusses this in detail. Harlen *et al.*'s 2010 report illustrates how progression in students' understanding of the principle that matter is comprised of atoms can be developed (Table 1.1).

Table 1.1 Progression in understanding the big idea 'all matter in the Universe is made of very small particles'

5–7

All the 'stuff' encountered in everyday life, including air, water and different kinds of solid substances, is called matter because it has mass and therefore weight on Earth and takes up space. Different materials are recognisable by their properties, some of which are used to classify them as being in the solid, liquid or gas state.

7–11

When some substances are combined, they form a new substance (or substances) with properties that are different from the original ones. Other substances mix without changing permanently and may be separated again. At room temperature, some substances are in the solid state, some in the liquid state and some in the gas state. The state of many substances can be changed by heating or cooling them. The amount of matter does not change when a solid melts or a liquid evaporates.

11–14

If a substance could be divided into smaller and smaller pieces it would be found to be made of tiny particles. These particles are not in a substance; they are the substance. All the particles of a particular substance are the same and different from those of other substances. The particles are not static but move in random directions. The speed at which they move is the temperature of the material. Differences between substances in solid, liquid or gaseous states can be explained in terms of the speed and range of particle movement and the separation and strength of attraction between neighbouring particles. The stronger the bonds in between particles are, the more energy has to be transferred to the substance to separate them. Energy is transferred to a substance to become a liquid from a solid (melting) or to become a gas from a liquid. This is why substances have different melting and boiling points.

All materials, anywhere in the Universe, living and non-living, are made of very large numbers of atoms. There are 118 different types of atoms known at the time of writing. Substances made of only one type of atom are called elements. Atoms of different elements combine together to form compounds. A chemical reaction involves rearrangement of atoms in the reacting substances to form new substances. The total amount of matter present in the reacting substances and the new substances remains the same. The properties of materials can be explained in terms of behaviour of the atoms and groups of atoms (molecules) of which they are made.

14–17

Atoms have an internal structure comprising a heavy nucleus, made of protons and neutrons, surrounded by electrons arranged in orbitals or shells. Electrons and protons are respectively negatively and positively charged. Atoms are electrically neutral, as the numbers of electrons and protons present in an atom are the same, so their charges balance exactly. When one or more electrons are removed or added to an atom, an imbalance of electrical charge occurs and the charged particle is called an ion. Loss of one or more electrons creates a positively charged ion, because there are more protons than electrons present. Gain of one or more electrons creates a negatively charged ion because there are more electrons than protons present.

Electrons move rapidly in matter, causing electric current and magnetic force. The availability of electrons to move freely within a structure causes chemical elements to conduct electricity. A small number of chemical elements can be magnetised. Some atoms have nuclei which are unstable, due to an imbalance in the numbers of protons and neutrons present. These atoms spontaneously emit small particles in a process called radioactivity. This process involves the release of large amounts energy in the form of radiation. The behaviour of matter on the scale of nuclei, atoms and molecules differs from that observed on the scale that human senses can normally experience.

Source: Harlen *et al.*, 2010

Chemical reactions involve rearrangements of reactant particles to form new substances

Atoms, molecules and ions can combine to make new types of matter with properties that differ from those of the original reagents, whether these are chemical elements or compounds. An event in which a new type of matter forms is a *chemical reaction*. In any chemical reaction, the amount or mass of matter present does not alter regardless of the physical state of the reagents and products. Chemists cannot change the amount or mass of matter present, only the type of matter. This principle is the *law of conservation of mass*, which states that matter (or mass) cannot be created or destroyed in a chemical reaction.

Students may ask why chemical reactions occur spontaneously, that is, without the addition of any additional energy in the form of heat or a change in pressure. The answer lies in the overall energetics of the change from reactants to products. A reaction will occur spontaneously if the overall change of energy is positive. However, a positive energy change does not necessarily mean an increase in temperature, which is easy to identify. A positive energy change can mean that a gas is produced from two liquids reacting together, or by a liquid being formed when two solids react together. The positive change in these cases means that the products are more disordered than the reactants. The energy change may not be immediately measurable by a thermometer, but each individual particle of gaseous or liquid product has slightly more energy than those of the reagents. The total effect is an energy increase. Chemists use the term *entropy* for disorder: entropy always increases when a chemical reaction occurs.

Chemists have devised many methods for undertaking chemical reactions. To decide if a chemical reaction has occurred or is occurring, chemists make observations. Qualitatively, these include physical state, smell, overall appearance and colour. Quantitatively, chemists measure, for example, temperature, gas pressure, mass, concentration, colour changes and pH (acidity and alkalinity). Chemists try to understand why chemical reactions do or do not occur and how to make product(s) they want efficiently and sustainably. The ability to control chemical reactions is essential.

The periodic table shows all known chemical elements in atomic number order

The periodic table lists all known chemical elements, arranging these in an orderly way. The location of an element in the periodic table indicates its chemical and physical properties (see Chapter 4). Familiarity with the periodic table is therefore a valuable starting point for understanding chemistry. The periodic table originated from chemists' observations of patterns and trends in the physical and chemical properties of elements. Chemists considered how elements could be arranged logically. Early attempts included 'triads' of three elements and 'octaves' of eight. By the mid-nineteenth century, scientists found that many newly discovered elements would not fit into these simple arrangements. A comprehensive system that covered all the chemical elements was sought. In 1859, Dimitri Mendeleev, a Russian chemist, proposed the periodic table. He organised elements by atomic weight, stacking those with similar chemical properties above one another in columns called *groups*. Mendeleev aligned the elements into rows called *periods*. Enough was known about chemical elements to reveal that properties changed gradually, leading to the notion of periodic trends. Crucially, Mendeleev left gaps for chemical elements that were yet to be discovered. As these were discovered, Mendeleev's predictions were found to be correct.

 ## Doing chemistry

Chemistry is a practical science. As a science, chemistry follows scientific principles. These include:

- → For every effect there is one or more cause.
- → Scientific explanations, theories and models are those that best fit the facts known at any one time.
- → Knowledge scientists produce is used in some technologies to create products that meet human needs.
- → Applications of science often have ethical, social, economic and political consequences.

In gaining understanding of chemical reactions and processes, chemists establish cause and effect. What chemists believe these causes and effects are has changed as scientific explanations based on techniques for measurement, analysis and observation have developed. In chemistry, application of these principles led to paradigm shifts in chemists' thinking and understanding of the behaviour of matter. This deepened and refined knowledge about substances and how to apply chemistry to improve the quality of human life.

A paradigm shift: overthrow of the phlogiston theory

Until the late eighteenth century, many scientists believed that a substance named phlogiston was emitted from substances on burning. At the time, chemists lacked understanding of differences between gases; since the gaseous atmosphere that surrounded them was known as air, all gases were called 'airs' and named according to whether or not they supported combustion and other crudely observable properties such as solubility in water. 'Dephlogisticated air' supported combustion, while 'phlogisticated air' did not: a substance burning in dephlogisticated air gave off phlogiston, changing the air to phlogisticated air. The theory gained wide support and the supposed existence of phlogiston dominated chemical reasoning and analysis. This was despite some experimental findings that the behaviour of a number of different airs suggested otherwise. In the 1780s, experiments conducted by the French chemists Antoine Lavoisier and his wife Marie-Anne Paulze Lavoisier used very careful modern (for the time) measurement techniques and analysis to show that dephlogisticated air was a pure gas they called *oxygène* (oxygen). Phlogisticated air became carbon dioxide. The Lavoisiers effectively debunked the phlogiston theory, causing a paradigm shift that allowed chemistry to replace primitive reasoning with precise understanding. They developed systematic measurement techniques and chemical language based on elements present in substances. Their work laid the foundations of modern chemistry.

Chemistry that makes life better: a brief history of aspirin

Chemistry involves application of chemical reactions and processes to achieve human ends. As noted earlier, its origins lie in alchemy, in which people sought methods for manipulating natural substances to achieve personal benefits. Outstanding global progress in treating major diseases via vaccination, drug development and analysis arises because scientists, including chemists, have devised and applied technologies to solve challenges threatening human life. An example is the development of the drug aspirin, which has powerful pain-relieving, temperature-reducing and anti-inflammatory properties.

1 The principles behind secondary chemistry teaching

The history of aspirin begins with the willow tree (various species of *Salix*), a recorded source of medicine since antiquity: the Assyrian, Sumerian and Egyptian civilisations (4000–1500 BCE) used willow extracts to relieve joint pains, fever and inflammation; from around 600–500 BCE, Chinese and Babylonian herbal practitioners used willow as a remedy for various illnesses; the Greek physician Hippocrates (400 BCE) administered willow leaf tea to women in childbirth, while Dioscorides (100 BCE) found willow effective for treating inflammation. As societies developed, accurate knowledge of the active agents in herbal medicines was sought, as dosages and efficacy were inconsistent. By the mid-eighteenth century, medical and scientific societies enabled systemic recording and publication of experimental methods and results. In 1763, the Royal Society published five years of data on administering willow-bark extract gathered by an English country clergyman named Edward Stone. His precise preparation instructions and evidence that over fifty people were cured of fevers and other ailments led to willow-bark extract becoming a main drug available to herbalists. For researchers, Stone's work provided reliable information that a chemical in willow bark acted as a potent drug. Advancements in analysis enabled German chemist Johann Buchner to extract bitter-tasting yellow crystals from willow in 1828. He named the substance salicin after the Latin name for willow, *Salix*. Roughly simultaneously, salicin was extracted from meadowsweet flowers by the Swiss pharmacist Johann Pagenstecher.

Next, chemists investigated the structure of salicin. This was published in 1853 by French chemist Charles Frédéric Gerhardt, who named the compound salicylic acid, and synthesised a variant, acetylsalicylic acid. In 1876, systematic trials of salicin showed it was effective in treating fever and joint inflammation in rheumatism. The German chemist Felix Hoffman, who worked for the Bayer pharmaceutical company, found a method of making the drug in a laboratory. He also showed that the irritant side effects of salicin were reduced when acetylsalicylic acid was used, and he synthesised this molecule too. In 1899, Bayer named acetylsalicylic acid 'aspirin', coining the word by combining 'a' (for acetyl) with 'spir' (from *Spirea ulmaria*, the Latin name for meadowsweet) and 'in' (a common suffix for drugs at that time).

Aspirin became the leading drug for treating many illnesses and was awarded the title of 'most frequently sold painkiller' by the Guinness Book of Records in 1950. Further studies led to an understanding of aspirin's mechanism of action within the body. Today, aspirin is recommended for treatment of and as a preventative medicine reducing incidence of strokes, heart attacks and various forms of cancer.

1.4 Practical chemistry

The book introduces the practice of scientific inquiry as undertaken by chemists. This includes:

→ Making observations and inferences based on real-time events.
→ Identifying patterns in data and exceptions to patterns.
→ Techniques for understanding substances and structures (chemical analysis).
→ Techniques for making new substances (chemical synthesis).
→ Examples of chemistry at work: the chemical industry and real-life chemists.

The nature of chemistry as a discipline means it is a hands-on, practical subject. While there are theoretical chemists, most people trained in chemistry gain experience in laboratories, manipulating equipment and chemicals. Experiencing chemical phenomena in a live setting generates motivation for teaching and learning theoretical ideas. In the UK, examination specifications for 14–16 and 16–18-year-olds require students to complete practical work, so hands-on experiences are essential. Chemistry experiments serve multiple purposes: they lead to knowledge and skill development, and require application of chemical principles, so creating a basis for advanced knowledge. Visualising reactions from virtual laboratories is no substitute for doing an experiment on the bench in real time. Even so, research evidence suggests practical work does not in itself teach chemical concepts, as students may follow instructions without knowing why they are doing so or the meaning of what they observe. Teachers should maximise the benefit (and expense) of practical work by giving careful consideration to expected learning outcomes.

Most chapters in this book feature chemical experiments, either teacher demonstrations or student-led activities, based on one or more of the principles of scientific inquiry listed above. Some activities produce immediately identifiable products that can be seen, smelt or tested qualitatively. Analysis skills progress to training students to make precise measurements of chemicals in, for example, titration reactions (Chapter 7). Other experiments introduce patterns of chemical behaviour displayed by sequences of elements in the periodic table (Chapter 4) or organic compounds with similar formulae (Chapter 10). Gradually, anomalies and exceptions to these patterns may be introduced. Post-16 students use advanced equipment and make a wide range of measurements, tests and observations. This methodology underpins chemists' development of productive and responsible applications. Open-ended, inquiry-based experiments raise questions that students investigate in a non-fixed way, devising their own method. These permit various outcomes from one starting point, so introducing the principles of research and encouraging creativity.

 Ethics in chemistry

Thankfully for the progress of science, and for the health and wellbeing of scientists, scandals involving lapses of ethics are relatively rare. Nonetheless, all scientists must adhere to ethical standards. They may be tempted to falsify data, perhaps in anticipation of a significant finding, so they can publish before a rival, gaining kudos and credibility. Disregarding health and safety regulations by habit or choice creates conditions ripe for serious accidents. Funds intended for research could be diverted elsewhere. Plagiarism transfers credit for work without acknowledgement. Junior members of a research team may be subjected to harassment, leading to stress. Working outside strict regulatory frameworks generates legal vulnerability. While all these are possible scenarios, in practice, the vast majority of scientists work reliably and honestly. For chemists, unexpected consequences of research probably cause the most significant ethical challenges. Thousands of new molecules are made annually, each part of a process of generating novel substances that may contribute positively to human development. However, some of these molecules may unintentionally have a negative impact on society or be deliberately turned to destructive uses. An example is discussed next.

Unintentional impact of a well-intentioned discovery: chlorofluorocarbons

Chlorofluorocarbons, or CFCs, were made innocently in a well-intentioned attempt to solve the problem of refrigeration. By the 1920s, electrically powered refrigerators were becoming a popular way to preserve food and other essential materials. Refrigeration relies on a refrigerant, a substance that changes state readily from liquid to gas and vice versa. Liquid refrigerant is pumped through pipes at the back of the refrigerator, absorbing heat from within the fridge and becoming a gas. As the gas leaves the refrigerator, a fan cools the pipes and the refrigerant re-condenses to liquid, transferring thermal energy to the atmosphere through a grille attached to the tubes. Initially, refrigerants were toxic and/or flammable substances such as ammonia, methane, propane, and sulfur dioxide. In the USA,

Frigidaire, a leading refrigerator manufacturer, sought a non-toxic, non-flammable alternative. Thomas Midgley Jr (1889–1944) obliged, synthesising the first CFC, dichlorodifluoromethane (CCl_2F_2), which was named freon. CFCs proved to be highly successful compounds ideally suited to their intended function. Midgley was feted for his contribution, winning international medals. By the 1970s, around one million tonnes of CFCs were produced annually for use as aerosol propellants, blowing agents for foams and packing materials, and as refrigerants. Their non-toxic nature led to CFCs being released into the atmosphere after use.

Meanwhile, in the 1960s, British scientist James Lovelock (1919–) noted build-up of atmospheric haze in certain weather conditions near his home. Curiosity led him to investigate the haze using his own invention, the Electron Capture Detector (ECD), a device sensitive to tiny amounts of synthetic chemicals. Lovelock detected CFCs even in places such as the west coast of Ireland and (in 1971) Antarctica where he expected to find 'clean' air. CFCs had dispersed across the globe. Lovelock presented his findings at a conference in 1972. His talk inspired American chemist, F. Sherwood (Sherry) Rowland to investigate atmospheric CFCs. With Mario Marina, Rowland published a seminal paper in 1974 showing that although CFC molecules are inert in the lower atmosphere, the troposphere, in the stratosphere, about 20 km above the surface of the Earth, interaction with UV-rays causes them to break up. Rowland and Molina calculated that a single chlorine atom from one CFC molecule initiates a chain reaction that destroys thousands of molecules of ozone, O_3, the vital substance that protects the Earth from excessive UV-radiation. They estimated that an immediate ban on CFC production would mean ozone depletion continuing for years, while maintaining production would exacerbate the loss. Rowland said this realisation sent 'a chill down his back'. He became increasingly troubled as industry experts disputed his claims, in part because results were difficult to replicate in tropospheric conditions.

Independent evidence supporting Rowland and Molina came from British scientists Joe Farman and colleagues working for the British Antarctic Survey (BAS). BAS data from 1957–58 onwards showed annual depletion of stratospheric ozone during the Antarctic summer when the Sun reappears. They suspected something was destroying ozone. In 1985 Farman and his team published a paper demonstrating that ozone over Antarctica was depleted by 40% each September (the end of the southern hemisphere summer) and showed significant overall decline since the 1960s. The Antarctic ozone hole was quickly recognised as an environmental disaster, prompting immediate change in human behaviour. By 1987, 56 countries had signed the Montreal Protocol, agreeing to halve CFC production, paving the way for a complete ban by 1996. Scientists expect Antarctic ozone to recover to 1980 levels by 2070.

1.6 Teaching and learning chemistry

Achieving conceptual change

Inevitably, students come to chemistry with naive knowledge which is often incorrect compared to formal chemical knowledge. The book gives many well-documented examples of naive knowledge in the section of each chapter entitled 'Students' prior knowledge and misconceptions'. Naive knowledge often (but not always) impedes learning of formal knowledge. Some naively held ideas are well-formulated into mental models that offer plausible explanations for phenomena and strongly resist change. It is worth noting that experienced scientists are as vulnerable to this failing as students. The phlogiston theory described above offers a good example: Joseph Priestley, a chemist often credited with discovering the gas that became known as oxygen, called it dephlogisticated air, and continued to believe in phlogiston despite strong evidence to the contrary. Similarly, students may persist in thinking that gases have no mass, solids 'disappear' on dissolving, matter is continuous not granular, and the periodic table is just a picture on the chemistry laboratory wall. Any learning that takes place is rote, for the convenience of passing an examination, and/or pleasing the teacher.

A major challenge for chemistry teachers is how to avoid being satisfied with students rote-learning formal chemical knowledge, instead helping them achieve deep understanding of at least the three big ideas discussed above. 'Deep' in this context means a major shift from naive to formal knowledge, not, for example, understanding quantum mechanics, mathematical details of entropy, or the intricacies of making new superheavy elements. Chemistry teachers must force students to abandon naive views, adopt chemists' ways of 'seeing' the particles comprising substances, realise what is meant by a chemical reaction, and understand the organising principles behind the periodic table and its impact on our understanding of matter. This formal knowledge creates an intellectually satisfying foundation for further study. The big question is how to achieve this knowledge and understanding.

Of course, this question cannot be fully answered in one paragraph. And it is not easily done. But a teacher who is serious about achieving more than surface learning can adopt four principles for their lessons.

The first principle is *not to make assumptions about what students know or understand*. Knowing what students know already is a powerful tool to achieve deep learning: establish students' prior knowledge, perhaps via a brief 'bell task' or introductory exercise (see White and Gunstone, 1992). Initially, this may feel like opening Pandora's box ('a source of great and unexpected troubles'), as realisation dawns that students' naive knowledge is complex, plausible, systematic and resistant to change. But students enjoy being asked to say what they think, and knowing their naive knowledge makes reconstruction to formal knowledge a possibility.

Second, *be prepared to ask hard questions*. The skill of asking and answering questions that probe understanding is essential. To prompt deep learning, find the question(s) that force(s) students to reconsider their naive views. These do not involve lazy description or guessing what is in the teacher's head. The questions must challenge long-held naive mental schemata. Finding the answers requires careful thought and time.

The third principle is to *provide support*. This means a variety of things. It includes continued reinforcement of new formal knowledge, such as confirmatory examples and situations that require application of the novel (to the students) reasoning. Support also means providing motivation for change, as students need reasons that are socially, attitudinally and emotionally acceptable in order to make the cognitive leap of accepting formal knowledge.

Finally, *take time*. This is often in short supply in science lessons: there is pressure to teach to a tight schedule, especially when examinations are looming. Faced with naive knowledge even after teaching, the temptation is to cut corners to formal knowledge, saying, 'You're supposed to know this' or, 'Don't think about it, just learn it.' Such grumpiness is tempting, but ultimately damaging to the process of achieving naive–formal knowledge shifts. It may seem hard to believe, but students whose formal knowledge is deep, not rote, have better examination results. Be patient.

Transferring between macro-, sub-micro- and symbolic scales: Johnstone's triangle

Chemistry explains macro-scale observable events in terms of sub-microscopic molecules, atoms, ions and electrons, representing them in chemists' language of symbols, formulae and equations. Chemists shift between macro, sub-microscopic and symbolic scales without reflection, often assuming others follow their reasoning automatically. Students focus on the macro-scale, the level of observable events, and cannot make shifts between levels rapidly, in part because their understanding of the sub-micro and symbolic levels is not intuitive. A challenge for teachers is how to support students first, in understanding the sub-micro- and symbolic levels, and, next in moving between three levels of observation without access to extremely expensive equipment that enables them to 'see' atoms for real. A tool proposed by Scottish chemistry educator Alex Johnstone (1991) is valuable. Johnstone illustrates the macro-, sub-micro and symbolic levels using the triangle shown in Figure 1.1.

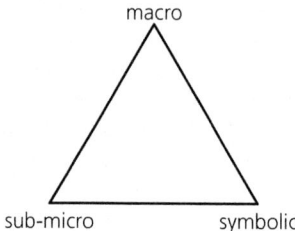

Figure 1.1 Johnstone's triangle

Source: Johnstone, 1991

The triangle scaffolds learning by prompting a need to be explicit when moving between macro-scale observations, sub-microscopic explanations and symbolic representations. The macro-scale is considered throughout the book when discussing observations and outcomes of experiments. Teaching students about particles, the sub-micro-scale, is considered in Chapter 3, and these ideas are applied throughout the book, beginning with modelling chemical reactions, discussed below. Symbolic representations of chemicals and reactions in formulae and equations is considered first in Chapter 2 and applied throughout the book.

Using models in chemistry teaching

Some molecular models have had enormous impact on scientific understanding. For example, Dorothy Hodgkin (1910–1994) deduced the structures of penicillin (in 1945) and vitamin B12 (in 1955), making models of both. In 1953, James Watson and Francis Crick elucidated the structure of DNA from photographs taken by Rosalind Franklin, building a three-dimensional model that changed understanding of genetics. In the 1980s, Harry Kroto made a paper model of C_{60} (carbon-60), based on an old football and futuristic buildings designed by American architect Richard Buckminster Fuller, which changed our understanding of carbon allotropes.

Chemists have also developed theoretical models of chemical and physical phenomena. For example, attempts to understand acidity led to models of an acid, beginning with the seventeenth-century notion that acids were made of needle-shaped molecules, the sharpness of which caused their sharp taste. The twentieth-century Lewis model (the most recent) regards an acid as a lone-pair acceptor (see Chapter 7). Similarly, atoms were originally believed to be indivisible. In the late nineteenth century, J.J. Thomson proposed a 'plum-pudding' model in which negatively charged electrons (the plums) were embedded in a positively charged atom (the pudding); in the twentieth century this changed to a planetary motion model in which a positive nucleus is surrounded by electrons in specific orbits related to their energy; this was superceded by the quantum mechanical model in which orbits become regions surrounding a nucleus in which there is a probabilistic chance of finding an electron. Models representing phenomena are important because they illustrate the impact of empirical science on our understanding, show inaccuracies and prompt assimilation of new knowledge.

Students' understanding of the sub-micro scale can be aided by modelling chemical reactions using hand-held manipulatives, large-scale molecular models and animated visualisations. The purpose of modelling is to enable students to 'see' microscopic processes and form explanations for chemical events, rather than relying on macro-scale observations. These explanations consider, for example:

→ the physical structures of reactants and products.
→ the relative stability or level of energy that particles have before and after a reaction.

In chemistry at post-16 and beyond, models can represent reaction mechanisms, illustrating stepwise interactions between atoms, molecules and ions.

Initially, students benefit from seeing diagrams and images of molecules and manipulating molecules to track and trace a chemical reaction. These models support understanding of how products form from reactants, connecting the symbol equation for the reaction with observable macroscopic events. Placing images of molecular models alongside the equation and photographs of the reaction itself can help. For example, the reaction between methane and oxygen can be represented as shown in Figure 1.2.

1 The principles behind secondary chemistry teaching

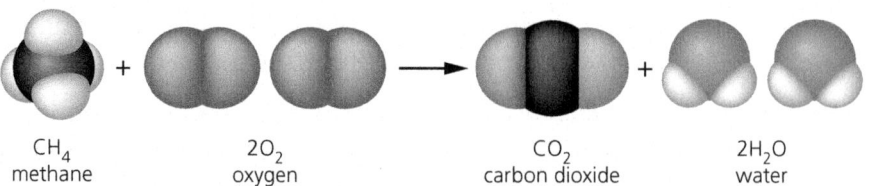

| CH$_4$ | 2O$_2$ | CO$_2$ | 2H$_2$O |
| methane | oxygen | carbon dioxide | water |

Figure 1.2 A pictorial representation to model a chemical reaction

In Figure 1.2, the image complements word and formula equations, showing the molecules involved. This model is space-filling, as no lines are drawn within molecules to represent chemical bonds. The shading shows that atomic cores are retained, as there are one carbon, four oxygen and four hydrogen atoms on both sides of the arrow. The model misleadingly gives the impression that atoms are genuinely unchanged: the carbon atom in methane 'relocates' and becomes the carbon atom in carbon dioxide. This is deceptive. Atoms are not full, intact atoms when bonded within molecules as electrons are shared across atomic boundaries. So it is important to make students aware this is a model, not reality.

Although models are extremely helpful in forming students' sub-micro reasoning related to chemical reactions, care is required, as students may develop misconceptions, some of which are listed below (misconceptions in italics).

- *Carbon atoms are black, hydrogen atoms are white, oxygen atoms are red, and so on.* In reality, atoms do not have these or any other colours. When observed in bulk, a substance may have a colour, but individual atoms are not coloured.
- *Atoms contain holes that need to be filled to make the atom 'happy'.* In practice, atoms do not have holes and are not solid. Nor do atoms have emotions so they cannot be happy or sad. Bonding occurs as a result of electrostatic attractions and energetics (see Chapter 6).
- *Bonds are physical objects, such as grey plastic sticks.* In reality, bonds are formed by electrostatic attractions between positive and negative particles (see Chapter 5).
- *Double bonds (bendy bonds) are made from two identical parts that are different from single bonds (straight bonds).* In practice, all bonds are formed by electrostatic attractions between particles.
- *Effort (energy) is required to make and break bonds.* Bond-breaking is endothermic, so requires energy, but bond-making is exothermic and emits energy to the environment. This can be shown when taking a model apart, as effort is needed. It is harder to use a physical model to show that energy is given out when bonds form.

Consider using magnetic models which will snap together, making clearer the exothermic nature of bond formation.

These possible misconceptions mean that, although using hand-held models is recommended, discussing their limitations is important. Animated visualisations of reactions can be helpful but, once again, care should be taken to discuss the accuracy of the representation. Illustrating how chemists visualise specific molecules, perhaps drawing on historical examples and showing images of atoms from scanning tunnelling microscopes, may prompt understanding and aid transition from macro to sub-microscale thinking. Ultimately, chemists only see particles in their minds' eyes: models are a valuable way of helping students achieve this.

Mathematics in chemistry

Chemists use mathematics to model chemical reactions and molecular structures, understand chemical concepts and solve problems. An understanding of mathematics is fundamental to the development of accurate knowledge of chemical concepts and processes. At university level and beyond, this leads to mathematical chemistry, a recognised subject in which mathematical approaches are applied to complex chemical problems, to model chemical phenomena and to design novel molecules. Mathematical chemists engage in research in molecular modelling, drug design, molecular engineering and theoretical chemistry. Developing the ability to apply mathematical knowledge to understand chemistry begins in many courses for 14–16-year-olds. Aspects of mathematics are introduced in various chemical topics, including chemical bonding (molecular structures), rates of reaction, acidity and alkalinity (titrations and pH measurement) and stoichiometry (mole calculations). Table 1.2 summarises mathematical requirements that are commonly expected of students aged 14 to 16. Post-16 and higher education courses feature increasingly advanced applications of mathematics.

Table 1.2 Mathematical requirements for chemistry for 14–16-year-olds

Arithmetic and numerical computation	Recognise and use expressions in decimal form
	Recognise and use expressions in standard form
	Use ratios, fractions and percentages
	Make estimates of the results of simple calculations
Geometry and trigonometry	Use angular measures in degrees
	Visualise and represent 2D and 3D forms including two dimensional representations of 3D objects
	Calculate areas of triangles and rectangles, surface areas and volumes of cubes

1 The principles behind secondary chemistry teaching

Handling data	Use an appropriate number of significant figures
	Find arithmetic means
	Construct and interpret frequency tables and diagrams, bar charts and histograms
	Understand the principles of sampling as applied to scientific data
	Understand simple probability
	Understand the terms mean, mode and median
	Use a scatter diagram to identify a correlation between two variables
	Make order of magnitude calculations
Algebra	Understand and use the symbols: $=, <, \ll, \gg, >, \infty, \approx$
	Change the subject of an equation
	Substitute numerical values into algebraic equations using appropriate units for physical quantities
Graphs	Translate information between graphical and numeric form
	Understand that $\{\{y = mx + c\}\}$ represents a linear relationship
	Plot two variables from experimental or other data
	Determine the slope and intercept of a linear graph
	Draw and use the slope of a tangent to a curve as a measure of rate of change
	Understand the physical significance of area between a curve and the x-axis and measure it by counting squares as appropriate

More specifically, mathematical principles and processes in chemistry occur in:

→ Measurement of volumes, mass, concentration, acidity and alkalinity.
→ Determination of factors controlling the rate of a chemical reaction, such as temperature, concentration and surface area.
→ Measurement of how much energy is produced or taken from the environment when a chemical reaction occurs.
→ Understanding molecular shapes in three dimensions.
→ Drawing graphs to represent and interpret data obtained by experiment.
→ Calculating reacting mass ratios and moles of product.
→ Using basic numerical functions such as calculation of percentages, appropriate units, standard form and decimal form.

Teachers who feel their mathematical skills would benefit from additional practice prior to teaching are advised to take time for this (see Resources). When teaching mathematical aspects of chemistry, several points must be considered. First, in mathematics, students may learn procedural steps by repetition without necessarily understanding why these steps are used, then apply these to solve identical problems. In chemistry, problems (including examination questions) often feature novel situations which are not an identical match to previous questions. Training students to realise they have the mathematical knowledge to solve a novel problem in chemistry is important. Second, students who are gifted mathematicians may have difficulty understanding how to apply mathematics in a chemical setting. Working successfully in an abstract setting does not necessarily enable common-sense application to a real-world setting. Also, cutting procedural steps and/or using mental arithmetic may cause such students to lose examination marks for not showing working. Third, the ability to estimate is valuable. Transferring numbers correctly to a calculator requires mental processes that need attention to detail and practice. Making a rough mental calculation to estimate an answer before starting is worthwhile. Understanding powers of ten, units and graphical data is helpful. Fourth, giving students opportunities to practice mathematical applications in chemistry is beneficial. Continued practice builds confidence and trains the ability to handle novel situations as well as to estimate answers to problems.

While practising, telling students 'not to worry about the mathematics, just put the numbers in the equation' should be avoided. This leads to rote learning, which means avoidance of conceptual understanding and failure to build confidence. Taking time to develop underpinning mathematical ideas, building practice towards solving a complex problem in simple steps is recommended.

Conclusion

Chemistry has global reach: everything we eat, touch, wear, see and use and everything of which we are comprised is chemical 'stuff'. Teaching chemistry needs to convey the relevance of the discipline to everyday life. This book attempts to do this in various ways, including:

- → Presenting historical and up-to-date examples of chemical discoveries that have significant impact on how we live.
- → Showing how the chemical industry has developed and continues to develop, making products that support our society.
- → Illustrating the environments in which chemists work.
- → Discussing ethical principles that underpin how chemistry is done.

Throughout the text, images, boxed information and stories illuminate chemistry through its applications and show people who work as chemists in a variety of settings. Teachers may use these to illustrate relevance, provide contextual settings and show the subject is up-to-date, broad and technologically advanced.

We hope that readers find this book useful from a number of perspectives. First, we expect that new teachers of chemistry, who may be experienced teachers of other sciences, will find information that supports, enables, challenges and encourages high-quality teaching. Second, we aim to convey the material required to teach chemistry to students aged up to about 16, using UK curricula as a guide. Consequently, chapters take topics to this level so may leave teachers with some unanswered questions. Reference materials and other sources may provide connections to sources of higher-level understanding as required. Third, our intention is that the book prompts teachers and their students to realise that chemistry is a vital, living and varied subject that touches us all in multiple ways. Understanding chemistry – even to a small extent – is life-enhancing.

 ## Resources

References and further reading

American Chemical Society (2017) *Chlorofluorocarbons and ozone depletion: A national historic chemical landmark.* www.acs.org/content/acs/en/education/whatischemistry/landmarks/cfcs-ozone.html

British Antarctic Survey (2017) *Science briefing: The ozone hole.* www.bas.ac.uk/wp-content/uploads/2017/04/Ozone-Hole-briefing_Apr17-1.pdf

Connelly, D. (2014) A history of aspirin. *The Pharmaceutical Journal.* https://pharmaceutical-journal.com/article/infographics/a-history-of-aspirin

Desborough, M.J.R. and Keeling, D.M. (2017) The aspirin story – from willow to wonder drug, *British Journal of Haematology* 177: 673–683. Available at https://onlinelibrary.wiley.com/doi/epdf/10.1111/bjh.14520

European Chemistry Industry Council (2019) *Chemical Industry Contributes $5.7 Trillion To Global GDP And Supports 120 Million Jobs, New Report Shows.* https://cefic.org/media-corner/newsroom/chemical-industry-contributes-5-7-trillion-to-global-gdp-and-supports-120-million-jobs-new-report-shows/

Harlen, W. et al. (2010) *Principles and big ideas of science education*, ASE. Available at www.ase.org.uk/documents/principles-and-big-ideas-of-science-education/

Harlen, W. et al. (2015) *Working with big ideas of science education*, InterAcademy Partnership. Available at www.ase.org.uk/documents/working-with-the-big-ideas-in-science-education/.

Johnstone, A.H. (1991) Why is chemistry difficult to learn? Things are seldom what they seem. *Journal of Computer Assisted Learning* 7: 75–83. Available at https://onlinelibrary.wiley.com/doi/abs/10.1111/j.1365-2729.1991.tb00230.x

Mahdi, J.G. (2010) Medicinal potential of willow: a chemical perspective of aspirin discovery. *Journal of Saudi Chemical Society* 14: 317–322. Available at https://www.sciencedirect.com/science/article/pii/S1319610310000578

Notman, N. (2018) Ethics in Chemistry. *Chemistry World* May 2018. Available at www.chemistryworld.com/features/ethics-in-chemistry/3008982.article?adredir=1

Oxford Economics (2019) *The global chemical industry: Catalyzing growth and addressing our world's sustainability challenges*, Oxford Economics. Available at https://www.oxfordeconomics.com/recent-releases/the-global-chemical-industry-catalyzing-growth-and-addressing-our-world-sustainability-challenges

Prather, M.J. and Blake, D.R. (2012) F. Sherwood Rowland (1927–2012): Obituary. *Nature* 484:168. www.nature.com/articles/484168a.pdf

White, R. and Gunstone, R. (1992) *Probing Understanding*, Routledge.

Websites

Further information about Thomas Midgley Jr: https://allthatsinteresting.com/thomas-midgley-jr

The Science Museum, London, UK has a good range of resources including this information relating to James Lovelock and his role in detecting atmospheric pollution: www.sciencemuseum.org.uk/objects-and-stories/chemistry/something-air-james-lovelock-and-atmospheric-pollution

Other resources

Molymod® molecular model kits (see molymod.com) are available from school equipment suppliers such as Better Equipped, Philip Harris, SciChem and TimStar.

Royal Society of Chemistry Maths skills course for teachers: www.rsc.org/cpd/resource/RES00001503/maths-skills

The language of mathematics in science by Richard Boohan (ASE) ISBN 9780863574559

Maths for Science by Alison Pyle (OUP) ISBN 9780199644964

2 Chemical change

Vanessa Kind

Introduction

This chapter introduces the principle of chemical change, which is the central concept on which chemistry relies. Understanding chemical change lays the foundation for understanding all other aspects of chemistry. This chapter provides activities and information to help develop secure understanding of chemical change. Students start learning chemistry with life experiences of chemical changes including combustion of fuels, cooking foods, taking medicines and using soap, bleach, washing powders and other stain removers for cleaning and washing. They may know that fertilisers support plant growth while weedkillers control unwanted plants. Students may be aware that chemical industries produce many synthetic materials and objects. However broad their experience, students may not recognise events as chemical changes, and their general knowledge is not a systematic understanding of chemistry. Introducing chemical change early in secondary education is recommended, alongside reinforcing the notion as students proceed to develop their knowledge

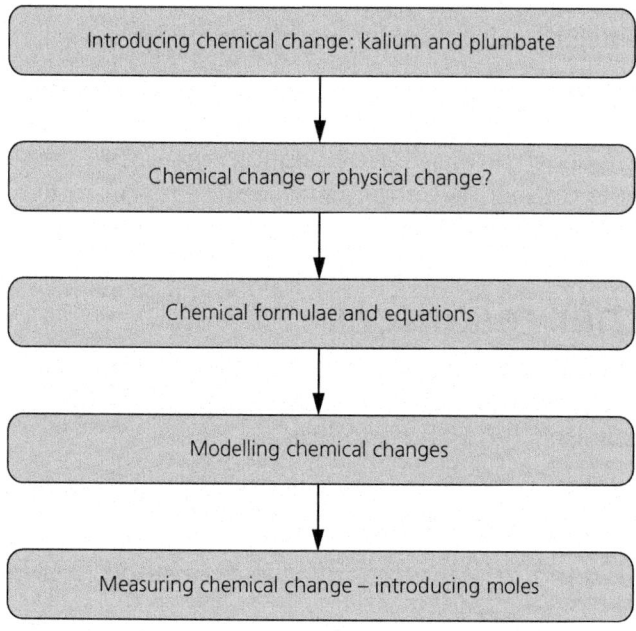

Figure 2.1 Chemical change: teaching sequence overview

of chemical bonding. This will aid the process of changing random experiences of the behaviour of substances into systematic knowledge. This chapter offers support for teaching the principles of a chemical change, promoting the understanding that new substances are formed in chemical reactions. A method for teaching the difference between a chemical and a physical change follows. Next, suggestions are provided about how to teach formulae and equations, which chemists use to represent chemical change. The final section offers a method for introducing students to the mole, chemists' unit for measuring substances.

Students' prior knowledge and misconceptions

Students cannot readily distinguish between a chemical and a physical change. One reason for this is that they are taught that chemical changes have consistent observable characteristics, such as production of a gas, a colour change or heat and light being emitted. Students will apply these to all changes, whether physical or chemical. This may lead to incorrect conclusions, as some chemical changes progress without any of these events while some physical changes do exhibit such characteristics. For example, sublimation, the change from a solid directly to a gas, may appear to students to be a chemical change because a gas is produced. Dissolving may be regarded (confusingly) as either a chemical or a physical change, as characteristics of both are observed on a macroscopic scale. This dual nature is discussed in detail later (Section 2.2). Students often interchange dissolving and melting (a physical change) because in both cases liquids are formed and the solid 'disappears'. They do not realise that melting is reversible and that the particles are unchanged in this process. Further difficulties arise when an apparent state change in fact masks a chemical change: research (Kind, 2004) shows that students may misunderstand what happens when a solid tablet, such as a vitamin C tablet, dissolves in water and produces a gas: they think the gas released on contact with water is already present in tiny bubbles in the tablet. They miss that the gas formed when two solid compounds in the tablet dissolved in water, then reacted in solution to produce the gas. Also, as many reactions presented to students are short-lived, they are unlikely to realise some chemical reactions take a long time (beyond the span of a lesson) to proceed.

2 Chemical change

Other misconceptions are commonplace. For example, students often think of specific chemicals as 'active agents' in a reaction. This may arise because marketing slogans refer to 'active cleaning power', 'best raising agent', 'germ-zappers' and so on. An example is the typical idea of an acid as a chemical destroyer that neutralises an alkali. A common perception, then, is that an active agent works on another substance. In truth, the active agent and the other substance react together: neither is any more active than the other.

Language used to describe chemicals can also create confusion. For example, 'pure' implies 'untampered with' or 'natural' in an everyday sense, leading to the belief that no chemical process has been involved in creating the pure substance which, in practice, could be a mixture of many substances (as is the case for things such as orange juice, butter or wool). Chemically speaking, 'pure' is used more specifically: to refer to one single substance that is the outcome of a process such as separating components of a mixture, or the result of the steps of a chemical synthesis that produce a single product. A contributory factor to this misconception is that students do not see, or consider the existence of, particles but view all matter as continuous.

Hence, although students may make accurate macroscopic observations of changes over the relatively short time span of a science lesson, they may not understand what causes the changes. Making the sub-microscopic events that generate the observed changes clear is essential. Visual images of particles such as molecular models and animations that are clear and unambiguous aid students' assimilation of new scientific ideas and clarify muddled thinking. A teacher's role is to cultivate accurate understanding of events.

Progression

Children begin formal learning about substances in primary education and, of course, learn informally about how substances behave through their lived experiences. In school, they may group substances by properties including physical state, density and ability to conduct electricity or act as a thermal insulator. Children will become aware of state changes such as melting, evaporation, freezing and condensation, but are unlikely to begin to label these prior to age 9–11. The notion of reversibility of state changes may be developed in this age group as children observe that water is able to freeze, melt and re-freeze.

Chemical changes are introduced in upper primary education as changes to materials. Children aged 9–11 may observe changes occurring when substances are burned in air and in an applied sense via baking and cooking items such as bread and cakes. The notion that new materials are produced and that these changes are not usually reversible is included.

On starting secondary education, curricula frequently present chemical changes to students as the rearrangement of atoms which can be represented using formulae and equations. Specific examples of chemical changes that students may meet include combustion, thermal decomposition, oxidation and displacement reactions. Acids and bases are introduced (see Chapter 7), together with the pH scale to measure acidity and alkalinity and the use of indicators to indicate if substances are acidic or alkaline. Study of the formation of salts in reactions between metals and acids and acids and alkalis is also commonplace at this point. The role of catalysts in increasing the rate of chemical change may be discussed.

In the middle years of secondary education, students aged 14–16 build on their prior learning to develop their range and depth of knowledge about chemical change. Their understanding of how to represent chemical changes using equations, formulae and chemical models also develops, as well as their conceptual understanding of changes occurring to particles in chemical bonding. The aim is to create a firm foundation for advanced knowledge.

Post-16, students extend their understanding of types of chemical change further, adding knowledge of reaction mechanisms, measurement of concentrations and yield of product, and tracking progress using techniques such as titration. They are likely to undertake a series of chemical changes to produce a product and learn how advanced analytical techniques can be applied.

 Chemical changes

Introducing chemical changes

In a chemical change (or reaction), atoms (or ions) are rearranged to form one or more new substances. Once reacted, the starting substance(s), the reactant(s), do not exist. This does not mean that material has disappeared: the total amount of material has not changed. Products form from the material originally present in the reactants. Burning a piece of magnesium ribbon in air, around 20% of which is oxygen, illustrates this. The starting substances are magnesium and oxygen gas. Magnesium is a metal with a shiny grey appearance. Magnesium ribbon is malleable, that is, it bends into shapes. Magnesium conducts electricity and has a melting point of 650 °C. Oxygen gas is colourless and odourless at room temperature and pressure. Oxygen has a boiling point of 54.36 K (about −219 °C). When magnesium ribbon is held in a Bunsen burner flame, it combusts in oxygen producing a white powder, magnesium oxide. Magnesium and oxygen do not exist after the reaction. Magnesium oxide was not present before the reaction. Magnesium oxide has very different properties from magnesium and oxygen: it is an insulator with a melting point of 2852 °C, a hygroscopic solid (that is, it absorbs water from the atmosphere) and is used in indigestion remedies. A chemical reaction has taken place in which two substances combined to form a new substance that was not present before. In this reaction, the atomic nuclei are unchanged, but electrons are transferred from magnesium atoms to oxygen atoms (see Chapter 5), forming the ions Mg^{2+} and O^{2-}. The ions bind together by electrostatic attraction, forming magnesium oxide, MgO. The ions are arranged in a lattice structure, not paired in molecules of MgO.

School science features chemical reactions that have an obvious change, such as producing a gas, exhibiting a colour change, or generating light or/and heat. These can provide variety and drama but, besides observing the change, students need to understand the chemical explanation for what has occurred: that one or more new substances have been produced by rearranging particles from the reactants. Without explicit explanations, this may not be obvious. This activity helps to make the point that a chemical reaction involves producing a new substance, and that the particles in the product differ from those of the starting substances, the reactants. Chemical reactions occur by transfer of electrons or hydrogen ions (proton transfer) or by sharing electrons between atomic nuclei (see Chapter 5).

Kalium and plumbate

Table 2.1 describes an activity adapted from der Vos and Verdonk (1985) and Barker (2002) that forces students to acknowledge a new substance forms in a chemical reaction. The activity can be a class experiment or a teacher demonstration. The reaction is between two white crystalline solids, potassium iodide, labelled kalium, and lead(II) nitrate, labelled plumbate. These false names minimise confusion that may arise from using scientific names. The solids react as follows:

potassium iodide + lead nitrate → lead iodide + potassium nitrate

$$2KI\,(s) + Pb(NO_3)_2(s) \rightarrow PbI_2(s) + 2KNO_3(s)$$

When the reactants combine at room temperature, lead(II) iodide is formed. As this is a bright yellow solid, a dramatic colour change is seen. Potassium nitrate is another white crystalline solid.

Preparation

In advance of the lesson, grind samples of potassium iodide and lead(II) nitrate in separate mortars using clean pestles. Place small samples in sealed containers, labelled kalium and plumbate respectively, for students to use.

If doing step 4, prepare sodium chloride by grinding it in a pestle and mortar and placing small samples in sealed containers labelled salt for students' use in the lesson.

Hazards

Lead compounds are harmful if inhaled or swallowed; doing so risks damage to an unborn child and possibly fertility. They may also cause damage to organs through prolonged or repeated exposure. Lead compounds are very toxic to aquatic life and have long-lasting effects. Wear eye protection. Wash hands thoroughly after the experiment. The reaction mixtures must be collected together at the end for recycling or waste disposal. See also http://science.cleapss.org.uk/Resource/SSS043-Lead-and-its-compounds.pdf

2 Chemical change

Table 2.1 Introducing the concept of a chemical reaction: kalium and plumbate

Step	Experiment	Questions	Possible student responses	Discussion points
1	**Kalium and plumbate** Shake the samples of kalium and plumbate separately. Open the containers, then tip the kalium into the plumbate. Close the container and shake again.	What do you notice about kalium and plumbate when shaken separately? Where did the yellow substance come from? Who put the yellow stuff in the container?	Kalium and plumbate stayed white, they became more powdery. I don't know, the yellow stuff just appeared. The white powder grains broke open and the yellow stuff came out. One powder was a bit yellow at the start and the colour got stronger when they were mixed.	The white powders react to form lead(II) iodide which is yellow, and potassium nitrate, which is white. The reaction occurs between the solids. Make it more difficult for students by not giving away the answer (above). Reject the incorrect answers: for example, if the powders contained the yellow stuff it would be seen in the first stage. Force students to say, 'No one put it there, it was formed when the powders were mixed.' This is the first step to recognising a new substance forms in a chemical reaction.
2	**Adding plumbate then kalium to water** Add a tiny amount of plumbate to the edge of a petri dish containing water. Then add a tiny amount of kalium to the dish at the opposite side, using a clean spatula.	What do you notice? How did the yellow substance form? What does this tell us about particles of kalium and plumbate?	A line of yellow solid forms in the dish. (The appearance of this line is often a surprise.) The yellow stuff formed because particles attracted each other across the dish. Particles must be tiny and can move in water.	A thin yellow line appears about halfway across the dish and thickens due to the same reaction as in Step 1 but here it happens in solution. Reinforce the substance is new, formed in a chemical reaction. Introduce the notion of scale: as we could not see particles moving, they must be very small. The attraction and particle movement points are addressed in the next step.

2.1 Chemical changes

Step	Experiment	Questions	Possible student responses	Discussion points
3	**Adding kalium then plumbate to water** Add a tiny amount of kalium to the edge of a petri dish containing water. Wait 3 minutes, then add a tiny amount of plumbate.	What do you notice? Why is this different from what happened in Step 2? What does this tell us about the particles of kalium and plumbate?	The yellow stuff forms immediately. The particles cannot attract each other across the dish. Kalium particles spread out in the dish. Plumbate reacts with kalium to make the yellow stuff where the particles meet.	The yellow substance (lead(II) iodide) forms immediately. Particles moved through the water, forming lead(II) iodide precipitate. Introduce the term diffusion (see Chapter 3) to describe particles moving in water.
4	**Salt and plumbate** (optional) Repeat Step 1 with samples of salt and plumbate. Tip one into the other. Repeat Step 2, using salt and plumbate.	What do you notice? After Step 1: has a chemical reaction occurred? After Step 2: now explain what happened in Step 1. What does this tell us about observations?	Nothing changed. The two substances stayed white both separately and together. I don't know if a chemical reaction occurred. A chemical reaction occurred because white stuff appeared in the dish. We cannot always see that a chemical reaction has occurred.	A white solid, lead(II) chloride, is produced. This is difficult to see in the container. When the reaction is repeated in water, a white precipitate forms in the dish. This confirms that a chemical reaction occurred. It tells us chemists need evidence from tests to confirm the products of a reaction.

2 Chemical change

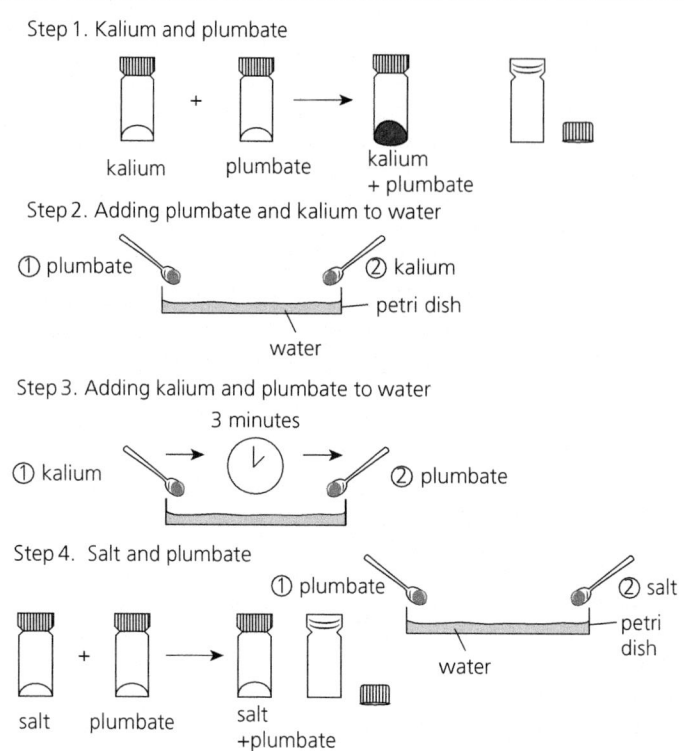

Figure 2.2 Kalium and plumbate activity

When students see the initial reaction, allow thinking time for them to understand the reason for the colour change. Emphasise the tiny size of the particles in Step 2. In Step 3, students are asked to wait several minutes before adding the second substance. This helps them realise time is required for chemical reactions to occur. In Step 4, different reactants are used and the chemical reaction is not immediately apparent, leading students to realise other tests will determine the outcome. Questions and discussion points are included in Table 2.1. Figure 2.2 shows the equipment and instructions.

Examples of chemical reactions

Table 2.2 summarises five chemical reactions. References to instructions are provided together with questions and discussion points. The reactions are briefly described.

Aluminium and iodine

In this teacher demonstration, two elements react completely to produce a new substance, aluminium iodide. After an initiation period, the chemical change from a grey solid and a purple solid to a white solid is dramatic, producing heat and light.

Iron and sulfur

This reaction can be a class experiment. The chemical change from a metallic grey solid and a yellow solid to a dark grey solid occurs with heat produced as a red glow. The metallic grey solid is attracted to a magnet. The grey solid produced is not attracted to a magnet.

Hydrogen and oxygen

In this demonstration, two well-known colourless gases are ignited in an enclosed space, a plastic bottle. The explosion fires the bottle around 10–15 m. Droplets of the product, liquid water, can be found in the bottle.

Candle wax and oxygen

Students can track the change in mass of a candle as it burns. Students often think the wick burns while the wax just melts. Observing the candle flame is worthwhile: the lower blue area is where combustion is occurring. The luminous yellow upper area comprises unburnt carbon particles from the wax that glow brightly in the heat produced by the reaction. These condense on a cold surface placed in the flame. Ask students how they could capture the products of combustion: water and carbon dioxide.

Hydrogen peroxide decomposition

This is a decomposition reaction that occurs slowly. In this demonstration, the rate of decomposition is enhanced by adding a catalyst. One product, oxygen, is produced as a gas, creating a foam. Water is the second product.

2 Chemical change

Table 2.2 Simple chemical reactions

Reaction	Description	Questions and discussion points	Reference
Aluminium and iodine Demonstration, fume cupboard required. 0.4 g iodine and 0.1 g aluminium are ground separately then mixed by pouring repeatedly from one container to the other. The mixture is a flammable solid that is harmful in contact with skin or if inhaled, and toxic to aquatic life. Flammable gases are released.	Two elements, aluminium powder and powdered iodine are mixed in a tin lid. A few drops of warm water are added. A spectacular exothermic reaction occurs producing clouds of iodine vapour and a white solid, aluminium iodide, in the tin lid. $2Al(s) + 3I_2(s) \rightarrow 2AlI_3(s)$	What is the purple vapour? *Excess iodine vapourises in the heat produced by the reaction. Iodine vapour is purple.* Why is water added? *Water acts as a catalyst. It wets the powdered reagents allowing the particles of aluminium and iodine to make contact. The water is vapourised in the reaction.* What happens to the aluminium and iodine? *Some iodine is vapourised. The aluminium reacts with the rest of the iodine to form aluminium iodide. The reactants do not exist after the reaction and the product is a new substance.*	https://edu.rsc.org/resources/reaction-between-aluminium-and-iodine/715.article
Iron and sulfur Demonstration or class practical. Eye protection essential. Student quantities are limited to 0.2 g Fe and S in an ignition tube, with the top plugged with mineral wool. If sulfur dioxide escapes, this may cause skin burns and eye damage, and it is toxic if inhaled.	Two elements, iron powder and sulfur powder, are mixed then heated. An exothermic reaction occurs producing black iron sulfide: $Fe(s) + S(s) \rightarrow FeS(s)$ Holding a magnet above a sheet of paper held over to the mixture before heating shows iron and sulfur can be separated. Iron sulfide is a new substance that is not attracted to a magnet.	What happens to the iron and sulfur? *They combine in a chemical reaction to produce iron sulfide, a black solid. The reactants do not exist after the reaction. The product is a new substance.* Why does the tube glow orange? *The reaction is exothermic. The glow is started by heating using a Bunsen flame. The glow continues as the iron and sulfur react together.* Why is a yellow vapour seen? *Some sulfur vapourises in the heat. This is seen in the tube. Some sulfur may crystallise at the mouth of the tube.*	https://edu.rsc.org/resources/iron-and-sulfur-reaction/713.article
Combustion of hydrogen: hydrogen–oxygen rocket Demonstration, preferably outdoors. Eye protection is essential. The explosion is very loud and may cause damage to hearing and buildings.	Small plastic fizzy drink bottles (the rockets) are filled with hydrogen gas and air, which are ignited electronically. An explosion occurs which fires the rocket several metres. The reaction occurring is: $2H_2(g) + O_2(g) \rightarrow 2H_2O(l)$	Why does the gas mixture explode? *The reaction is exothermic producing a large amount of heat. The heat vapourises the water to water vapour which expands massively in a very short period of time, creating the explosion.*	https://edu.rsc.org/resources/a-hydrogen-powered-rocket/1705.article Barker and Hadi-Talab, (2005)

2.1 Chemical changes

Reaction	Description	Questions and discussion points	Reference
	Changing the amounts of each gas will change the outcome: a stoichiometric mixture of 2:1 hydrogen: oxygen produces the most dramatic results.	What happens to the hydrogen and oxygen? *The hydrogen and oxygen react to produce water. They do not exist after the reaction.* Why does the size of the explosion change? *The equation shows that hydrogen and oxygen react in a 2:1 ratio. This mixture produces the most dramatic explosion.*	
Burning a candle Demonstration or class practical. Use tea lights if this is a class practical. Use a large pillar candle if a demonstration is preferred. Dim lighting in the room.	Candle wax is paraffin wax, a hydrocarbon comprised of only carbon and hydrogen atoms. When a candle burns, this reaction occurs at the base of the wick where a blue flame can be seen: $C_{25}H_{52}(l) + 28O_2(g) \rightarrow 15CO_2(g) + 26H_2O(g) + 10C(s)$ Carbon particles (soot) absorb heat and light, creating a luminous yellow flame above the wick. The temperature at the centre of the flame is about 1000 °C. This is a combustion reaction.	What is burning? *The candle wax is burning in oxygen. This is a chemical reaction. The wick does not burn.* What is the flame? *Students may say heat, burning, bacteria, wick, candle or oxygen. It is actually carbon particles from the wax.* Why does the candle mass decrease? *Students may say wax evaporates; turns into carbon or ash; burns away. In fact, the wax combusts in oxygen. It does not exist after the reaction. The products are carbon dioxide and water, which are gases at room temperature and pressure.*	Barker (2002) Unit 6
Decomposition of hydrogen peroxide: elephant toothpaste Demonstration or class investigation. Use only 20 volumes of hydrogen peroxide in the class investigation. Catalysts must be checked for potential hazards. Eye protection is essential.	Hydrogen peroxide decomposes to form water and oxygen gas $2H_2O_2(l) \rightarrow 2H_2O(l) + O_2(g)$ The reaction happens very slowly at room temperature. Adding a catalyst increases the rate of decomposition. In the demonstration, adding washing up liquid creates a foam, which is 'toothpaste'. The investigation studies the effect of changing catalysts on rate of decomposition.	Can we prove the gas is oxygen? *Yes, by collecting a sample of gas and testing with a lighted splint, which should relight.* What does the catalyst do? *The catalyst provides a surface on which hydrogen peroxide molecules can collide. More collisions are successful with the catalyst than without. Decomposition occurs very slowly without the catalyst.* What happens to the hydrogen peroxide? *Hydrogen peroxide molecules decompose, that is, split into water and oxygen. The hydrogen peroxide does not exist after the reaction.*	http://science.cleapss.org.uk/resource-info/sra011-spectacular-decomposition-of-hydrogen-peroxide-to-produce-a-foam-catalysed-by-potassium-iodide-elephant-s-toothpaste.aspx https://www.york.ac.uk/chemistry/schools/chemrev/projects/peroxide/ Barker and Hadi-Talab (2005)

 Chemical change or physical change?

School science often introduces students to the notion of chemical and physical changes. In general, a chemical change involves a reaction resulting in the formation of an entirely new substance (or substances), while a physical change results in the same substance (or substances) being present in a different phase or state. This sounds straightforward but is actually surprisingly difficult to establish consistently in practice. An initial example that can be discussed is the states of 'hydrogen monoxide', which are ice, water and steam.

$$H_2O(s) \rightarrow H_2O(l) \rightarrow H_2O(g)$$
$$\text{ice} \rightarrow \text{water} \rightarrow \text{steam}$$

One argument, which chemists would not accept but that some students may believe, is that the properties of ice, water and steam are so different that they can be regarded as unique substances, so chemical rather than physical changes occur when one is formed from another. Importantly, the molecular formula of water is unchanged in all three, so the changes are not chemical in the same way that in the reaction between magnesium and oxygen to produce magnesium oxide (see above). Understanding this difference is essential.

However, many events do not easily fit this distinction. Additional criteria that attempt to distinguish between chemical and physical changes are shown in Table 2.3. Of these, formation of one or more new substances with a change in chemical formulae is the most distinctive characteristic of a chemical change. Any new substance(s) will have different physical properties to those of the original reactants, including melting and boiling points, structure, conductivity and general appearance. Students can observe substances before and after an event to assess if a physical or a chemical change has occurred, noting if these properties are the same or have changed.

The remaining criteria are harder to apply consistently. For example, not all chemical changes involve large energy changes and/or they may occur slowly, over extended time periods. Rusting is one such change: iron reacts so slowly with oxygen, it is very hard to determine whether energy is transferred or not. The notion of reversibility is more reliable, as physical changes tend to be relatively easy to reverse. Even so, appearances tend to alter when a physical change is reversed, so dependence on this criterion alone may be unconvincing. The chocolate and egg activity below illustrates this.

2.2 Chemical change or physical change?

Table 2.3 Criteria for distinguishing between chemical and physical changes

Chemical change	Physical change	Determined by
A new substance (or substances) is (are) produced.	The same substance is present before and after the change.	Knowledge of the chemical formulae and molecular structure of the substances.
Often involves a large change in energy level from reactants to products.	Energy change is small-scale and gradual.	Measuring or observing change in energy – heat and light being produced implies a chemical change.
Irreversible under normal laboratory conditions.	Reversible under normal laboratory conditions.	Testing for reversibility: a gas may be condensed and a liquid boiled. Note that, theoretically, all chemical changes can be reversed but the conditions for reversal are not usually room temperature and pressure.
Involves breaking bonds within molecules.	Involves breaking bonds between molecules.	Tracking a change by examining the particles present at different stages. This is very difficult, if not impossible, in a school laboratory but can be done by professional scientists.

Dissolving can be classified as a physical or a chemical change depending on circumstances. Many examples in which dissolving occurs are easily reversed, so can be considered physical changes. Examples include: pigments in ink dissolved in water being recovered by evaporation; components of crude oil (liquids dissolved together in a complex mixture) separated by fractional distillation (see Chapter 10); a liquid dissolved in a second liquid, such as ethanol and water, separated by heating then condensing the vapour of the liquid with the lower boiling point (see Chapter 9); a gas dissolved in a liquid, such as oxygen in water, released by gentle heating; sugar dissolved in a drink recovered by evaporation. In these examples, the chemical identity of the substances involved is unchanged in their dissolved and separated states, so the situations meet another criterion of physical changes. However, when an ionic solid (see Chapter 5), such as sodium chloride (NaCl) dissolves in water, it does not dissolve as molecules of, in this case, NaCl. Ions in the solid dissociate, that is, they separate from each other. Sodium chloride crystals dissociate into sodium and chloride ions (Na^+ and Cl^-). The solid vanishes from view because individual ions are too small to see. This may lead some students to believe that the mass of a solution is less than the mass of the solid (solute) and the solvent (water). Individual dissociated ions are surrounded by water molecules, so aqueous sodium chloride (sodium chloride dissolved in water) does not have the same structure as solid sodium chloride. Thus, this is a chemical change. If the water surrounding the ions is evaporated, the water molecules change state to steam, leaving the ions as solid matter, having reformed a crystal structure. The evaporated salt structure is often not as precise as the original, in part because some water molecules become incorporated in the new structure as 'water of crystallisation'.

2 Chemical change

Students are often taught separation techniques leading to recovering a salt by evaporation. The notion of describing this process as making pure salt from rock salt needs care: the rock salt is pure rock salt, as only rock salt is present in the sample. The salt produced in a separation experiment is unlikely to be pure sodium chloride. This is because the techniques used in a school laboratory are not sufficiently sophisticated to remove every impurity from the rock salt to leave 100% sodium chloride. This is hard to explain convincingly to students.

Some chemical reactions are readily reversible (see Chapter 6). This means it is not always true that a chemical change cannot be reversed. So using this as an absolute principle is problematic.

Understanding the fourth criterion, bond-breaking, requires secure understanding of chemical bonding (see Chapter 5). Even then, there are exceptions to the general rule as some elements, such as carbon and phosphorus, exist in different forms, known as allotropes. Changing from one allotrope into another requires bonds to be broken and formed, creating new substances with very different properties – diamond, graphite and buckminsterfullerene in the case of carbon. Melting and boiling metals requires breaking of metallic bonds, which are a form of intramolecular bond. Discussion of this criterion could get very detailed, requiring advanced chemistry to progress.

For these reasons, teaching chemical and physical changes is potentially challenging. Perhaps as a consequence, some chemistry educators prefer to ignore the topic. However, it is common in science curricula and so requires attention. The best advice is to ask students to consider broad generalisations, then illustrate exceptions that break the 'rules'. This allows for establishment of a framework within which variation is acceptable. The following activities provide introductions and revision opportunities.

KEY ACTIVITY

The chocolate and egg experiment

This experiment provides good evidence of a difference between a chemical and physical change. Samples of grated chocolate (no tasting!) and liquid albumin (egg white) are placed in separate test tubes which are then heated in a water bath. The water bath is a beaker of cold water heated slowly using a Bunsen burner. Full details of this class experiment are available at https://edu.rsc.org/resources/chocolate-and-egg-experiment/441.article.

Students observe that solid chocolate melts and transparent jelly-like albumin cooks to an opaque white solid. When the test tubes are removed from the hot water and cooled to room temperature, chocolate solidifies while albumin remains a solid white substance.

2.2 Chemical change or physical change?

Freshly grated chocolate does not solidify to its pre-melted structure. This potential source of confusion can be eliminated by using chocolate that has been melted and solidified before the lesson. When students re-melt melted chocolate, it does solidify back to the same structure. Note that previously melted chocolate takes longer to become liquid than un-melted chocolate.

Questions to ask
- What do you notice about the physical states of the chocolate and egg white before heating?
- What happens when the egg white and chocolate are heated?
- What happens when the egg white and chocolate are cooled?
- Explain why the chocolate and egg white behave differently.

Once students have grasped the basic principle shown in the practical, develop this by discussing additional examples, such as those listed below. For each example, ask students if they think it is a physical change (P) or a chemical change (C), with a reason. Do not reveal the correct answer (here shown next to each example) until they have given theirs. Be prepared to discuss their viewpoint. Focus on whether or not the change could be reversed and, if so, how; and if the change produces a new substance with different particles to those present before the change occurred.

Frying an egg	C
A banana going black	C
Dissolving sugar in a hot drink	P
Washing up using detergent	C
Combining two paints to make a paint with a new colour	P
Making toast	C
Adding water to a fruit juice drink	P
Using fuel when driving a car	C
Cleaning teeth using toothpaste	C

Changes in chemistry

KEY ACTIVITY

This worksheet activity, available from https://edu.rsc.org/resources/changes-in-chemistry/1085.article, complements the chocolate and egg experiment above. It is a diagnostic tool that supports students learning about, or revision of, chemical and physical changes. The tool includes questions that utilise a particle model of matter, focusing attention on particles present before and after a change. The tool is self-explanatory, including questions and discussion points.

2 Chemical change

> **Cross-disciplinary**
>
> Many delicious aromas and flavours in cooked food are due to chemical changes known as Maillard reactions, named after Frenchman Louis Camille Maillard, who, in 1912, published a paper about reactions between sugar, a carbohydrate, and amino acids. These occur, for example, when a slice of bread is heated to around 150 °C. At this temperature, proteins and sugars react, producing molecules including glucosamines that darken and harden on the surface, creating the burned texture and characteristic aroma of toast (from Latin *torrere*, which means 'to burn'). If the reaction continues too long, bitter-tasting acrylamides form.

2.3 Chemical formulae and equations

Chemical equations represent chemical and physical changes using formulae and symbols. These enable chemists worldwide to communicate consistently and efficiently. Chemical formulae represent specific substances, while chemical equations summarise what are often complex events. A chemical equation can be considered as a model of the reaction represented (see Chapter 1). Like all models, chemical equations have limitations. Students may find writing chemical equations challenging, as working memory is required to combine understanding of chemical concepts and recall of formulae with precision and knowledge of basic mathematics, particularly ratios. Persistence in supporting students as they learn how to write chemical formulae and equations is important as it is an essential part of chemistry and offers students a sense of 'being a chemist'. The ability to write chemical equations aids problem-solving and analysis skills, as well as application and interpretation of observations, logic and deduction.

Introducing chemical formulae

Chemical formulae act as a chemists' code which gives precision to colloquial names such as Milk of Magnesia, Epsom salts, bleach, oil of wintergreen and cream of tartar. The level of precision required demands recall of the symbols of chemical elements and meanings of various terms such as hydroxide, oxide and chloride. This activity helps introduce students to the formulae of non-metal compounds, which is a good starting point.

2.3   Chemical formulae and equations

> **KEY ACTIVITY**
>
> ## Chemical formulae and equations
>
> A Royal Society of Chemistry resource, available from https://edu.rsc.org/download?ac=517716, introduces writing chemical formulae for non-metal compounds. Misconceptions and advice about formative assessment are included. The activity is self-contained with a student worksheet and take-home points.
>
> Note that although the names of some chemicals allow their formulae to be deduced, some formulae simply have to be learned, as their names give no clues. Examples include ammonia, sulfuric acid and water. In general, the best advice to help students learn chemical formulae is to keep practising. Include formulae and equations routinely in chemistry lessons where possible, reinforcing connections to macroscopic, observable events occurring in the reactions students see.

Principles of chemical equations

A chemical equation shows the reactant(s) and product(s) of a chemical or physical change:

$$\text{reactant(s)} \rightarrow \text{product(s)}$$

The arrow symbol ($\rightarrow$) means 'change(s) to'. Reactant(s) are the substances present at the start. Product(s) are the substances present after the change. For example:

$$\text{water} \rightarrow \text{water vapour}$$

In this word equation, the arrow represents the change from liquid to gaseous water.

The word 'equation' implies being equal. Students sometimes write an equals sign (=) in a chemical equation, as they would in mathematics:

$$\text{reactant(s)} = \text{product(s)}$$

$$\text{water} = \text{water vapour}$$

An equals sign is incorrect as reactants and products are not the same, and the symbol does not mean that the amounts of reactants equals the amounts of products, so its use in chemical equations should be discouraged.

The law of conservation of mass states that matter cannot be created or destroyed, so the total mass of product(s) after the change must be the same as the total mass of reactant(s) present before the change. Similarly, the numbers of sub-atomic particles – that is, electrons, neutrons and protons (see Chapter 4) – do not change. Electrons, protons and neutrons cannot be made or lost in a reaction. The arrow means that electrons, protons and neutrons are organised differently in product(s) and reactant(s).

Care is required when representing energy in chemical equations. The law of conservation of energy states that energy cannot be created or destroyed; it can only be transferred from one form to another. In chemistry, the total energy of a *system* – that is, reactant(s) and product(s) – cannot change (Chapter 6). Secondary-aged students do not distinguish clearly between matter and energy. They may not understand that energy is transferred because they think incorrectly that energy (or heat) is a substance (see Chapter 4 in *Teaching Secondary Physics*). It is tempting to write '+ energy' or include 'heat' as a product or reactant but doing so may reinforce the misconception. To make explicit that energy is *not* a substance, write the energy term in a different colour, set apart from the equation; and, once students are secure, write a formal enthalpy change term alongside the equation as $\Delta H = x \text{ kJ mol}^{-1}$.

Word and symbol equations

Word equations are often taught first, as these represent changes without involving mathematics and chemical formulae. However, word equations present difficulties as they do not always make what is happening completely obvious. This inhibits students understanding the shift between macro-, sub-microscopic and symbolic representations (see Figure 1.1). Research suggests students may struggle to complete word equations from which only one word is missing, as they can only proceed by recall rather than application of logic.

Symbol equations represent reactant(s) and product(s) using chemical formulae. A recommendation is to teach symbol equations early, to reinforce macro–micro–symbol representations. The change of state:

$$\text{water} \rightarrow \text{water vapour}$$

is represented in symbols as:

$$H_2O(l) \rightarrow H_2O(g)$$

Students may recall the molecular formula for water (H_2O) and learn that (l) and (g) represent liquid and gas states of matter. Word and symbol equations reinforce each other. Presenting both supports students' learning by giving the same information in multiple forms.

Balancing equations

A chemical equation must balance: that is, numbers of particles of reactants and products are equal. The reaction between hydrogen and oxygen gases illustrates the steps students should learn to balance equations. The word equation is:

$$\text{hydrogen} + \text{oxygen} \rightarrow \text{water}$$

This tells us water is the product but offers limited information. The symbol equation is:

$$H_2 + O_2 \rightarrow H_2O$$

This says: 'one molecule of hydrogen and one molecule of oxygen change to one molecule of water'. Although correct in principle, the equation is not balanced, as one oxygen atom is missing. One hydrogen molecule comprises two hydrogen atoms, as indicated by subscript 2 after the symbol for hydrogen, H. Both hydrogen atoms are present in the product, one molecule of water, H_2O. The other reactant, an oxygen molecule, also comprises two atoms, this time of oxygen, but only one is present in the water molecule. No other product was formed. The law of conservation of mass states atoms cannot be destroyed. This symbol equation does not show every atom in the reactants is present in the product.

Students often solve this by conveniently 'forgetting' a missing atom or adding it to a product. Forgetting the second oxygen atom leaves the equation unbalanced. Adding an oxygen atom to water makes H_2O_2. This fundamentally changes water to another compound as H_2O_2 is the formula for hydrogen peroxide. This is not correct, because *a chemical formula cannot change*. The equation must be balanced without changing formulae or losing an atom. Two ways of balancing this equation are:

1. Use stoichiometric coefficients
   The word 'stoichiometric' (pronounced with a hard 'k'– stoy-keyo-metric) combines two Greek words: *stoicheion* meaning 'element' and *metron* meaning 'measure'. Stoichiometric coefficients give precise information about a reaction. In this example, the number of hydrogen molecules and water molecules can be doubled by placing a large 2 in front of H_2 and H_2O. The equation already shows one oxygen molecule. By convention, a large 1 is never shown:
   $$2H_2 + O_2 \rightarrow 2H_2O$$
   The large 2s and the (invisible) 1 for oxygen are *stoichiometric coefficients*. The large 2s mean the equation reads, 'two molecules of hydrogen and one molecule of oxygen react to form two molecules of water'. Counting atoms, the equation shows four atoms of hydrogen (two in each molecule) react with two atoms of oxygen to make two molecules of water, which together comprise four hydrogen atoms and two oxygen atoms. This equation is 'balanced'.

2. **Use an empirical approach**
   Hydrogen and oxygen are both gases composed of diatomic molecules – that is, molecules comprising two atoms. The equation shows one molecule of hydrogen (that is, two atoms) reacts with one atom of oxygen. One atom of oxygen is half of one molecule, so can be shown as $½O_2$:
   $H_2 + ½O_2 \rightarrow H_2O$
   This equation shows reacting ratios and *empirical* formulae. This avoidance of stoichiometric coefficients may seem easier. However, it introduces the notion of 'half a molecule', which is hard to grasp.

When teaching symbol equations, the first method is preferred until students have a secure understanding of what chemical equations represent. Continued practice is required to raise confidence and build secure knowledge. Freely available online resources support students' learning. This resource, https://edu.rsc.org/download?ac=140227 introduces the principle of writing chemical equations from a piece of text.

> ### Enrichment
>
> Notation for chemical equations was first developed in the seventeenth century. French chemist Jean Beguin is acknowledged as the first scientist to draw a reaction diagram (in 1615) showing changes occurring between compounds known today as mercury(II) chloride ($HgCl_2$) and antimony sulfide (Sb_2S_3). In 1757, Scottish chemist William Cullen pioneered use of the arrow, $\rightarrow$, to symbolise 'affinity force', an early explanation of chemical change. In 1775, the Swedish chemist Torbern Bergman used the arrow to signify change from reactants to products. By the mid-nineteenth century, chemists had adopted symbols for chemical elements to use when writing chemical equations.

2.4 Modelling chemical changes

Modelling was introduced in Section 1.6. Here, the focus is on applying molecular models to aid students' understanding of chemical reactions. Students benefit from seeing diagrams and images of molecules and manipulating molecules to track and trace what happens in a chemical reaction. These support understanding of the formation of products from reactants, connecting the symbolic equation for the reaction with observable, macroscopic events. Placing images of molecular models alongside the equation and

2.4 Modelling chemical changes

photographs of the reaction itself can help. For example, the reaction between methane and oxygen can be represented as shown in Figure 2.3.

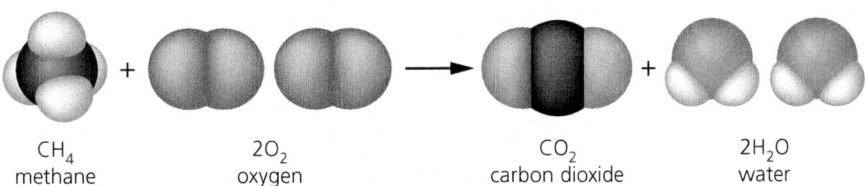

CH$_4$
methane

2O$_2$
oxygen

CO$_2$
carbon dioxide

2H$_2$O
water

Figure 2.3 A pictorial representation of the reaction between methane and oxygen

The image complements word and formulae equations, showing the molecules involved and the meaning of the stoichiometric coefficients. This model is space-filling, as no lines are drawn within molecules to represent chemical bonds. The shading shows that atomic cores are retained, as there are one carbon, four oxygen and four hydrogen atoms on both sides of the 'change to' arrow. The model misleadingly gives the impression that the atoms involved are genuinely unchanged: for example, the carbon atom in methane 'relocates' and becomes the carbon atom in carbon dioxide. This is a deceptive simplification: atoms are not full, intact atoms when bonded within molecules as electrons are shared across atomic boundaries. So it is important to make students aware this is a model, not reality.

Models can be constructed from Molymod® kits (see Section 2.8). Besides allowing students to track reactants becoming products in a reaction, Molymod illustrates molecular shapes by providing grey connectors to represent bonds which join atoms that have appropriate numbers of holes at correctly spaced points. This is a sophisticated version of ball-and-stick models. Hence, a Molymod water molecule is V-shaped while a methane (CH$_4$) molecule is tetrahedral. For some molecules, such as oxygen (O$_2$), long flexible connectors are needed to get the right shape, due to positioning of holes in the atoms. Although Molymod models are useful manipulatives to start students thinking about molecules, reactions and molecular shapes, they share some limitations of space-filling models, especially in terms of the colours of atoms and relative atom sizes, and may generate some misunderstandings (see Chapter 1), so these should be used with appropriate caveats.

To supplement hand-held models, animations that show movement of particles in chemical reactions are valuable. Good examples are available via the Australian website http://vischem.com.au/index.html

2 Chemical change

KEY ACTIVITY

Introducing molecular models

This activity, which can be freely downloaded from www.tes.com/teaching-resource/balancing-chemical-equations-molymod-activity-6409214, offers an introduction to using Molymod kits.

2.5 Measuring chemical change

Professional chemists measure amounts of substances accurately. This avoids waste by maximising the amount of product(s) formed while minimising excess reagents. The need for a measuring unit for substances arises because particles are extremely small. We cannot count or find the mass of 100, 1000 or 1 000 000 particles under normal conditions. To convey this, imagine that cooking a meal required counting exact numbers of beans or grains of rice, flour, salt or sugar: try counting 1000 rice grains. Discuss why this is practically impossible: the grains are too small to count accurately with ease. Instead, measurements are made using scales of volume and mass, calibrated in grams, kilograms, litres and millilitres. We count objects in multiples such as dozen (12), score or quire (20), gross (144) and ream (500). Cooking relies on correct proportions of ingredients: a useful illustration could be the same cake made several times, using differing ratios of flour, sugar, fat and eggs. Proportions and amounts must be accurate. Cakes made with incorrect ratios are (usually) inedible.

Similarly, chemists need a convenient, reliable measuring unit that enables chemicals to be measured in fixed proportions. The chosen unit is the *mole* (mol), an abbreviation of the German word for molecule, which is *molekül*. German chemist Wilhelm Ostwald (1853–1932) suggested the mole in 1894, defining it as 'the relative atomic or molecular mass of a substance weighed in grams'. Atomic mass (symbol: A_r) and molecular mass (M_r) are explained below.

An essential piece of information is that *one mole of any substance contains exactly the same number of particles*. This number is the Avogadro constant, referred to as Avogadro's number, named after Italian scientist Amadeo Avogadro (1776–1854). Particles are very small, so the number of particles in one mole is huge: it is defined by the International Union of Pure and Applied Chemistry (IUPAC) as $6.02214076 \times 10^{23}$. One mole of any substance contains this number of particles. If this seems strange, recall that one dozen means twelve regardless of the item.

Students often find learning the mole challenging as understanding why Avogadro's number applies to a mole of any substance, as well as its colossal size, can be difficult. They must connect a chemical reaction with an equation representing it and know the numbers of moles of reactant(s) and product(s) and the relative atomic or molecular masses for each substance involved. The topic requires understanding of basic mathematical operators, ratios, rearrangement of mathematical equations and standard form. This places demands on working memory and secure understanding of underpinning chemical concepts is a prerequisite.

Introducing moles

This is a three-stage process, starting by introducing reacting mass ratios, progressing to understanding that one mole of any substance contains Avogadro's number of particles, then consolidating this understanding using the iron and sulfur reaction.

Reacting mass ratios

In chemical reactions, reactants and products are in constant ratios. As explained above, hydrogen and oxygen react in a 2:1 molecular ratio when gases at room temperature and pressure are ignited to produce water, never hydrogen peroxide.

Recall that the periodic table (Chapter 4) arranges elements in atomic number order. The atomic number gives the number of protons present in atomic nuclei. All atoms, except those of hydrogen, also have one or more neutrons in their nuclei. Neutrons and protons have very similar masses. In measuring atomic mass, protons and neutrons are assigned an atomic mass unit (a.m.u.) of 1. Electrons have very little mass. The mass of an atom is therefore almost entirely the mass of its nucleus. The mass of an atom in atomic mass units is referred to as the atomic mass, symbol A_r. Hydrogen and oxygen have an A_r value of 1 and 16 respectively: an oxygen atom is 16× heavier than a hydrogen atom. The molecular mass, symbol M_r, is used for a molecule. To find an M_r value, add together the A_r value for each element present in a molecule. For water, for example, the M_r value is 18 (H: 2 × 1 = 2; O: 1 × 16 = 16; 2 + 16 = 18). This means the mass of one mole of water is 18 g.

2 Chemical change

In chemical reactions, substances react in fixed ratios. The stoichiometric coefficients of 2 for hydrogen and 1 for oxygen in the chemical equation for the reaction:

$$2H_2 + O_2 \rightarrow 2H_2O$$

(see above) correspond to *numbers of moles* of hydrogen and oxygen reacting. Previously we read the equation as 'two molecules of hydrogen react with one molecule of oxygen to produce two molecules of water'. We can also read the equation as, 'two moles of hydrogen react with one mole of oxygen to produce two moles of water'. This works because one mole represents a fixed number of particles. In practice, this means two moles of hydrogen, 4 g, react with one mole of oxygen, 32 g; or 2 g of hydrogen, (one mole) reacts completely with 16 g of oxygen (half of one mole); or 8 g hydrogen reacts with 64 g oxygen, and so on. This ratio is fixed and cannot be changed. The reacting mass ratio for the reaction between them is 2:16, which simplifies to 1:8, or 0.125 as a decimal. If masses of hydrogen and oxygen react in a ratio outside this fixed arrangement, an excess of one or more reagents will remain. This principle is true for all reactions. Table 2.4 introduces examples for use with students. Understanding reacting mass ratios is a valuable step preceding introduction of the mole as a counting unit. There is no need to mention the mole when working through these examples. However, students can be invited to comment on the numbers, comparing these to the A_r values given in the periodic table as appropriate. Questions and solutions are included in the table.

2.5 Measuring chemical change

Table 2.4 Introducing reacting mass ratios

Reaction	Equation	Questions	Solutions
Magnesium and oxygen	$Mg + \tfrac{1}{2}O_2 \rightarrow MgO$ M_r values 24 16 $24 + 16 = 40$ Ratio Mg:O $24:16 = \mathbf{1.5}$	How much oxygen would react with 3 g, 6 g, 12 g, 24 g, 600 g and 3 kg of magnesium? What would be produced when 35 g magnesium reacts with 16 g oxygen?	Divide each mass of magnesium by 1.5. Answers are: 2 g, 4 g, 8 g, 16 g, 400 g, 2 kg 16 g of oxygen will react with 24 g of magnesium to give 40 g magnesium oxide, leaving (35 −24) = 11 g magnesium unreacted (in excess).
Hydrogen and oxygen	$H_2 + \tfrac{1}{2}O_2 \rightarrow H_2O$ M_r values $(2 \times 1) = 2$ 16 $(2 \times 1) + 16 = 18$ Ratio H:O $2:16 = \mathbf{0.125}$	How much hydrogen would react with 1 g, 4 g, 8 g, 16 g, 32 g, 320 g and 1.6 kg of oxygen? What would be produced when 40 g oxygen reacts with 4 g hydrogen?	Multiply each mass of oxygen by 0.125. Answers are: 0.125 g, 0.5 g, 1 g, 2 g, 4 g, 40 g, 200 g 4 g of hydrogen will react with 32 g of magnesium to give 40 g water with 8 g excess oxygen.
Aluminium and iodine	$Al(s) + 1\tfrac{1}{2}I_2(s) \rightarrow AlI_3(s)$ M_r values 27 $(3 \times 127) = 381$ $27 + (3 \times 127) = 408$ Ratio Al:I $27:381 = \mathbf{0.071}$ Note that not all ratios are neat numbers	How much iodine would react with 3 g, 9 g, 27 g, 54 g, 108 g, 540 g and 9 kg of aluminium? What would be produced when 500 g iodine reacts with 30 g aluminium?	Divide each mass of aluminium by 0.071. Answers are: 42.3 g, 127 g, 381 g, 762 g, 1.521 kg, 7.6 kg, 126.8 kg 30 g Al will react with 30 ÷ 0.071 = 423 g I_2 to produce 30 + 423 = 453 g AlI_3 with 77 g iodine in excess.
Iron and sulfur	$Fe + S \rightarrow FeS$ M_r values 56 32 88 Ratio Fe:S $56:32 = \mathbf{1.75}$	How much iron would react with 4 g, 8 g, 16 g, 32 g, 128 g, 256 g and 8 kg sulfur? What would be produced when 112 g iron and 80 g sulfur react?	Multiply by 1.75. Answers are: 7 g, 14 g, 28 g, 56 g, 224 g, 448 g, 14 kg. 176 g iron sulfide with 16 g sulfur in excess

KEY ACTIVITY

One mole of any substance contains Avogadro's number of particles

Table 2.5 gives a method for introducing the mole as the chemist's unit to measure amount of substance, leading to showing that one mole of any substance contains the same number of particles. Follow the steps in whole class discussion. The sequence enables students to connect measuring atomic or molecular masses in grams with the concept of a mole, realising that one mole of any substance contains Avogadro's number of particles. At Step 8, the best outcome is achieved by giving students plenty of thinking time. Do not be tempted to give the answer. If progress is slow, start again from Step 1. Eventually, one or two students will realise the answer to the crucial question, then others will want to understand too. Invite students who first understood this point to give a peer-to-peer explanation. Once students are secure in their understanding that one mole of any substance contains Avogadro's number of particles of that substance, usually rounded to 6.02×10^{23}, reinforce this with further examples. For example, one mole of water, 18 g, contains 6.02×10^{23} molecules of water and $3 \times 6.02 \times 10^{23}$ atoms; one mole of sodium chloride (NaCl) contains one mole of sodium ions (Na^+) and one mole of chloride ions (Cl^-); one mole of calcium fluoride (CaF_2) contains one mole of calcium ions (Ca^+) and two moles of fluoride ions (F^-), that is, 1.2×10^{24} F^- ions.

2.5 Measuring chemical change

Table 2.5 Introducing the mole as a unit of measurement of substance

Step	Instruction	Points to note
1	Make two columns on a board, one headed, for example, Cu and the other S. Write the atomic mass values for each element under the symbol (see between Steps 4 and 5 below)	The copper–sulfur pair is convenient because these numbers (64 and 32) are relatively straightforward to follow if the copper A_r value is rounded to 64 from 63.5. Also, the copper–sulfur reaction can be demonstrated relatively easily. Other element pairs that have atomic mass values in simple whole number ratios are: calcium and bromine (40:80, ratio 1:2); magnesium and carbon (24:12 ratio 2:1); titanium and carbon (48:12, ratio 4:1).
2	Say that these mass values are measured in atomic mass units. Ask what the ratio of the two values is. The answer for these elements is 2:1. Write this in a third column headed 'Ratio of masses'.	
3	Imagine we now have 100 atoms – add 100 under the atomic mass values. Ask for the ratio of masses of 100 atoms. The answer is still 2:1. Write this in the ratios column.	Reinforce the ratio is constant regardless of the number of atoms present. If the numbers of atoms of both elements are the same, the ratio of mass values is the same. Repetition of this point is important so students can understand that the ratio is independent of the number of atoms.
4	Repeat step 3 for increasingly large numbers of atoms, for example, 1000, 10 000, 1 000 000. Write the numbers of atoms in columns for the two elements. Ask for the ratio each time. It remains 2:1.	

Cu	S	Ratio of masses
A_r = 64 a.m.u.	A_r = 32 a.m.u.	
100 atoms	100 atoms	2:1
...	...	2:1
		...
1 000 000 atoms	1 000 000 atoms	2:1
64 g	32 g	2:1

Step	Instruction	Points to note
5	Ask why the ratio cannot be changed. The answer is that atoms of elements always have the same mass values, so the ratio is fixed.	This helps students realise that atomic masses are in fixed proportions.
6	Ask if it is possible to measure the mass of 1 000 000 atoms. The answer is yes, but atoms are tiny, so this number would not register on a normal balance weighing in grams.	In practice, measuring this number of atoms is possible with extremely expensive technology. Reinforce this shows atoms are extremely small.

2 Chemical change

Step	Instruction	Points to note
7	Explain that we can weigh out the atomic mass values in grams, that is, 64 g copper and 32 g sulfur. Ask for the ratio of these two amounts. The answer is 2:1.	This is the point at which the switch from focusing on atoms to weighing amounts is made. Understanding why the ratio 2:1 applies to these mass values is essential before moving to step 8.
8	Ask the crucial question: What can you say about the number of atoms in these two masses? The answer is that the number of atoms must be the same.	Allow plenty of thinking time. It is tempting to give students the answer if nothing seems forthcoming. However the sequence works best when students realise this for themselves. The sequence may need repeating several times to help students realise this point.
9	Repeat that the ratio is fixed at 2:1. The number of atoms of each element was the same before (as shown on the board) so this must also be true for 64 g copper and 32 g sulfur.	Reinforce once again that the ratio is 2:1 because the atomic mass values are in this ratio, which cannot be changed.
10	Introduce the term 'one mole' for an atomic mass value measured in grams. Using moles, we can measure amounts of substances easily.	Background information about the mole is helpful here (see text). The mole is a unit of measurement, like a dozen (12), score (20), gross (144), or a ream of paper (500 sheets).
11	The number of atoms in one mole must be very large. Ask how we know that this number must be very large. Answer: because when we were talking about millions of atoms, we still could not measure this number in grams.	
12	Introduce the number 6.02×10^{23} as the number of atoms in one mole explaining that this is called Avogadro's number or Avogadro's constant. Illustrate the size of this number.	Images of atoms taken with scanning tunnelling microscopes and of the size of Avogadro's number are useful here (see Enrichment box, below).
13	Say that the same number of particles, Avogadro's number, is present in one mole of any substance. This is because, for example, the mass of a calcium atom is always 40 a.m.u., that of magnesium atoms is 24 a.m.u., oxygen atoms 16 a.m.u. and so on. Weighing out these amounts in grams means each sample must contain the same number of particles.	This arises because the masses of atoms of all elements are in fixed ratios relative to each other.

2.5 Measuring chemical change

> **Enrichment**
>
> One mole of drink cans would cover the Earth's surface to a depth of 200 miles; one mole of marbles would cover the UK and Ireland to a depth of 1500 km.

KEY ACTIVITY

Understanding moles

This activity helps to consolidate understanding that when reactions occur, fixed numbers of moles of reagents produce fixed numbers of moles of products. The reaction between iron and sulfur provides the context. Instructions for carrying out the reaction between iron and sulfur are available at https://edu.rsc.org/resources/iron-and-sulfur-reaction/713.article

The activity can be run as a demonstration or class practical. Two reaction mixtures are required, with a clean tube for each:

1 **Reaction mixture A** (excess sulfur): for example, 112 g/56 g/28 g/14 g iron and 80 g/40 g/20 g/10 g sulfur
2 **Reaction mixture B** (stoichiometric amounts of iron and sulfur): for example, 112 g/56 g/28 g/14 g iron and 64 g/32 g/16 g/8 g sulfur

Heat the reaction mixtures in order, A then B. Reaction mixture A produces significantly more sulfur vapour than B. Sulfur vapour can be absorbed by a mineral wool plug at the mouth of the tube. Should the vapour ignite, cover the mouth of the tube with a damp cloth to absorb sulfur dioxide. In both cases sulfur melts, the reaction glows orange and a black substance, iron(II) sulfide, is produced.

The equation for the reaction is:

Words: iron + sulfur $\rightarrow$ iron sulfide

Symbols: Fe + S $\rightarrow$ FeS

For each mixture, ask students to:

- observe closely what happens in the tube
- note what was left over after the reaction.

Moles and reacting mass ratios for reaction mixtures A and B

Write the numbers of moles and masses reacting using the equation and compositions of the reaction mixtures. For example:

2 Chemical change

Equation:	Fe	+ S	→	FeS
Moles:	1	1		1
Mass of 1 mole:	56 g	32 g		88 g
Reaction mixture A:	56 g	40 g		88 g + 8 g unreacted sulfur
Reaction mixture B:	28 g	16 g		44 g

How many moles of iron and sulfur reacted?

Using the sample masses of iron and sulfur given above:

1 **Reaction mixture A**: 1 mole of iron and 1 mole of sulfur reacted to give 1 mole iron sulfide, with 8 g sulfur in excess.
2 **Reaction mixture B**: 0.5. mole of iron and 0.5 mole sulfur reacted to give 0.5. mole iron sulfide.

The equation shows iron and sulfur react in a mole ratio of 1:1 or 1 ÷ 1 = 1.0. This means that for every mole of iron and sulfur reacted, one mole of iron sulfide is produced. Reaction mixture A contains more sulfur than this ratio, so excess sulfur remains. From the A_r values, an iron atom and a sulfur atom have a mass ratio of 56:32. This simplifies to 7:4, so the reacting mass ratio is 7:4 or 7 ÷ 4 = 1.75.

Students need to understand that the mole and reacting mass ratios cannot be changed. They may be tempted to add unreacted sulfur to the mass of iron sulfide, giving a mass of 96 g iron sulfide for reaction mixture A. Connect this to the observation that excess sulfur is produced when A is heated. This shows that only a fixed amount of sulfur will react with the fixed amount of iron.

Careers

Many careers in chemistry apply knowledge of chemical changes. For example, pharmaceutical chemists create new substances that are crucial to developing new medicines. Initial steps in drug development involve interaction with doctors and other scientists whose research and observations lead to a potential route or mechanism within a disease that could be reversed or inhibited using a chemical. Chemists design, synthesise and purify new molecules for testing. To develop new drugs, chemists apply combinatorial technology, a process which defines a specific molecular target and designs a new structure to fit this. A range of compounds (called a library) may be made, which are tested using a technique called high throughput screening to identify the most promising leads for follow up. This results in a significant number of new drugs each year. In 2019, these included treatments for various cancers, malaria and multiple sclerosis.

Chemical engineers design and develop industrial processes that lead to new products. They focus on changing the chemical, biochemical and/or physical state of substances using cost-effective processes that minimise waste and follow appropriate health and safety procedures with minimal environmental impact. They may work for energy and power generators, in water treatment plants, be developing nanotechnology, carry out research in biomedical sciences or be employed in industries including those producing pharmaceuticals, oil and gas, food, beverages, wine and cosmetics. A chemical engineer applies understanding of the chemical change(s) and reaction conditions such as temperature, pressure, solubility, viscosity and humidity to design and run a production facility. Current research combining synthetic chemistry with molecular biology and process engineering is creating new biocatalysts that can help increase productivity in chemical and pharmaceutical industries.

Flavour scientists understand chemical reactions that occur in cooking and how these impact smell and taste. They may also develop flavours and extracts using natural and synthetic chemicals. These are added to food and drinks to enhance flavour and texture. Food companies commission flavour scientists to create specific taste combinations for their products. For example, vanilla is a popular flavour worldwide. The molecule responsible for the characteristic vanilla taste is vanillin, formula $C_8H_8O_3$, a compound present in the vanilla bean. Natural vanilla is very expensive, so using synthetic equivalents is popular. Flavour scientists synthesise vanillin in formats adapted for foods such as custard (vanilla sauce), yoghurt, sponge cakes, ice cream, meringues and crème brulée (baked vanilla custard).

2.6 Resources

References and further reading

Barker, V. (2002) *Building success in GCSE science: Chemistry*, Folens.

Barker, V. and Hadi-Talab, R. (2005) 'Demonstrating chemistry: Part 2' *School Science Review* 86 (317): 95–106.

Cheng, M.M.W. and Gilbert, J.K. (2017) 'Modelling students' visualisation of chemical reaction', *International Journal of Science Education* 39 (9): 1173–1193. https://doi.org/10.1080/09500693.2017.1319989

2 Chemical change

Kind, V. (2004) *Beyond appearances: Students' misconceptions about basic chemical ideas.* Available at https://edu.rsc.org/resources/beyond-appearances/2202.article

Taber, K.S. (2002) *Chemical misconceptions: Prevention, diagnosis and cure* (Volumes 1 and 2), Royal Society of Chemistry.

Tasker, R. and Dalton, R. (2008) 'Visualizing the Molecular World – Design, Evaluation, and Use of Animations' pp. 103–131 in *Visualisation: Theory and Practice in Science Education*, Gilbert, J.K., Reiner, M. and Nakleh, M. Dordrecht (editors), Springer.

de Vos, W. and Verdonk, A. (1985) 'A new road to reactions part 1', *Journal of Chemical Education* 62: 238–240.

de Vos, W. and Verdonk, A. (1985) 'A new road to reactions part 2', *Journal of Chemical Education* 62: 648–649.

de Vos, W. and Verdonk, A. (1985) 'A new road to reactions part 3', *Journal of Chemical Education* 63: 972–974.

de Winter, J and Hardman, M. (2021) Teaching Secondary Physics (3rd edition), Hodder Education.

Websites

This is a concept cartoon that can be used to elicit students' ideas about electrons and chemical reactions: https://edu.rsc.org/download?ac=15463

High-quality particle-based animations of physical changes and some chemical reactions are available via this resource: https://phet.colorado.edu/en/simulations/states-of-matter-basics

Celebrate Mole Day with students on 23 October from 6.02 am to 6.02 pm. A new theme is featured each year: www.moleday.org/

This introductory level resource shows students how to use Molymod® kits to represent molecular models: www.tes.com/teaching-resource/balancing-chemical-equations-molymod-activity-6409214#

Resources on Chemical Reactions are available at the BEST evidence in science teaching project at the University of York Science Education Group: www.york.ac.uk/education/research/uyseg/research-projects/bestevidencescienceteaching/ and from: www.stem.org.uk/best/chemistry-earth-science/big-idea-chemical-reactions

This class experiment introduces the difference between a chemical and a physical change using chocolate and egg white: https://edu.rsc.org/resources/chocolate-and-egg-experiment/441.article

Royal Society of Chemistry (RSC) resource introducing chemical formulae for non-metal compounds: https://edu.rsc.org/download?ac=140710

RSC resource illustrating the principle of writing chemical equations from a piece of text: https://edu.rsc.org/download?ac=140227

RSC How to teach chemical equations: https://edu.rsc.org/cpd/chemical-formulas-and-equations/3010001.article

RSC Iron and sulfur experiment: https://edu.rsc.org/resources/iron-and-sulfur-reaction/713.article

RSC Changes in Chemistry resource: https://edu.rsc.org/resources/changes-in-chemistry/1085.article

Other resources

Molymod® molecular model kits (see molymod.com) are available from school equipment suppliers such as Better Equipped, Philip Harris, SciChem and TimStar.

Particle theory

Maurice M.W. Cheng and Vanessa Kind

Introduction

Chemistry involves substances changing in one of three ways: changing state, mixing, or changing chemically. For example, when water boils, it changes state to gaseous water (steam); when sugar dissolves in water, sugar and water form a mixture which tastes sweet; when a candle burns, a chemical change occurs involving wax and oxygen and emitting light and heat. Explaining these changes accurately relies on understanding that all matter is made of particles. This presents challenges to learners because our daily language and thinking naturally focus on macroscopic observations of substances, not particles.

We cannot 'see' particles directly, yet we need to know they are present to enable correct and accurate explanations for phenomena. Moving from macroscopic observations to particle ideas to explain observable phenomena represents an entirely new way of thinking (see Chapter 1). Taking this cognitive leap is essential for students to make progress in learning chemistry. This chapter discusses approaches for teaching the particle model of matter. Once established, students can learn that 'particles' encompasses atoms, ions and molecules as well as neutrons, protons and electrons (among other sub-atomic species). Understanding the particle theory of matter is the essential concept that underpins chemistry. The aim of this chapter is to help teachers navigate students' formation of this idea.

Students' prior knowledge and misconceptions

'Seeing is believing' is a maxim commonly applied to daily events. Students' experiences of natural phenomena rely on their senses. Children develop a naive view that matter is continuous (not made of discrete particles) based on their experiences with 'stuff', including iron/steel, water, ice, snow, meat, vegetables, stone, sand, jelly, paper, paint, glue, wood, cork, cloth and wool. Research suggests that reliance on sensory (macroscopic) information about matter persists to the age of about 14 (Kind, 2004). Children do not use particle ideas spontaneously to explain the behaviour of stuff (matter), as the continuous model of matter, implied (and confirmed) by sensory information works perfectly well. Without the concept of substance and particles, sensory-based or common-sense reasoning is likely to lead to misconceptions such as those shown below in italics.

- *Particles can change shape by exploding, expanding or contracting, and take on the properties of the substance itself.* Students may explain changes of state in terms of macroscopic properties; for example, ice particles melt to form water particles which evaporate to form steam particles. This primitive reasoning is incorrect.
- *Particles are static and held in position by bonds or forces.* Students may over-stretch the application of the concept of bonds between particles to arrive at a static model. In fact, particles are in continuous random motion, even in solids. The amount of movement depends on the temperature of the substance.
- *The space between (gas) particles is filled with air, bacteria, dust or 'something'.* The space between particles is an empty vacuum. This is a very difficult idea to understand because it lacks supporting sensory evidence. Diagrams of gases significantly under-estimate distances between particles.
- *Gases do not have weight and/or do not occupy space.* Gases are made of particles, so they have weight and occupy space. We experience air pressure pushing down on us every day. In gases particles are quite far apart from each other but occupy space nonetheless. Air is invisible to the eye, because gases in the air do not have intense colours under normal atmospheric conditions. Some gases can be perceived by their colour, including nitrogen dioxide, chlorine, and iodine vapour. It is easier to see that these gases occupy space.
- *Water, ice and steam have different physical properties, so are not the same substance.* Water, ice and steam are the same substance, made of water particles. In water, ice and steam, the water particles have differing amounts of energy and spatial distributions. The word steam is also used to describe the fine mist made of condensed tiny water droplets dispersed in space. This usage helps support the misconception that steam is visible.
- *Boiling a substance always requires a temperature that is above room temperature and pressure.* Some substances, such as oxygen and nitrogen, are gases at room temperature. Their boiling points are below room temperature. These substances have boiled at temperatures far below room temperature.
- *When solids dissolve in water, they disappear. The mass of the solution is less than the combined mass of the solid (solute) and the water.* When a solid dissolves via a process known as dissolution, particles of solid disperse among water particles. The solid seems to disappear because individual particles are too small to be seen

by the human eye. The mass of the solution is the same as the total mass of the solute and the mass of the water. The law of conservation of mass applies: this states that matter cannot cease to exist.

Although our senses may tell us otherwise, substances are made from particles. The challenge is how to force our cognitive processes to accept this.

Progression

In primary (elementary) school, children begin by distinguishing everyday materials – such as wood, plastic, glass, metals, water and rock – then characterise them using physical properties including hardness, absorbency, shininess, stretchiness and opacity. They group objects by these properties. Children also learn that a variety of objects can be made from one material: for example, metal is used for coins, cars, food cans, cutlery and lampposts, while wood is found in matches, bowls, cutlery and used for paper-making. Solids, liquids and gases are introduced from the age of about 7 with the notion that materials change when heated and cooled. Water in the water cycle is introduced as an example. Processes leading to transitions between states are labelled using terms such as evaporation, condensation, melting and solidifying. Advanced properties, including solubility, conductivity and magnetism, are introduced at around the age of 10 alongside separation techniques such as filtering, sieving and evaporating. The aim at this point is to develop a systematic understanding of materials based on a range of properties.

The particle model of matter is often introduced when students are aged 11–14, in the context of explaining three states of matter: solid, liquid and gas. This is reinforced with information about atoms, differences between particles in elements, compounds and mixtures, and an introduction to chemical symbols and formulae (see Chapter 2). Purity is introduced as a concept relating to substances that have one type of particle. Students start to consider how to identify pure substances and to understand the principles of chemical reactions. These ideas are developed further by 14–16-year-olds as they learn about atoms, ions, molecules and how these form via chemical bonds (see Chapter 5). Post-16, students' knowledge of particles deepens to include complex ions, molecular shapes, and analysis techniques such as mass spectrometry.

Johnson (2012) notes that introducing particles via states of matter is likely to cause challenges for students and teachers because this emphasises three separate types of matter, namely, solid, liquid and gas. A physical state of matter depends on melting point, boiling point and surrounding temperature (ignoring pressure). State changes do not change the identity of the matter. Using these three states and labelling changes between them creates the impression that this is all there is, so leaves students adrift for particulate explanations of, for example, solids that pour (sand, flour and sugar); semi-solids such as gels and pastes; the seeming continuity of gases such as air, nitrogen, oxygen and carbon dioxide; and complex mixtures such as emulsions and foams. Particle ideas are applied to explain specific circumstances, usually related to water. Thus, reliance on the curriculum alone to help students learn about particles means that misconceptions such as those listed above are highly likely to be held.

3.1 A simple particle model and the concept of substance

Teaching students the concept of *substance* offers a secure route to establishing the particulate nature of matter. The principle is that a substance is made of only one kind of particle. A simple particle model expressed using the concept of substance includes these ideas:

→ A single substance is made of a collection of identical particles. There is nothing but empty space between particles.
→ Particles are extremely small so are invisible to human eyes, even via microscopes available in schools. To provide perspective, magnifying a drop of water to the size of planet Earth, a single water particle would have the size of a ball with a diameter of 80 cm.
→ Under normal conditions, particles can neither be created nor destroyed in physical changes of state that we observe in our daily lives. This means particles are not destroyed during, for example, melting, dissolving or evaporating.
→ The behaviours and appearances of a substance differ from those of particles. For example, water and ice look different and have contrasting properties. But both are made solely of water particles.
→ Particles are not static, but constantly moving. In a piece of metal that appears to be static, particles are vibrating. Particle movement depends on the temperature of the substance.
→ Particles are held together by bonding. Bonding relies on electrostatic attractions. Bond strength between particles varies depending on the type of particle and temperature. It is not always true that bonds between gas or liquid particles are weaker than those between particles of a solid, so this should not be taught as a fixed rule. For example, bonds between solid oxygen molecules are not 'stronger' than those between particles of iron in the liquid state.

Teaching that intermolecular bond strength is a consequence of the state of a substance, using language such as 'because X is a solid, the bonds are ...' is unhelpful.

A common teaching approach illustrates three states of matter, solid, liquid and gas using water as the only example, explaining transitions by changing arrangements of water particles. This implies that *all* matter can exist in these three states. In fact, the properties of water are unusual in that the transition temperatures between states means they are convenient for science lessons. Some substances – wood, fabrics, paper, plastics and some foods, for example – undergo chemical reactions on heating but not state changes. Other substances are gases at room temperature, giving the impression that they can never be solid or liquid. Others, including gels, pastes and granular solids such as sand, seem to share characteristics of solids and liquids. Students need a rationale that explains these while also arising from knowing that all matter is particulate in nature. Focusing on substances with individual (and overlapping) characteristics, all of which are made of particles, is a better and ultimately more satisfying route than teaching that everything is either a solid, liquid or gas. This is discussed further below. To introduce particles as the foundation for understanding substances, start by getting students to think on an atomic scale, the nanoscale.

KEY ACTIVITY

Seeing on a nanoscale

The concept of substance enables explanations that focus on state changes and emphasises that the nature of a substance and the identity of particles are unchanged. Introducing particle size is a good starting point, as this illustrates the limits of our vision. The video Powers of Ten (1977), available on YouTube at www.youtube.com/watch?v=0fKBhvDjuy0, is useful for this. The film starts with an image of a family picnicking by Lake Michigan in the USA. The image is enlarged stepwise by powers of ten. At about 100 metres, streets, cars and boats can be seen, with the family visible but very small. At 1000 m, large streets and buildings can be determined, but people, cars and boats cannot be seen. At 10^6 m, the camera shows the whole USA, showing the location of the original image beside Lake Michigan marked in a small square. The scale extends to give a sense of the vastness of space. The story then returns to the starting image, this time reducing the scale by powers of ten to reveal micro- and nanoscale details. A cell can be seen at about 10^{-4} m. Students may think that atoms and cells are similarly sized, so emphasise that this is not true. DNA becomes 'visible' at 10^{-6} m; atoms in DNA are 'seen' at 10^{-9} m. Atoms are about 10^5 times smaller than cells. The diameter of an atom ranges from 0.1 nm to 0.5 nm (nm = nanometres, 1 nm = 10^{-9} m).

> **Questions to ask**
> - What can be seen on the Earth's surface from above, and at what distances?
> - What are we unable to see when we are above the Earth's surface, and at what distance?
> - What is the smallest size of object that our eyes can see without magnification?
> - How big is a cell?
> - Which is smaller, an atom or a cell? How do we know?
> - How big is an atom in a cell?
>
> To reinforce these ideas, the activity 'Zooming in' from the Best Evidence Science Teaching resource pack, available at www.stem.org.uk/best/chemistry-earth-science/big-idea-particles-and-structure is helpful.

 How do we know matter is made of tiny particles we can't see?

The notion that matter is made of tiny particles was originally known as atomism and first proposed by the Greek philosopher Democritus (460–370 BCE) from Miletus, a wealthy city in ancient Greece (now ruined), located on the southern coast of modern day Turkey. Atomism was disregarded by European scientists for centuries, although it was retained in Islamic science. In Western Europe, the idea resurfaced incrementally through the eighteenth and nineteenth centuries. In this period, attempts were being made to understand substances and chemical reactions. Scientists had not established the relationship between masses of reactants consumed and that of the products produced in chemical reactions; and stoichiometry (see Chapter 2) had not been developed. The role of oxygen in combustion was not understood. Scientists knew that when a metal burned, the product seemed heavier than the original metal but found this hard to justify, in part because data about reactions were inconsistent. Nevertheless, scientists sought rationales for their observations, devising theories such as the phlogiston theory (see Chapter 1).

Eventually, the invention of sophisticated mechanical balances and heating equipment allowed chemists to accurately measure mass changes occurring during chemical reactions (Brock, 2016). Antoine and Marie-Anne Lavoisier carried out precise experiments with mercury calx, now known as mercury(II) oxide, HgO. (*Calx* is an old word for a substance formed when a mineral or metal ore is heated.) Mercury calx decomposes on heating to give mercury (called quicksilver at the time) and a gas, now known as oxygen. The Lavoisiers showed that when mercury calx was heated in a sealed system, the products and reactant had exactly the same mass. They also demonstrated that on recombining to form mercury calx, the masses of the product and reactants were once again the same. Antoine Lavoisier summarised this finding in 1774 in the law of conservation of mass. Lavoisier's publications, including *Reflections on Phlogiston* (see Chapter 1) and, in 1789, the first modern chemistry textbook *Elementary Treatise on Chemistry*, transformed chemists' thinking. He clarified the idea of a chemical element as a substance that could not be broken down into a simpler substance by chemical methods and explained the formation of chemical compounds. Lavoisier's reasoning and evidence were convincing. However, particles were not mentioned. For Lavoisier, lack of observable evidence meant that atoms could not be dealt with in the realm of science but were a philosophical idea.

However, Joseph Proust (1754–1826) then proposed the law of definite proportion. This states that 'all pure samples of a particular chemical compound contain the same elements combined in the same proportion by weight'. This means, for example, that pure water always contains hydrogen and oxygen in the mass ratio of 1:8. Proust's publications of sound experimental evidence led to gradual acceptance of the law. Why substances behaved this way and detail of the composition of substances remained unresolved puzzles. At this point, Lavoisier's concept of elements was the most fundamental idea regarding substances, and this was a function of the limit of available analysis techniques.

Also in this era, John Dalton (1766–1844) proposed the law of multiple proportions (Dalton's law). This states that 'when two elements combine to form more than one compound, the ratios of the weight of an element that combine with a fixed weight of another element are of *small whole numbers*' (italics added). Dalton's law is based on investigations of the compositions of methane (CH_4) and ethylene (ethyne, C_2H_2). The phrase 'small whole numbers' implied something fundamental lay behind the fixed ratios. Subsequently, Dalton suggested the laws of conservation of mass, definite proportion and multiple proportions could only be understood by applying atomism. Unlike Lavoisier, Dalton was undeterred by the lack of direct observation of particles. He reasoned that atoms were the only sensible justification. Dalton's atomic model (1803) proposed that:

3.2 How do we know matter is made of tiny particles we can't see?

- All matter is made of very minute particles.
- The particles that make up elements are called atoms. They are indivisible and indestructible in chemical reactions.
- All atoms of a given element are identical and that each atom of a given element has the same weight.
- The weight of atoms of different elements is different.
- When elements combine to form compounds, the atoms of elements combine in a fixed ratio for that compound.

Atomism, or 'particle theory' became an accepted explanation for scientific evidence that generated three seemingly unrelated laws about substances. That atoms are indivisible and indestructible, explained the law of conservation of mass; the formation of a fixed atom ratio when elements react to form a compound explained the law of definite proportion; and the small whole number ratios in Dalton's law are explained by the existence of particles and differences in particle sizes, measured in atomic weights. For nineteenth-century scientists, particle theory represented a major achievement. Nothing has yet shown the theory is false. But this first step was not taken easily. Similarly, students may not find it easy to think of matter as particulate rather than continuous.

KEY ACTIVITY

The development of particle theory

This activity helps students understand that scientific theories develop in leaps, and that scientists develop new theories based on weaknesses in existing ones. Theories arise because humans seek explanations for phenomena. Introduce the notion of a theory as a model for a phenomenon, using, for example, a model car, aeroplane or train, a toy dog, a doll, or a computer game. Each model bears a resemblance to reality but represents a model-maker's representation of a 'real' thing. Discuss the features of a model that match the real item and the model's limitations. Theories are scientists' representations of phenomena and/or explanations for experimental data. Some, all or no parts of a theory will be proven correct as scientists continue to work on the phenomenon: theories are subject to change.

Table 3.1 shows models (or theories) about atomic structure starting with John Dalton's 'billiard ball' in 1805. The table extends beyond the Rutherford atom required in curricula for 14–16 s. It is valuable for students to see that scientists have progressed understanding of atomic structure beyond this. They may be aware of 'big-science' experiments such as the Large Hadron Collider particle accelerator. Post-16 chemistry courses require understanding of spectra, a feature of the Bohr model. All these models have uses: a billiard-ball model is valuable when considering changes of state; the Rutherford model helps understand arrangement of electrons.

3 Particle theory

Table 3.1 Particle theory: models describing atomic structure

Model	Date proposed	Proposer	Description	Supporting evidence	Contradictory evidence
Billiard ball	1805	John Dalton (1766–1844)	Atoms are solid, spherical balls of matter that cannot be divided or destroyed. All atoms of an element are identical.	Chemical elements always combine in a fixed mass ratio.	Data illustrating the existence of subatomic particles. Data illustrating that some atomic nuclei contain extra protons, hence existence of isotopes.
Saturnian	1904	Hantaro Nagaoka (1865–1950)	Atoms have massive nuclei around which electrons revolve, like Saturn's rings around the planet, held by electrostatic forces.	Prediction confirmed by Rutherford's evidence.	Data showing relative sizes of nucleus and subatomic particles. Calculations regarding electron movement.
Plum pudding	1900	Joseph John Thomson (1856–1940)	Atoms comprise negative 'corpuscles' surrounded by a 'soup' or cloud of positive charge.	Thomson's 1897 demonstration that electrons come from atoms.	1909 Geiger–Marsden gold-foil experiment.
Rutherford	1911	Ernest Rutherford (1871–1937)	Central positive charge surrounded by orbiting electron cloud.	Gold-foil experiment. Thomson's evidence for electrons.	Emission spectra for atoms showing lines suggesting that electrons were ordered within the atom. If this model were right, electrons would spiral into the nucleus – but they don't.
Bohr	1913	Niels Bohr (1885–1962)	Central positive nucleus. Electrons moving in circular orbits. Orbits ordered by energy level.	Explained basic features of the atomic emission spectrum for hydrogen. Used new mathematics, called quantum mechanics, removing reliance on Newton's theories.	The model could not explain: emission spectra for larger atoms the effects of magnetic fields on line spectra finer lines in spectra different intensities of lines in spectra

3.2 How do we know matter is made of tiny particles we can't see?

Model	Date proposed	Proposer	Description	Supporting evidence	Contradictory evidence
Sommerfeld	approximately 1922	Arnold Sommerfeld (1868–1951)	Central positive nucleus. Electrons moving in elliptical orbits.	Explained more complicated emission spectra.	The maths needed to support this proved very difficult. Evidence that electrons are waves as well as particles from Louis de Broglie (pronounced 'de Broy'). Evidence that electrons spin from Wolfgang Pauli.
Schrodinger	1925	Erwin Schrodinger (1887–1961)	Central positive nucleus. Electrons move in differently shaped 3D areas, not orbits.	Uses quantum mechanics to provide the maths, rather than Newton's system. Explains emission spectra features that the Bohr and Sommerfeld models can't.	None yet found – this is our best model so far, but work has continued on the nature of subatomic particles and the forces within the atom and nucleus.

Source: Kind and Kind, 2008

3 Particle theory

Information in the table can be presented in different ways, for example:

- As a card sort – each box on the table can be presented as a separate card. Students must match the lines across the table. This may be easier using the first four lines of the table (Billiard ball to Rutherford).
- Working in groups, each takes one line of the table and prepares a presentation of the information. This is an opportunity to carry out further research on one model. Presentations could include images of the scientists, their laboratories, their experiments and challenges they faced. Focus on what each model explains about particle behaviour.

Questions to ask

- What does the billiard ball model say an atom is like?
- What does the Rutherford model say an atom is like?
- Which do you think is the better model? Explain why.
- Who was the first scientist to show that an atom is not completely solid? What was the evidence for this?
- Use information in the table to draw a timeline showing changes in our understanding about particles.
- Why are there so many theories about atoms?

Enrichment

Mercury, known for over 5000 years, is an unusual metal, because it is a shiny grey liquid at room temperature. Its symbol is Hg, from *hydrargyrum* which combines Latinised versions of Greek words for 'water' and 'silver', as it was once known as quicksilver (meaning 'runny silver') based on its resemblance to solid silver. Our eyes are deceived into thinking silver and mercury are two forms of the same substance when, in fact, they are each unique substances comprising particles which are responsible for the metals' properties. Mercury is obtained by roasting the toxic red ore cinnabar, which is rich in mercury sulfide, HgS.

Cinnabar is found worldwide, most notably in Almadén, at Ciudad Real, in Spain. *Almadén* means 'the mine' in Arabic. Spain was ruled by Arabs between the eighth and fifteenth centuries CE. In around 936 CE, before the toxicity of mercury became known (and in the absence of risk assessments), caliph Abd al-Rahman III built a beautiful palace, an *alcázar*, near Córdoba. He designed a pool of mercury, a liquid mirror, sited to reflect shafts of sunlight around the room. Guests dabbled their hands in the cool metal, creating mercury waves, which bounced sunbeams in a glitterball effect. Such ornamental mercury pools were a feature of Islamic luxury living in this period.

 # Introducing the simple particle model

Changing substances: using simulations and models

Students will be aware that some substances change physical state as temperature changes. The next step is to prompt explanations for these changes using a particle model. The physical state of a substance is determined by the surrounding temperature. In the solid state, particles vibrate but the substance retains a relatively rigid structure. Increasing the temperature increases particle movement (kinetic energy) allowing them to move around each other: this is the liquid state. In gases, particles are further apart due to their greater kinetic energy. Online simulations presenting particles as billiard balls provide insights by relating temperature to transition points and showing increased particle movement. Online versions (see Section 3.7) include PhET interactive simulations (https://phet.colorado.edu/sims/html/states-of-matter-basics/latest/states-of-matter-basics_en.html) which allow manipulation of temperature and illustrate particle movement in neon, argon, oxygen and water. A kinetic particle theory simulator (see Other resources) has a base that vibrates increasingly, causing model atoms to move. It is very important to explain at this point that the particles themselves do not alter. This means particles do not expand, contract, shrink, explode or break apart when substances change state. Run simulations for a range of substances, so that students understand that, for example, oxygen also has a solid state and that diamonds can be gaseous.

KEY ACTIVITY

Representing particle movement with static models is challenging. Nevertheless, spatial distribution and bonding between particles may be represented by Bunchems® (see Other resources). Each Bunchem ball has hooks that allow a collection to represent a solid (Figure 3.1) or a liquid (Figure 3.2). As Bunchems do not need sticks to hold the balls together, this avoids leading students to think that a bond is a physical entity (see Chapter 5). The balls are easy to handle. A limitation to discuss with students is that particles do not have hooks to hold them together.

Questions to ask
- What causes a substance to change?
- Why do changes occur at different temperatures for each substance? What does this tell us about particles?
- What happens to particles when substances change?

3 Particle theory

Figure 3.1 A model representing solid, showing 'particles' bonded to each other in a regular arrangement

Figure 3.2 A model representing liquid, in which 'particles' bond closely but are irregularly arranged.

Changing states: a substance-based approach to melting and freezing/solidifying

Once students understand that particles are present in every substance and that changes occur due to temperature variations which increase or decrease particle movement, they can measure these changes. Figure 3.3 shows an experimental set-up that permits students to measure the melting points of substances within the range 30–80 °C.

KEY ACTIVITY

Substances that can be melted include wax, chocolate, ice, dodecanoic acid (lauric acid, used in soaps), ice cream, butter, margarine.

Discuss what happens on melting. The sample size is not relevant, other than noting that, for example, an iceberg takes longer to melt than a small ice cube. Melting does not change the particles of a substance: melted wax is still wax, melted chocolate is chocolate, butter is butter. Continuing to heat beyond the temperature needed to change the substance from a solid to a liquid may cause further physical and chemical changes. For example, under very gentle heating, butter melts to a yellow liquid. Continued heating separates butter into water, milk solids and fats (lipids), and will eventually burn the milk solids. The separation of butter is used to make *ghee* (clarified butter) which consists of butter fats only and is used in Asian dishes to create an intense, nutty butter flavour.

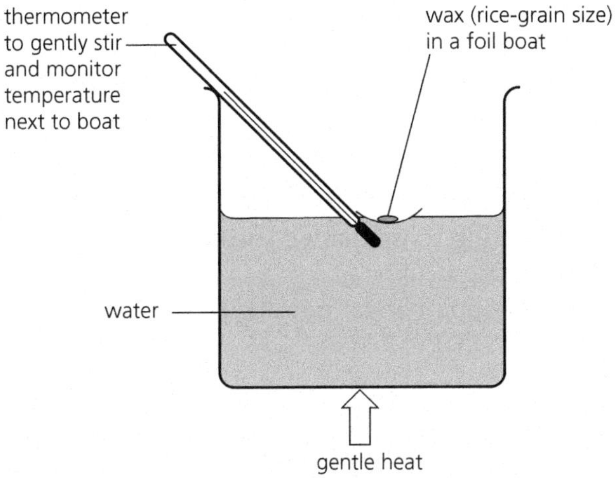

3.3 Introducing the simple particle model

Figure 3.3 Finding the melting point of candle wax

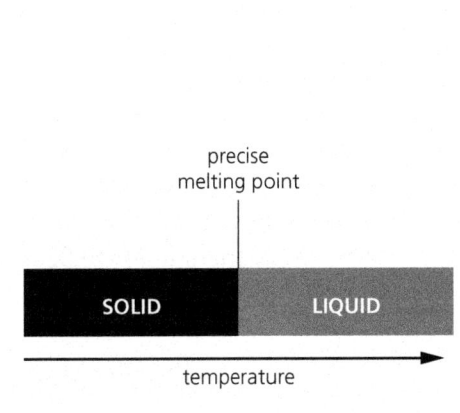

Figure 3.4 A substance has a precise melting point

When considering the results, remind students to apply the model from the previous step when thinking about what happens to particles: they move more, and interactions between them loosen. Differences in size and type of particles and the strength of the interactions between them cause melting points to vary.

Note that when substances solidify, they do not look exactly the same as the starting substance. It is next to impossible to arrange particles back into identical starting arrangements. Also, some substances from the list above, such as chocolate, butter and ice cream, contain more than one type of particle (see Chapter 4 and Chapter 9), so interactions are varied causing more changes to the structure. But the particles themselves are, nonetheless, unchanged. Consider how to use molecular models to represent this (see above).

A substance can be defined as a sample of matter that has a precise melting point, an exact temperature at which it switches between solid and liquid. On heating, a solid substance melts at this temperature. On cooling, a liquid substance solidifies at this temperature (see Chapter 9).

Students may think freezing temperatures have to be below 0 °C. Reinforce that the freezing point of a substance is the same as the melting point but is measured on cooling a liquid.

Demonstrate melting lead (this happens at 328 °C), tin (at 232 °C) and solder, an alloy of tin and lead, which melts at about 220 °C, to illustrate the melting points of metals. Instructions are available at https://edu.rsc.org/resources/solid-mixtures-a-lead-and-tin-solder/447.article. Note that the solder must be rosin-free.

Demonstrate the melting point of sodium chloride (801°C): instructions are available at https://spark.iop.org/effect-heating-common-salt-and-paraffin-wax#gref (**Safety notes**: Wear eye protection. If the electrolysis of molten sodium chloride is carried out for more

3 Particle theory

than a few seconds, this activity must be carried out in a fume cupboard. Beware of the risk of igniting the paraffin wax.)

Questions to ask
- What are the temperatures at which these substances change from solid to liquid?
- Why do some substances have higher melting points than others?
- What would happen if we carried on heating these melted substances?
- What are the freezing points of these substances?
- What is the difference between a melting point and a freezing point?

> **Enrichment**
>
> Chocolate is a solid food that melts in the mouth to a smooth, sweet-tasting liquid. The main component of chocolate is cocoa butter, obtained from cocoa beans, which has a melting point of between 34 °C and 38 °C, much the same as human body temperature. Cocoa beans come from cocoa trees, cultivated from around 600 BCE by Aztecs in Mexico and Incas in Peru. Cocoa beans were so highly prized they were used as currency, as well as to produce the drink *chocolatl*. *Chocolatl* was a bitter, acidic-tasting, fatty drink. Honey, vanilla and spices were added to improve the flavour. In the seventeenth century, chocolate became a popular, if expensive, drink in European chocolate houses. The Dutch company Van Houten invented a cocoa press that reduced the fat content of cocoa butter by creating cakes of cocoa that were milled to make cocoa powder, which dispersed better in milk or hot water, improving the drink. The next challenge was marketing the leftover cocoa butter. In 1847, Joseph Fry, from Bristol, UK, mixed it with milled sugar and cocoa nibs to make the first chocolate bars for eating. His lead was followed by John Cadbury, Joseph Rowntree and Milton S. Hershey.

3.4 Gases: boiling and condensing

The gaseous state

As most gases are not visible, the concept that gases occupy space and have mass appears counterintuitive to students. Having established that solids and liquids are made from the same particles, invite students to predict what would happen when a melted substance is heated. Record and discuss their speculations, noting that these may include misconceptions (see above) such as that the particles might expand, explode or vanish (all incorrect); or move apart because they will have too much kinetic energy to stay as a liquid (correct).

KEY ACTIVITY

A helpful demonstration on which to base this discussion involves injecting a small drop of water (about 0.05 cm³) into a (heat-resistant) glass gas syringe heated in an oven to 150 °C as shown in Figure 3.5. Full details of this experiment are available at www.stem.org.uk/best/chemistry-earth-science/big-idea-particles-and-structure. (**Safety note**: Set up the experiment behind safety screens and wear eye protection.)

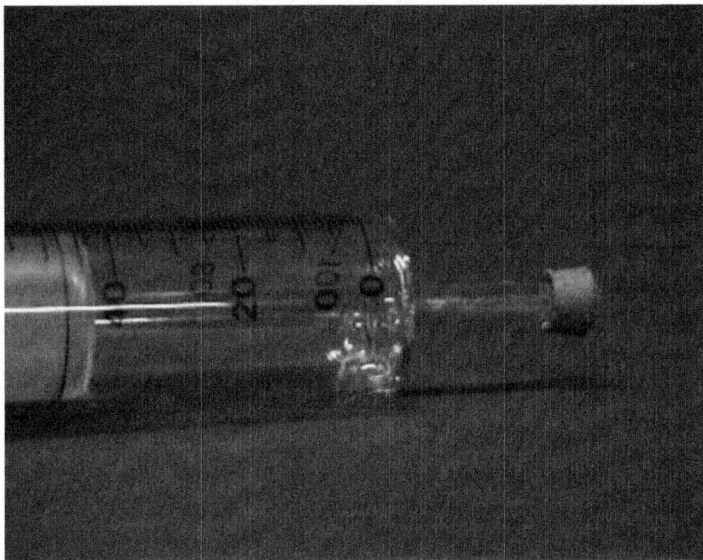

Figure 3.5 Water changing to the gas state

The water droplet expands immediately on meeting the hot glass to a volume of around 100 cm³, moving the plunger dramatically. This is much bigger than the volume students usually predict. A volume of 0.05 cm³ contains about 1.7×10^{21} water particles, each of which moves much more rapidly when heated by the heat energy in the glass, creating massive expansion. This creates pressure that forces the plunger to move. The plunger stops moving when pressure inside and air pressure outside the syringe are equal. There is nothing (a vacuum) in between the particles of gaseous water. Students may suggest bacteria or air are in between the particles, using this faulty idea to explain the extraordinarily large volume of gas relative to the small amount of water injected. Challenge this by asking how these got there when clearly only water was introduced into the syringe which was previously empty. Also remind students about the relative sizes of cells and particles (see above). An image representing the gas particles is shown below.

3 Particle theory

Gas state
The particles do not hold themselves together. Particles are apart, moving freely in all diections. There is a lot of empty space (nothing) between the particles.

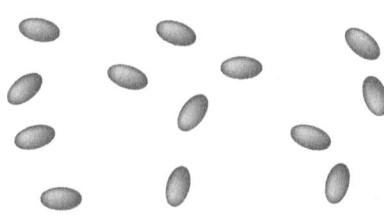

Figure 3.6 The gas state

Practice the demonstration, as dexterity is required to hold the hot gas syringe steady in heatproof gloves while simultaneously injecting a precise amount of water through the sealed tip of the syringe. Have more than one hot syringe available for a repeat. Prior to the demonstration, explain what will be done, inviting students to predict what they think will happen (see the resource sheet from the link above) with an explanation. Carry out the demonstration, then revisit their responses. Examine the syringe carefully, discussing what happened to the liquid droplet. Allow the syringe to cool, observing that water condenses and the plunger can be gradually pushed back to 0 cm³ gradually.

Questions to ask
- How many water particles do you think are present in the droplet?
- Was your prediction correct? If not, explain what was different and why.
- What has happened to the liquid water?
- Why has this happened?
- Why does the plunger stop moving?
- What does this tell us about water particles?
- What is in between the water particles in the liquid state?
- What is in between the water particles in the gas state?

Additional activities that support this experiment are available from the link above. 'Particle explanations – gas state' directly addresses points made in the gas-syringe demonstration; 'Empty Space' reinforces the notion that there is empty space between particles; and 'A particle model for the gas state' introduces an image of particles in the gaseous state.

A similarly dramatic demonstration is the fountain experiment. Full instructions are available here: https://edu.rsc.org/resources/ammonia-fountain-experiment/979.article

A round-bottomed flask filled with ammonia gas is inverted above a trough of water. A narrow-gauge tube and a syringe containing water are fitted to a double-holed stopper and placed in the mouth of the flask. A few drops of phenolphthalein (or Universal Indicator) are added to the water. A drop of water is injected into the flask via the syringe. A massive pressure drop occurs, which causes water to surge up the tube, spraying into the flask. A colour change (colourless to pink with phenolphthalein, or green to purple with Universal Indicator) occurs. The explanation for the effect is that ammonia gas is highly soluble in water. As it dissolves, the pressure inside the flask rapidly reduces compared to the air pressure outside. Air pressure forces water from the trough up through the tube, creating the fountain.

Boiling points

The phenomenon of evaporating water can be extended to introduce the concept of boiling point. Start by asking students to observe water boiling in a transparent glass beaker, noting when bubbles appear and the appearance of the bubbles (see https://www.stem.org.uk/best/chemistry-earth-science/big-idea-particles-and-structure, 'Explaining Bubbles'). In fact, two sets of bubbles form as water is heated: very small bubbles of dissolved gases (oxygen, carbon dioxide) are released almost immediately. At 100 °C, large bubbles form which break on the surface releasing gaseous water into the atmosphere. These bubbles contain water molecules, formula H_2O. Students may think they contain a mixture of hydrogen and oxygen, air, carbon dioxide, heat, or water molecules. The notion that water splits up on heating into hydrogen and oxygen is often part of a plausible model for evaporation and condensation in which students argue that during condensation, hydrogen and oxygen molecules recombine to form water. Time with molecular models (such as those made from Bunchems®, referred to above) to help establish that water molecules do not split up on heating is well-spent.

Figure 3.7 illustrates what happens when water is boiled in a kettle. Once students have observed water boiling in a beaker, observe the mouth of a kettle while water is boiling. Notice the clear zone comprising gas state water (steam) and the mist of condensed water droplets further away. Steam is, strictly, water in the gas state but the word is also often used incorrectly to mean the heated mist of fine droplets that forms as water vapour condenses in cool air. Recall that our eyes cannot see particles because they are extremely small, revising their size. Represent gaseous water using molecular models, as before.

3 Particle theory

Figure 3.7 The gaseous state of water from a kettle

KEY ACTIVITY

Students can measure the boiling point of water using standard equipment (beaker, Bunsen burner, heat-resistant mat, eye protection, gauze mat and thermometer). Boiling points of other substances such as ethanol (78 °C) and propanone (56 °C) can be measured by heating samples on a hot plate. (**Safety notes**: ethanol and propanone are highly flammable. Ethanol is harmful if swallowed and may cause damage to organs; propanone may cause serious eye irritation, drowsiness or dizziness by inhalation; repeated exposure may cause skin dryness and cracking.)

As previously with the solid/liquid transition, focus discussion on the switching temperature at which the liquid becomes a gas on heating, and, on cooling, gas condenses to become liquid. Reinforce that when a liquid boils, temperature is steady. An explanation for this is that time is required to raise the energy levels of the particles in the liquid to enable the change from liquid to gas. The change does not happen all at once. The experiment with the syringe is an exception, as the temperature (150 °C) of the glass was far above the switching temperature of water. The diagram shown earlier (Figure 3.4) can now be extended to show that a substance may have three states.

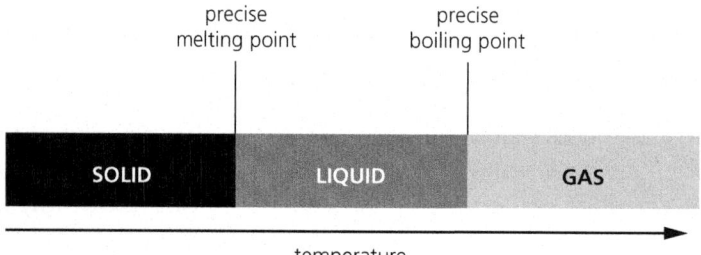

Figure 3.8 A substance and its three states

3.4 Gases: boiling and condensing

Students should recognise that substances have different boiling points. Discuss substances such as carbon dioxide, oxygen and nitrogen, which have boiling points of −78 °C, −183 °C and −196 °C respectively. Many substances boil at very low temperatures. Revisit the PhET simulations (see above) to support this discussion. Figure 3.9 represents information and relationships visually to help. Note that room temperature is an arbitrary measure for convenience, not a fundamental constant. The chart shows how cold liquid nitrogen is!

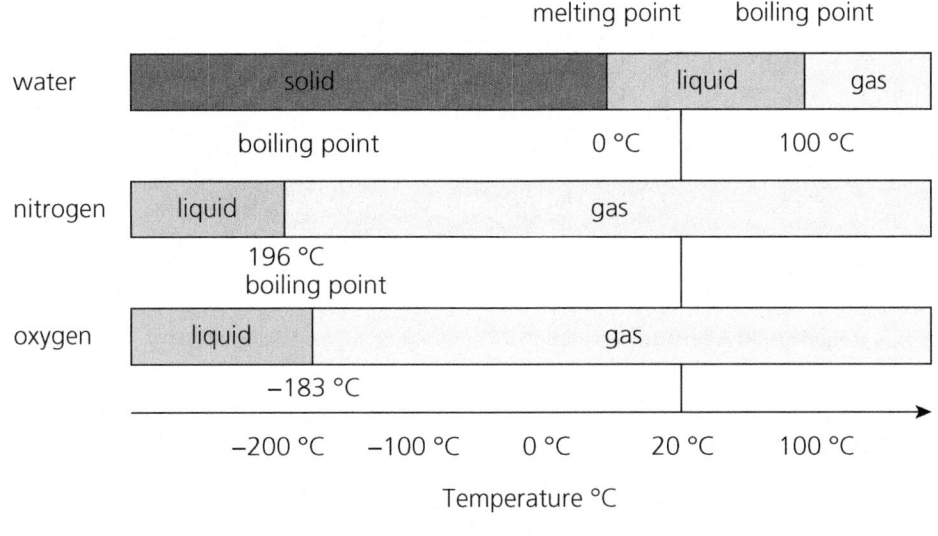

Figure 3.9 States of substances at room temperature and their melting/boiling points

Evaporation

For water, the same liquid–gas state change occurs via evaporation in everyday life, but at temperatures below 100 °C. This occurs, for example, when washing dries, the level of a pond drops during dry weather, or a puddle is there one day and gone the next. It is helpful to explain this in terms of the energy levels of particles. Figure 3.10 is an illustration for water.

Water particles with high energy close to the surface move into the air above the surface. Escaped water particles mix with air particles and are carried away. Meanwhile, high-energy air particles close to the liquid surface collide with other water particles. Energy is transferred between them, giving more water particles sufficient energy to escape into the air. This process continues until all the water particles mix with the air. A breeze helps to carry away particles; still air causes

3 Particle theory

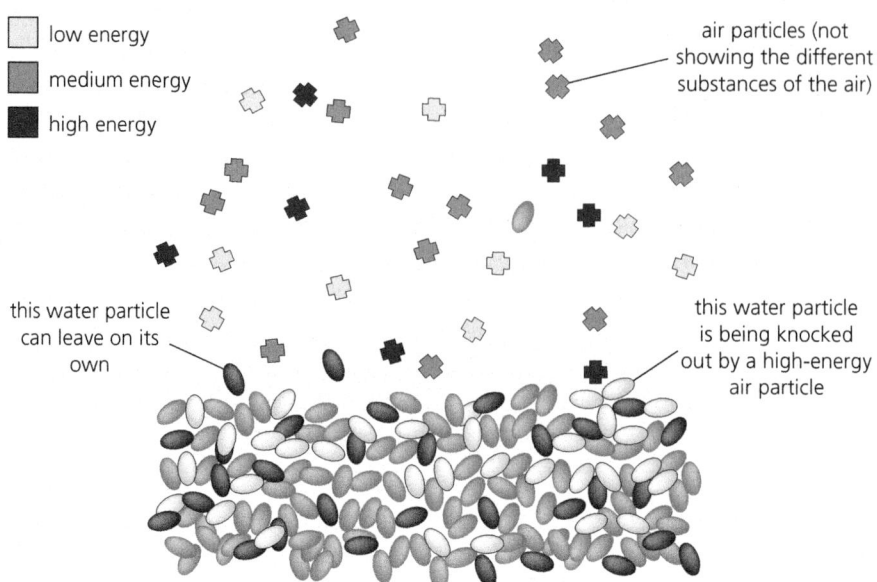

Figure 3.10 A particle explanation for evaporation below boiling point

them to remain. A drop in temperature below the boiling point causes condensation. Connections to weather conditions can be made through this discussion: for example, clouds, rain, snow, mist and fog. A liquid does not need to be boiling to evaporate: these processes can occur at temperatures below the boiling point of a substance because they rely on the relative energy levels of particles of the substance and air.

What are gases?

Gases can be introduced as substances that have boiling points that are below room temperature. They have already boiled. One reason for this is that the bonds between particles of these substances are very weak, so their kinetic energy at room temperature is sufficient to keep them apart, at the distance characteristic of gases. Most gases we encounter are clear, transparent and invisible to the eye. To repeat, there is nothing in between the particles of gases. This is important to reinforce, as students often think incorrectly that smoke and mists are gases, whereas in fact smoke comprises solid particles moving in air and

3.4 Gases: boiling and condensing

mists are droplets of liquids in air. The You Tube video available at www.youtube.com/watch?v=QLrofyj6a2s shows balloons containing five separate noble gases, comparing their densities relative to air. Wait for xenon – it's a true lead balloon! This illustrates that gases have mass and density, and that these can vary. Samples of gases such as nitrogen, oxygen and methane, look identical but they behave very differently chemically. To illustrate this, sit three (covered) gas jars, one each of nitrogen, oxygen and methane, then introduce a lighted splint into them in turn: it will be extinguished in nitrogen; burn more brightly in oxygen; and combust the methane. (**Safety note**: Wear eye protection; oxygen may cause or intensify fire and is an oxidiser; methane is extremely flammable.)

Introduce air as a mixture of substances that are all above their boiling points in the conditions on the Earth's surface. The percentages of gases in the air can be converted to particle numbers to reinforce the point that they are particulate. In a sample of 10 000 air particles, 7800 are nitrogen, 2100 oxygen, 93 argon, 4 are carbon dioxide and 3 are various others. Note that the proportion of carbon dioxide is very low, so any increase represents a significant increase.

Everyday applications of gases may also be useful in illustrating their properties. Nitrogen gas is used to cushion-pack crisps and salads, and to make packaging materials. Puffed-up bags release gas with a small gasp on opening, showing they are filled under pressure. The gas preserves the shape and texture of the food item and prevents oxidation of fresh leaves. Carbon dioxide is soluble in water, particularly under pressure. It is pumped into soft drinks, adding a pleasant degree of acidity to the drink, and a fizz on release, creating a pleasurable mouth feel. Discuss why fizzy drinks go flat: carbon dioxide is denser than air and diffuses into the gap above the liquid when the drink is poured out and the bottle top replaced. A fizz keeper will pump air into the space but the effect is of short duration, as carbon dioxide is simply released at a slightly slower rate.

Questions to ask
→ How do we know what gases are made of?
→ Why are most gases invisible?
→ Why are some substances gases at room temperature?

Enrichment

William Ramsay, a professor of chemistry at University College London, was working with Lord Rayleigh at the Cavendish Laboratory in Cambridge in the 1890s. They noticed that nitrogen gas obtained from minerals was lighter than atmospheric nitrogen. Ramsay burned magnesium in atmospheric nitrogen, finding that a small amount of gas remained. Spectroscopy (see Chapter 9) revealed this gas to be a new element, which Ramsay and Rayleigh named argon, a name derived from Greek for 'idle'. Ramsay realised argon might be one of a group of gases. He discovered four more: helium, named after the Sun; krypton, meaning 'hidden'; neon, for 'new element'; and xenon, meaning 'stranger'. The gases all seemed curiously chemically inactive, so became known as inert gases. With his assistant, Morris Travers, Ramsay experimented on the new gases, finding with amazement that neon produced what Travers described as a 'blaze of crimson light' when electricity passed through it. This property was picked up by French inventor Georges Claude, who made a neon lamp for the 1902 Paris motor show. The astonishing colour of the vapour, produced without a filament, created a magical effect, making neon perfect for advertising. It became known as liquid fire, symbolising modernity in the new century.

 Particle models for gas, liquid and solid

Note that in reality many (most) substances do not fit the description of solid, liquid or gas neatly, so teaching about substances as if they can only exist in one of these three states is not recommended. A major difficulty is that teaching 'solid, liquid and gas' presents these three states as if they are separate 'species' of matter. Chemists think about substances and their states. Room temperature is an arbitrary number without fundamental significance. Also, these three terms do not explain much about substances, even though they conveniently enable simple classification of some substances. The effect of this is that students may acquire persistent misconceptions, especially about gases, as these tend to remain mysterious, rather than realising they are simply substances. In reality, as the preceding discussion has shown, the state of a substance is a feature of particle movement, which is a function of the kinetic energy that particles possess. Nevertheless, science curricula require students to classify substances as solids, liquids or gases, for which the activity available at www.rsc.org/education/teachers/resources/aflchem/resources/20/index.htm may help. The

3.5 Particle models for gas, liquid and solid

activity requires a display of samples of substances that meet the criteria of solids, liquids and gases and offers scope for demonstrating changes of state. (**Safety note**: For Experiment 7, use surgical spirit not laboratory alcohol.)

Students match images on 'particle cards' (printed from the resource) with substances, learning language that describes organisation of particles. The activity addresses possible confusion between, for example, 'sand particle' and particles of silicon dioxide, which make up the grain itself. The important points to make are that particles:

→ are present in all types of matter
→ move, and the amount of movement is related to bonds between particles and their (kinetic) energy levels; this can be controlled by changing the temperature
→ have nothing in between them
→ stay the same, regardless of the physical state.

Take care to avoid the suggestion that bonds between particles are a consequence of the state ('because X is solid') rather than a factor in determining state along with particle energy. Students may be aided by learning the same topic in physics and chemistry lessons simultaneously. This allows demonstrations to be interwoven and timed appropriately, enabling teachers to agree upon a shared language to explain observed phenomena.

For particle diagrams, Johnson (2012) recommends using shapes other than circles to represent particles. Using ovals, as in Figure 3.11, below, helps to illustrate disorder in the liquid state. Circles tend to be drawn too far apart. Circles can be reserved for atoms, which creates a distinction between particles of a substance and particles that make up the particles of a substance. An image of a gas in the same style is shown in Figure 3.6 above.

Solid state
The particles hold themselves close together.
Each particle is in a fixed position.
The particles are vibrating.
There is an ordered arrangement.

Liquid state
The particles hold themselves close together.
The particles move around from place to place, randomly.

Figure 3.11 The same substance in the solid and liquid states

3 Particle theory

To reinforce these ideas, and to promote critical thinking about particle diagrams, the BEST resource pack (www.stem.org.uk/best/chemistry-earth-science/big-idea-particles-and-structure) includes an activity 'Particle diagrams – Liquid state'. Additional resources are listed in Section 3.7.

3.6 Diffusion and dissolving

Diffusion and dissolving are two contexts that can be used to help strengthen students' development of a particle model of matter, connecting the macro- and sub-microscopic levels (see Chapter 1).

Diffusion

Diffusion occurs when particles from two (or more) substances mix gradually. In most everyday examples, the substances are gases or liquids, but it is possible for substances to diffuse from or into a solid. For example, diffusion of various molecules occurs across cell membranes; drugs can be delivered via patches stuck on the skin; metals can diffuse between alloys at high temperatures. Diffusion means particles of a substance move and mix with those of one or more other substances. Simple demonstrations of diffusion include the following:

KEY ACTIVITY

Perfume diffusion

Place a few drops of perfume on a watch glass at the front of a class. Ask students to raise their hands when they can smell it. A wave of hands appears, as perfume particles diffuse through the air. Perfume is volatile, containing one or more liquids that boil at temperatures slightly above room temperature. Skin temperature is around 33–37 °C. Discuss why it takes time to smell the perfume. This is because perfume particles mix gradually, diffusing through the air particles. Smell is caused by particles interacting with olfactory receptors in the nose. Smells are not separate 'stuff'. If possible, test if perfume diffuses faster when everyone is waving their arms around and windows are open than when they are sitting still in a closed room. Air currents carry particles faster than they can diffuse.

Ammonia and hydrogen chloride

This demonstration involves clamping a clean, dry, 0.5–1 m long, 2 cm diameter glass tube horizontally in a fume cupboard. Working quickly, wearing gloves and eye protection, about 2 cm³ of concentrated hydrochloric acid (corrosive) is dripped on one cotton wool ball fixed into a rubber stopper, then placed into one end of the tube. About

3.6 Diffusion and dissolving

2 cm³ 880 ammonia solution (corrosive, dangerous for the environment) is dripped on a second cotton wool ball, also fixed into a stopper, and placed into the other end. Start a timer. Watch for a white ring of ammonium chloride forming in the tube, closer to the hydrochloric acid end. This was not present before, so could only form because gases emitted from the solutions diffused then reacted. Ammonia molecules diffuse faster than hydrogen chloride molecules. Ask students to explain the formation of the ring and why it is not exactly halfway along the tube. Full instructions are available at https://edu.rsc.org/resources/diffusion-of-gases-ammonia-and-hydrogen-chloride/682.article

Microscale diffusion of chlorine

Full instructions for this activity are available at https://edu.rsc.org/cpd/states-of-matter-and-particle-theory/3010239.article The experiment involves half a petri dish placed over a grid of sixteen spots drawn on paper which is laminated to create a plastic sheet. On fifteen spots, students drop potassium iodide solution and starch. These appear white, due to the paper underneath. On one spot, at a corner of the grid, students drop bleach solution (irritant), followed by dilute hydrochloric acid (irritant). Eye protection must be worn. The half of a petri dish is placed over the sheet. A small amount of chlorine is formed in a reaction between the hydrochloric acid and the bleach. This diffuses across the sheet under the dish. The liquids on the fifteen other spots change colour to dark blue as the chlorine diffuses through the air under the dish. Ask students to explain the colour change, and why this takes time to occur.

Dissolving

Dissolving is a process in which solids (known when in solution as solutes; the liquid part of a solution is the solvent) often seem to disappear. We cannot see solute particles because they are too small. We can see the starting solid because this is a structure made from many particles and the collection is big enough to see. When a solid, for example sodium chloride, is added to a solvent such as water and stirred, water particles collide with solute particles on the edge of the salt crystals. These get knocked off, mixing with water particles. Gradually, all solute particles separate from the crystal structure, forming a solution. The salt particles are dispersed through the water. Students may think that the mass of the dissolved substance no longer counts when considering the mass of the solution. Reasons include that the substance has disappeared; it is suspended in water; or has reacted with water and been given off as a gas. Discuss these ideas and demonstrate that mass is conserved by using a balance to show that the mass of the solute and solvent before and after dissolving is unchanged.

3 Particle theory

Students may confuse dissolving with forming suspensions. To illustrate the difference, use calcium carbonate (chalk), which has very low solubility in water. Powdered chalk spreads through water, which becomes white. Students may think of this as a solution. Leave the chalk–water mixture to settle, showing that the chalk forms a layer at the bottom of the vessel. Discuss the difference between dissolving and forming a suspension. A key criterion is that a solution is a clear, transparent liquid. Confirm this by dissolving a copper(II) sulfate crystal (skin irritant; cause serious eye damage; harmful if swallowed; very toxic to aquatic life) in water. Pass solutions and suspensions through a filter funnel to reinforce the difference. In each discussion, use particle ideas.

Discuss solutions and suspensions in everyday life. Examples include hot and cold drinks: milk is a suspension; black/green tea and black coffee are solutions; cola, lemonade are solutions. Paints are a suspension of solid pigments in liquids; mud is a suspension of soil particles in water; smoke is solid ash particles in air; fog is droplets of water in air.

Careers

Material scientists develop new materials based on research into properties and interactions of particles. Nanotechnology and nanoscience investigates and manipulates materials on the atomic or molecular scale (around 1–100 nanometres, 10^{-7}–10^{-9} m). Nanotechnology has led to the development of storage components such as microchips and hard drives in computers and mobile phones. Scientists make successively smaller microchips. The basic units of computer language are ones and zeros. The smallest possible entities that can represent 1 or 0 are *with* and *without* an atom respectively. This example illustrates the implication of particles being the most basic unit in technological development. Nanotechnology has also led to advances in everyday materials such as fabrics that resist bacterial growth and staining; the development of lightweight materials for vehicles and aircraft, so reducing fuel use; and diagnostic tests for and treatment of diseases. The particle theory is the key concept for these advancements.

3.7 Resources

References and further reading

Adadan, E., Irving, K.E. and Trundle, K.C. (2009) Impacts of multi-representational instruction on high school students' conceptual understandings of the particulate nature of matter. *International Journal of Science Education* 31(13): 1743–1775.

Aldersey-Williams, H. (2011) *Periodic tales: The curious lives of the elements*, Viking Books.

Brock, W. (2016) *The history of chemistry: A very short introduction*, Oxford University Press.

Cheng, M.M.W. (2018) Students' visualisation of chemical reactions – insights into the particle model and the atomic model. *Chemical Education Research and Practice* 18(1): 227–239.

Johnson, P. (2012) Introducing particle theory in K.S. Taber (editor), *Teaching secondary chemistry* (2nd edition, pages 49–73), Hodder Education.

Kind, V. (2004) *Beyond appearances: students' misconceptions of basic chemical ideas*. Available at https://edu.rsc.org/resources/beyond-appearances/2202.article

Kind, V. and Kind, P.M. (2008) *Teaching secondary science: How science works*, Hodder Education.

Lavoisier, A. (2019) *Oxygen, acids and water: Eight chapters from the Elementary Treatise on Chemistry*, Green Lion Press.

Rudmann, S., Collins, S., Ellard, K. and Matchett, B. (2017) *Investigating science in focus (Year 11)*, Cengage Nelson.

Taber, K.S. (2002). *Chemical misconceptions – prevention, diagnosis and cure: Volume 1: Theoretical background*, Royal Society of Chemistry.

Taber, K.S. (2002). *Chemical misconceptions – prevention, diagnosis and cure: Volume 2: Classroom resources*, Royal Society of Chemistry.

Websites

Nitrogen dioxide diffusing in air. Nitrogen dioxide (formula: NO_2) is a toxic brown gas. Its colour means it is useful for demonstrating gas diffusion. Full instructions are available at: https://edu.rsc.org/download?ac=12149

A boy and his atom: The world's smallest movie. This 93-second film made by IBM illustrates that atoms can be moved individually using a scanning tunnelling microscope: www.research.ibm.com/articles/madewithatoms.shtml

The Science History Institute has very informative pages about John Dalton and Antoine Lavoisier: www.sciencehistory.org/historical-profile/john-dalton and www.sciencehistory.org/historical-profile/antoine-laurent-lavoisier

An image search using 'history of the atom' reveals a range of images of different atomic models to support Table 3.1.

Instructions and safety advice are available via the CLEAPSS website, www.cleapss.org.uk, including CLEAPSS guide L195 'Safer chemicals, safer reactions' (http://science.cleapss.org.uk/resource-info/1195-safer-chemicals-safer-reactions.aspx) which includes advice relevant to activities described in this chapter.

Other resources

Bunchems® are available from various online suppliers in a variety of kits.

Kinetic Particle Theory Simulator: https://www.philipharris.co.uk/product/chemistry/materials-and-their-properties/states-of-matter/kinetic-theory-model/b8h25365

States of Matter: Basics – Atoms | Molecules | States of Matter - PhET Interactive Simulations (colorado.edu)

4 The periodic table

Vicky Wong

Introduction

The periodic table is the main organising tool in chemistry, laying out all known chemical elements systematically in a way which is recognised worldwide. The origins of the periodic table we use today reflect research undertaken by chemists in many countries, so this topic provides an opportunity to emphasise the international and collaborative nature of science. This chapter explains how the periodic table was devised and assumptions made in its development. Various versions of the periodic table are available. These feature the chemical elements organised in different ways. Chemists will select a version of the table that best suits their needs. For school purposes, the most common organisation is by atomic number. Figure 4.1 is a frequently used version. The chapter begins by discussing the origins of the periodic table, then introduces the arrangement of elements in this version and the structure of the atom.

Understanding the rationale behind the periodic table requires understanding the concept of a chemical element. This chapter suggests how to teach this concept including the distinction between an element and a compound. The periodic table introduces students to the chemistry of groups, the shared properties and chemical reactions of chemical elements in columns of the periodic table. Examples of the chemistry of three groups of chemical elements, namely the alkali metals (Group 1), the halogens (Group 7) and the noble gases (Group 0), is described with consideration of patterns in properties within the groups. Atomic structure and, in particular, the organisation of electrons within atoms helps students to understand why the periodic table has groups and periods, and how these explain element properties. The atomic model is therefore an underpinning concept that explains the periodic table's structure.

Group 1	Group 2											Group 3	Group 4	Group 5	Group 6	Group 7	Group 0
							1 **H** hydrogen 1										4 **He** helium 2
7 **Li** lithium 3	9 **Be** beryllium 4											11 **B** boron 5	12 **C** carbon 6	14 **N** nitrogen 7	16 **O** oxygen 8	19 **F** fluorine 9	20 **Ne** neon 10
23 **Na** sodium 11	24 **Mg** magnesium 12											27 **Al** aluminium 13	28 **Si** silicon 14	31 **P** phosphorus 15	32 **S** sulfur 16	35.5 **Cl** chlorine 17	40 **Ar** argon 18
39 **K** potassium 19	40 **Ca** calcium 20	45 **Sc** scandium 21	48 **Ti** titanium 22	51 **V** vanadium 23	52 **Cr** chromium 24	55 **Mn** manganese 25	56 **Fe** iron 26	59 **Co** cobalt 27	59 **Ni** nickel 28	63.5 **Cu** copper 29	65 **Zn** zinc 30	70 **Ga** gallium 31	73 **Ge** germanium 32	75 **As** arsenic 33	79 **Se** selenium 34	80 **Br** bromine 35	84 **Kr** krypton 36
85 **Rb** rubidium 37	88 **Sr** strontium 38	89 **Y** yttrium 39	91 **Zr** zirconium 40	93 **Nb** niobium 41	96 **Mo** molybdenum 42	[98] **Tc** technetium 43	101 **Ru** ruthenium 44	103 **Rh** rhodium 45	106 **Pd** palladium 46	108 **Ag** silver 47	112 **Cd** cadmium 48	115 **In** indium 49	119 **Sn** tin 50	122 **Sb** antimony 51	128 **Te** tellurium 52	127 **I** iodine 53	131 **Xe** xenon 54
133 **Cs** caesium 55	137 **Ba** barium 56	139 **La*** lanthanum 57	178 **Hf** hafnium 72	181 **Ta** tantalum 73	184 **W** tungsten 74	186 **Re** rhenium 75	190 **Os** osmium 76	192 **Ir** iridium 77	195 **Pt** platinum 78	197 **Au** gold 79	201 **Hg** mercury 80	204 **Ti** thallium 81	207 **Pb** lead 82	209 **Bi** bismuth 83	[209] **Po** polonium 84	[210] **At** astatine 85	[222] **Rn** randon 86
[223] **Fr** francium 87	[226] **Ra** radium 88	[227] **Ac**** actinium 89	[261] **Rf** rutherfordium 104	[262] **Db** dubnium 105	[266] **Sg** seaborgium 106	[264] **Bh** bohrium 107	[277] **Hs** hassium 108	[268] **Mt** meitnerium 109	[271] **Ds** darmstadtium 110	[272] **Rg** roentgenium 111	[285] **Cn** copernicium 112	[284] **Nh** nihonium 113	[289] **Fl** flerorium 114	[288] **Mc** moscovium 115	[293] **Lv** livermorium 116	[294] **Ts** Tennessine 117	[294] **Og** Oganesson 118
Key relative atomic mass **atomic symbol** name atomic (proton) number			*	140 **Ce** cerium 58	141 **Pr** praseodymium 59	144 **Nd** neodymium 60	(145) **Pm** promethium 61	150 **Sm** samarium 62	152 **Eu** europium 63	157 **Gd** gadolinium 64	159 **Tb** terbium 65	162 **Dy** dysprosium 66	165 **Ho** holmium 67	167 **Er** erbium 68	169 **Tm** thulium 69	173 **Yb** ytterbium 70	175 **Lu** lutetium 71
			**	232 **Th** thorium 90	231 **Pa** protactinium 91	238 **U** uranium 92	237 **Np** neptunium 93	(244) **Pu** plutonium 94	(243) **Am** americium 95	(247) **Cm** curium 96	(247) **Bk** berkelium 97	(251) **Cf** californium 98	(252) **Es** einsteinium 99	(257) **Fm** fermium 100	(258) **Md** mendelevium 101	(259) **No** nobelium 102	(260) **Lr** lawrencium 103

☐ Metal ■ Non-metal ☐ Difficult to classify

Figure 4.1 The periodic table

Students' prior knowledge and misconceptions

Children meet chemical elements within their experiences of 'stuff' (see Chapters 1 and 3). They are likely to be aware of metals such as gold, silver, copper, aluminium and iron, as well as non-metals oxygen and carbon (as graphite and diamond). They may have seen coloured advertising signs made by passing electricity through the noble gases, including neon (red-orange), argon (blue) and krypton (green); and helium-filled balloons at parties and celebrations. Older children and students may know that semiconductors in electronics include silicon; vehicle parts and other metal objects are chromium-plated for durability and shine; and aircraft and replacement hip joints feature titanium, a light and strong metal. These experiences mean that although students may have good levels of knowledge and awareness about substances that chemists recognise as chemical elements, they may not appreciate that these substances are examples of only around 100 special and distinctive substances known as chemical elements, available on Earth in finite quantities. They are also unlikely to understand how these are organised.

Students will know about and use many simple chemical compounds, such as salt (sodium chloride), water and baking soda (sodium hydrogencarbonate), as well as, for example, complex compounds in sugar and other carbohydrates, vitamins, proteins such as albumin (in egg white), and medicines. Besides air, they may know of and/or regularly use mixtures including cosmetics, paints, plant fertilisers and metal alloys. However, they will probably be unaware that these are compounds and lack awareness of differences between elements, compounds and mixtures. Students will tend to group and distinguish substances by their macroscopic properties only. A standpoint that some students may take is that the elements are earth, air, fire and water. This definition, proposed over two thousand years ago in ancient Greece, is not accepted scientific fact, but persists in astrology and other pseudo-sciences.

Research on students' misconceptions in this area is limited. Evidence suggests that students struggle to recognise periodic trends in data. Scaffolding by the teacher is required to support students appreciating periodic trends from data about chemical elements. Focusing on macroscopic phenomena – in other words, the observable properties of elements – provides a starting point but an understanding of atomic structure is required to explain the trends and properties observed.

4 The periodic table

Progression

In pre-school, children aged 3–5 place items (perhaps buttons, toy cars, leaves, or parts of a toy tea set) in groups based on properties including colour, size or shape. In primary school, grouping by properties becomes more sophisticated, for example, according to whether substances are solids, liquids or gases, and by physical properties such as magnetism, hardness, solubility, transparency and conductivity. In addition, children learn to classify living organisms, use simple keys to identify a range of animals, and distinguish between types of flowering and non-flowering plants.

Students aged 11–14 learn about the particle model of matter and use it to explain the different properties of solids, liquids and gases. They make observations about chemical elements, and establish differences between mixtures, elements and compounds. These ideas provide a foundation for understanding how the chemical and physical properties of elements vary, and thus the principles underpinning modern versions of the periodic table. Students learn positions and properties of metals and non-metals, then predict reaction patterns by referring to the periodic table. Properties of metal and non-metal oxides are often taught to this age group (see Chapter 7).

At the age of 14–16, students learn the chemistry of elements in groups of the periodic table. They use the position of the element relative to others in the group to predict and explain an element's properties. They learn that the modern periodic table is arranged by atomic number and to relate the number of electron shells and the number of electrons in the outer shell of an atom to the position of an element in the periodic table.

Post-16, students explore the link between atomic structure and the periodic table in greater depth and detail. Chemical and physical properties of elements in additional groups are studied, together with some transition metals (the lines of ten elements in the centre of the table). This knowledge provides a foundation for university-level inorganic chemistry, which features detailed consideration of the chemical and physical properties of an extensive range of elements, periodic changes, and the reactions between elements and organic compounds in organometallic chemistry.

Teaching sequence overview

→ Introducing the periodic table.
→ What is an element?
→ Grouping elements based on properties: Groups 1, 7 and 0.
→ Atomic structure and electron configurations.

 # Introducing the periodic table

Historical overview

Perhaps more than any other topic in chemistry, the periodic table is the study of ideas and models. The periodic table we use today is a product of successive ideas and models. The development of the periodic table coincided with discoveries of new elements and understanding the internal structure of the atom (see Section 4.5).

Chemical elements, such as gold, diamonds (carbon), silver and iron have been used in civilisations since ancient antiquity. As science developed, new elements were identified in increasing numbers. In the eighteenth century, scientists identified oxygen, nitrogen, phosphorus and platinum among others. The industrial revolution that began in Europe from around 1800 led to new chemical techniques that led to discoveries of many more elements.

Several scientists attempted to order the elements, or identify an underlying structure, but none of these early arrangements withstood new additions. By 1863, 56 elements were known, with new elements being added at the rate of about one each year. Dmitri Mendeleev (1834–1907), a Russian chemist and physicist, made the most significant contribution to the development of the periodic table. He is sometimes referred to as the father of the periodic table. Mendeleev made two assumptions which made his method of ordering the elements superior to previous ones. Firstly, he realised that not all elements had been discovered, so he left spaces for new elements where he thought they might fit. Secondly, although he initially assumed the elements should be placed in order of atomic weight, he knew these values could be erroneous. So, if elements seemed to fit better elsewhere on the basis of their properties Mendeleev located them accordingly. Mendeleev noted that chemical and physical properties change across the rows of the table and that patterns of change occur across all rows. This observation is known as the periodic law, and the rows are called periods.

Figure 4.2 shows Mendeleev's original periodic table. Mendeleev's basic organisation survives, but the periodic table has developed over the intervening 150+ years. Even today, the periodic table is not fixed but evolving. Periodic tables in classrooms may be out of date due to the recent discoveries and naming of new elements.

The modern periodic table (Figure 4.1) arranges elements by atomic number, not atomic mass. Atomic number (Z number, see Enrichment, below) is the number of protons in an atom. When Mendeleev devised his table, the internal structure of atoms, including

4 The periodic table

Reihen	Gruppo I. — R²O	Gruppo II. — RO	Gruppo III. — R²O³	Gruppo IV. RH⁴ RO²	Gruppo V. RH³ R²O⁵	Gruppo VI. RH² RO³	Gruppo VII. RH R²O⁷	Gruppo VIII. — RO⁴
1	H=1							
2	Li=7	Be=9,4	B=11	C=12	N=14	O=16	F=19	—
3	Na=23	Mg=24	Al=27,8	Si=28	P=31	S=32	Cl=35,5	—
4	K=39	Ca=40	—=44	Ti=48	V=51	Cr=52	Mn=55	Fe=56, Co=59, Ni=59, Cu=63.
5	(Cu=63)	Zn=65	—=68	—=72	As=75	Se=78	Br=80	
6	Rb=85	Sr=87	?Yt=88	Zr=90	Nb=94	Mo=96	—=100	Ru=104, Rh=104, Pd=106, Ag=108.
7	(Ag=108)	Cd=112	In=113	Sn=118	Sb=122	Te=125	J=127	
8	Cs=133	Ba=137	?Di=138	?Ce=140	—	—	—	—
9	(—)	—	—	—	—	—	—	—
10	—	—	?Er=178	?La=180	Ta=182	W=184	—	Os=195, Ir=197, Pt=198, Au=199.
11	(Au=199)	Hg=200	Tl=204	Pb=207	Bi=208	—	—	
12	—	—	—	Th=231	—	U=240	—	—

Figure 4.2 Mendeleev's periodic table. Note the spaces under aluminium (Al) and silicon (Si).

the existence of protons, was unknown. In fact, at the time, some chemists contested the existence of atoms, making the accuracy of Mendeleev's table even more astonishing. By ignoring atomic weight values that did not fit known patterns of chemical properties, Mendeleev built a periodic table that was useful to chemists.

Mendeleev's genius was realising the predictive power of the periodic table. He predicted properties of undiscovered elements from patterns in properties of known elements. For example, he predicted the existence of an element below silicon (see Figures 4.1 and 4.2), which he called eka-silicon (*eka* means 'one' in Sanskrit). This element was isolated in 1886 and named germanium (symbol: Ge). Germanium's properties closely align with those predicted by Mendeleev. Other predictions were for eka-aluminium, found in 1875 and named gallium (Ga); and eka-boron found in 1875 and named scandium (Sc).

Discoveries of elements and developments in the structure of the periodic table continued. The gases helium (from the Greek word *helios* meaning 'the Sun') and argon (from the Greek word ἀργόν meaning 'idle' or 'lazy') were discovered in 1894 and 1895 respectively by the Scottish chemist William Ramsay. Similarities in their chemical and physical properties led Ramsay to suspect that these elements were from a new group of the periodic table. Ramsay's experiments led to isolation of neon, krypton and xenon in 1898 (see Enrichment, Chapter 3). They were added as the noble or inert gases in 1902 (Group 0 in Figure 4.1).

At the time of writing, 118 chemical elements have been identified, of which 92 are naturally occurring. Numbers 93–118 are synthetic elements made by fusing atoms of other elements together. They are named after scientists involved in creating them or places where they were made. For example, element number 106, first synthesised in 1974, is named seaborgium (Sg), after American nuclear scientist Glenn T. Seaborg. By the late 1990s, techniques had developed that allowed researchers to produce seaborgium at the rate of one atom an hour. The short half-life of the seaborgium atoms made finding out about their chemistry very difficult. Nevertheless, properties could be predicted using the periodic table. Seaborgium was expected to be similar to the elements above it (molybdenum (Mo) and tungsten (W)), so would have a valency of 6 and react with oxygen and chlorine to form an oxychloride with the formulae SgO_2Cl_2. Using fewer than ten atoms, predictions made about the chemistry of seaborgium proved correct.

In 2015, a team led by Japanese chemist Kōsuke Morita was the first from Asia to be credited with the synthesis of an element, number 113, later named nihonium (from the Japanese name for Japan, *Nihon*). Elements 113–118 complete the last line of the periodic table. Any further elements would require addition of a new period.

The chemical elements can be organised in alternative representations (see Key activity: Other arrangements of the chemical elements, below).

> **Enrichment**
>
> Mendeleev's periodic table needed one crucial piece of information to make it correct. This was provided by Henry Moseley (1887–1915), a physicist at the University of Oxford. Moseley's experiments used what was then a new technique, X-ray spectroscopy, to investigate how atomic number related to the charge of an atomic nucleus. He showed that the number of protons was the defining difference between chemical elements. This led to atomic number being adopted as the organising principle for the periodic table, as this resolved anomalies created by ordering elements by atomic weight. Atomic number is represented by Z, an abbreviation of *Zahl* which is German for 'number'. A soldier in the First World War, Moseley was killed at Gallipoli in Turkey.

Understanding the structure of the periodic table

KEY ACTIVITY

Finding chemical elements in the periodic table

It is very helpful to develop students' familiarity with the location of elements in the periodic table. These activities aid development of this knowledge. Students should also be able to label groups and periods in the periodic table. Note that period 1 contains just hydrogen and helium.

- A class periodic table could be constructed if students each research 1–3 elements and present the information on a standard-sized piece of paper. These can be joined together to make a very large periodic table for display. To support this, students can explore the interactive periodic table produced by the Royal Society of Chemistry (see Section 4.6).
- To locate elements in the periodic table, play element bingo. Use a periodic table with either names or symbols removed. Call out names (assuming these have been removed) so students have to match symbols to the names. Play the other way round, calling symbols to which students should match element names. The first student to get, say, ten elements should shout 'Mendeleev!' These resources will help:
  - Periodic table with symbols missing: https://14823.stem.org.uk/page/modules/iveperiodictable/Periodic%20Table%201.pdf
  - Periodic table with names missing: https://14823.stem.org.uk/page/modules/iveperiodictable/PeriodicTable3.pdf

- Alternatively, students could use a 3 × 3 grid with symbols (or names) on so they just have to gain 3 or 9 to finish. For students who are likely to find this challenging, focus on the elements they need to know (usually elements 1–20, or/and hydrogen and the elements of groups 1, 2, 6, 7, and 0).
- Favourite elements. This resource comprises short videos of well-known people talking about their favourite chemical elements. Use these to introduce specific elements. Students can find the elements on the table and investigate them further. They could also make their own videos and podcasts of their favourites. https://www.stem.org.uk/resources/collection/3952/my-favourite-element-suitable-home-teaching

Additional suggestions to familiarise students with the elements in the periodic table are provided in Section 4.6.

Introducing groups of the periodic table

KEY ACTIVITY

This video introduces some chemical elements: www.stem.org.uk/resources/elibrary/resource/30887/periodic-table-ferocious-elements-suitable-home-teaching. Answers to the questions are in the video.

Questions to ask
- How many chemical elements are there? How many are metals?
- Where are the elements, gold, silver and copper in the periodic table?
- Why are gold, silver and copper used for jewellery?
- What are the trends in reactivity of the Group 1 and Group 7 elements?
- Name a chemical element that is extracted from its compounds.

The metal–non-metal boundary

KEY ACTIVITY

Most chemical elements are metals. Non-metals tend to be on the right-hand side of the periodic table. This activity shows students the 'staircase' between the metals and non-metals.

Starting at boron, 'walk' diagonally to silicon (Si), then to arsenic (As), tellurium (Te) and finally astatine (At). A diagonal line drawn between each of these elements and the one below them gives a staircase that separates metals and non-metals.

To help recognise the metal–non-metal boundary, they can learn this staircase by writing a walk with made up names. These might start with Boron Beach and end with Astatine Avenue. Students must say where to, for example, turn right onto Silicon Street, look out for number 33 Arsenic Alley, take another right turn into Tellurium Terrace and finally look for 85 Astatine Avenue. They can draw the route onto a periodic table to mark the boundary. Students learn that metals are 'under the stairs'. As long as they know where the staircase begins, they can identify an element as a metal or non-metal.

4 The periodic table

> **KEY ACTIVITY**
>
> ## Other arrangements of the chemical elements
>
> Scientists have devised ways of arranging the chemical elements that highlight alignments, properties, characteristics and of course, use many languages. These variants of the periodic table are easy to find using the internet. A selection current at the time of writing is provided, together with suggested questions that students can answer. These questions can change depending on the arrangements presented. Alternatively, students can choose one variant of the periodic table for themselves and give a presentation describing their choice.
>
> A common feature is that element symbols, atomic numbers and mass numbers are the same in every version of the table. Groups in the standard periodic table are also aligned in these versions.
>
> - The chemical elements vary in abundance. Some are very rare. This version of the periodic table shows the abundance of the chemical elements: https://cdn.mos.cms.futurecdn.net/beXS4QFQdwVSpuJKzH2sre-320-80.jpg
> - A non-conventional arrangement of the chemical elements which adopts a circular arrangement is available at https://images.newscientist.com/wp-content/uploads/2019/02/25140237/table.jpg
> - A race-track periodic table is available at https://66.media.tumblr.com/2db188198073f0cc2a7a1babff19faf8/tumblr_n3tbz5rlKk1s3r80lo3_1280.jpg
> - This version places groups as if they are spokes of a wheel: https://upload.wikimedia.org/wikipedia/commons/5/55/Alternative_circular_periodic_table.png
> - This periodic table is in Russian: www.sliderbase.com/images/referats/151b/(3).PNG
> - This periodic table is in German: https://upload.wikimedia.org/wikipedia/commons/thumb/b/bf/Periodic_table_%28German%29_EN.svg/1473px-Periodic_table_%28German%29_EN.svg.png
>
> **Questions to ask**
> - What are the common features of these arrangements of the periodic table?
> - Why are some chemical elements more abundant than others?
> - Why are non-standard periodic tables useful?

 ## What is an element?

In chemistry, the term element refers to a discrete number of substances (118 at the time of writing). Figure 9.2 shows elements as a sub-group of substances distinct from compounds and separate from mixtures. Students need to understand what chemists mean by the term element. Once grasped, they can learn the distinction between elements and compounds.

Figure 4.3 applies two sides of Johnstone's triangle (see Figure 1.1) to the notion of an element. The figure shows three overlapping meanings for element based on it having particular chemical properties (macroscopic); as a substance comprising a type of atom with a specific number of protons (sub-microscopic); and as something common to itself and its compounds (abstract).

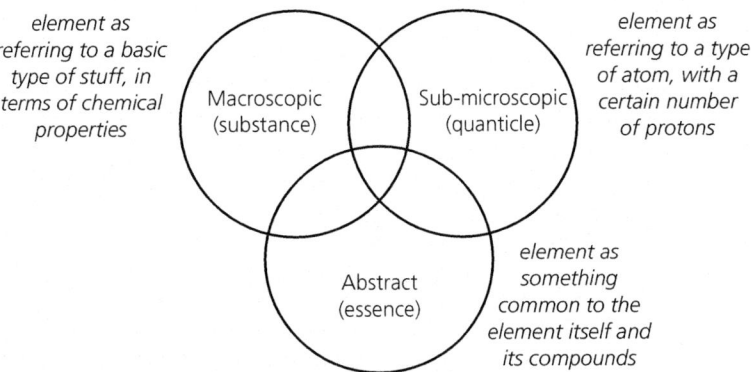

Figure 4.3 Three facets of the concept of element

Explaining exactly what an element is can be complicated. Macroscopically, an element is a substance that cannot be converted to a simpler substance by chemical means, unlike a compound: elements are a fundamental type of 'stuff'. Sub-microscopically, based on its atoms, an element is a substance in the periodic table, with a fixed position based on its unique atomic number. It is important that students realise the role of models in explaining science ideas. As their knowledge grows and new ideas are met, they can transition to the use of advanced models for explanations. For 11–13-year-olds, who have not been taught about isotopes or the structure of the atom, the definition of an element is usually that an element is *a substance or chemical where all the atoms are the same*. This is further complicated in that they have often not yet met the idea of atoms either. Once the structure of the atom and the existence of isotopes are introduced, the definition of an element becomes *a substance where all the atoms have the same atomic number or number of protons*. When introducing increasingly sophisticated models in chemistry, emphasise to students that they have learned a more advanced model, *not* that the previous definition was 'wrong'. Similarly, scientists revisit definitions and theories as their experimental results dictate.

The definition of an element enables students to contrast this with those for compounds and mixtures. A compound is two or more elements chemically joined together, so the original chemical properties of the elements no longer exist: a compound is a new substance. A mixture comprises two or more substances which are not chemically joined, or which can be separated. These definitions also alter as students continue studying chemistry; for example, some chemical reactions which form compounds are reversible, and some mixtures are very difficult to separate. Chemical change is discussed in Chapter 2.

A further complication is that some elements exist in more than one form, as allotropes. Graphite, diamond, graphene and buckminsterfullerene (C_{60}) are elemental forms of carbon (allotropes) yet have contrasting physical properties. Oxygen (O_2) and ozone (O_3) are both gaseous forms of atoms with atomic nuclei containing eight protons but have different chemical properties. Students must accept that two or more substances are the same element on the basis that they are made of the same type of atoms, which cannot be seen. Carbon allotropes seem to be more easily accepted as the same element than oxygen and ozone. For oxygen and ozone, that the differences are in chemical properties make the relationship harder to believe. Allotropes illustrate that the language and concepts chemists use to describe substances have limitations and are imperfect.

KEY ACTIVITY

Students' prior knowledge about the chemical elements

Start by finding out what students know about the chemical elements and the periodic table. These questions might be useful:

- What chemical elements do you know about?
- What symbols are used to represent these elements?
- Why do chemists use symbols to represent chemical elements?
- How many chemical elements have been discovered?
- How do you know if a substance is a chemical element?
- Name a scientist who has discovered a specific chemical element. What evidence led to the discovery?
- Which chemical elements are found as native elements, that is, directly from the ground, or via mining?
- If a chemical element is not found in the ground or via mining, where does it come from?

To date, 118 chemical elements are known. The newest elements, atomic numbers 104–118, are the superheavy elements (SHEs), all radioactive. Elements 113–118 complete the seventh row (period) of the periodic table (see Figure 4.1). Detailed information about discovery of the chemical elements can be found at https://en.wikipedia.org/wiki/Timeline_of_chemical_element_discoveries

4.2　What is an element?

Students may have heard of Marie Curie (who discovered both radium and polonium in 1898) and possibly Joseph Priestley (with Wilhelm Scheele, he is credited with discovering oxygen in 1774). Students may think Einstein discovered the chemical element that bears his name, einsteinium. Einstein was not involved in its discovery. He died in 1955 before the new element was named.

Relatively inert precious metals, such as gold, silver, copper and platinum, are found native, that is, in their elemental metal state. Some elements, such as the SHEs, are artificially prepared in experiments in cyclotrons ('atom smashers'). Others are extracted from minerals in the Earth's crust. Students are likely to have heard of gases in the atmosphere (oxygen, nitrogen, argon) and other gases such as hydrogen and helium. They may know about calcium (in milk, bones), aluminium (foil), carbon (graphite, diamonds), iron, zinc, copper, gold and silver. Metal alloys, such as bronze, brass and pewter, are not elements. Water, earth, fire and air were regarded as elements in ancient Greece and for many years afterwards, but this is no longer accepted science.

To extend this activity, students can investigate the discovery of one or more elements in greater depth and present their findings to other class members. Images of scientists involved, their laboratories and biographical details are widely available on the internet. Emphasise that knowledge of the chemical elements draws on centuries of international scientific effort. Facts we take for granted, such as the composition of air, were once unknown, and finding this out was at the cutting edge of scientific discovery.

KEY ACTIVITY

Introducing the chemical elements

This activity introduces chemical elements as pure substances, or fundamental types of stuff and invites students to classify them based on macroscopic properties. The activity familiarises students with chemical elements and their places on the periodic table.

Have samples of chemical elements, or objects made from single elements, available around the room labelled with the symbol, molecular/atomic model, boiling point, melting point and atomic number. Avoid elements stored under liquid, such as the Group 1 elements and phosphorus. Elements that could be included are: iron, copper, iodine, aluminium, carbon, sulfur, zinc, chromium, nickel, gold, silver, platinum, zirconium, argon, nitrogen, oxygen, helium, tin, lead, magnesium, titanium, manganese, carbon, neon, argon, silicon.

Ask students to classify/group the elements as logically as they can, based on the information available.

4 The periodic table

> **Questions to ask**
> - What do they notice about the elements?
> - What tests might they do to establish how to classify the elements?
> - On what basis do they classify/group the elements?
> - Why did they choose these criteria?
>
> There is no correct or expected answer. Students should think about classification and sorting elements logically, so that they are prepared for examining the periodic table in greater depth. An essential point is that elements all contain one type of particle (setting aside isotopes for the moment). Support this with particle diagrams. Suitable sources to support this are listed in Section 4.6.
>
> Revisit that scientists considered how to classify chemical elements for many years before the periodic table was proposed, and that the table has developed through the work of scientists worldwide over time. Reinforce the definition of chemical element at macroscopic and sub-microscopic levels: elements are substances with specific properties, comprising atoms with the same number of protons in the nucleus.

 Elements and compounds

Having established the concept of an element, students need to understand the difference between an element and a compound. This builds on particle theory (Chapter 3) and leads towards understanding chemical change (Chapter 2). Comparing elements with substances that are not elements, in other words, mixtures (discussed in Chapter 9) and compounds is important to establish the concept of an element. Students benefit from opportunities to investigate how elements change as they mix and react to form compounds. Examples of reactions between chemical elements are provided in Chapter 2. Students can carry out the reaction between iron and sulfur (see below) safely. A common misconception is that elements are unchanged in compounds: if this were true, adding table salt (sodium chloride) to food would mean eating a highly reactive metal (sodium) and a very toxic gas (chlorine). Chemical and physical changes are explored further in Chapter 2. Studying elements, mixtures and compounds leads to use of chemical symbols and formulae (see Chapter 2) and consideration of chemical bonding (see Chapter 5). A chemical formula tells us that elements are 'there' but 'not there', as explained in Chapter 2. The activity suggested below demonstrates this point.

4.3 Elements and compounds

KEY ACTIVITY

From elements to a compound: the reaction between iron and sulfur

There are few examples of direct reactions of two elements to form a compound that students can perform safely, but the iron and sulfur reaction is one such. Full practical details are available at https://edu.rsc.org/resources/iron-and-sulfur-reaction/713.article

When heated, iron and sulfur form a new substance, iron sulfide, a compound made from elements iron and sulfur. Iron and sulfur are not present as elemental iron and sulfur in iron sulfide. A word equation is:

$$\text{iron} + \text{sulfur} \rightarrow \text{iron sulfide}$$
$$\text{element} \quad \text{element} \quad \text{compound}$$

Introduce iron and sulfur as elements. Locate these in the periodic table. Students will explore what happens when they are heated and react to form a compound. Before heating, give students an opportunity to investigate the properties of iron and sulfur separately and mixed. Table 4.1 is a suggested results table. Adding the elements to water in a test tube can be quite messy so you may prefer to demonstrate this.

Table 4.1 Iron and sulfur: elements and a compound

	Appearance	Test with a magnet	Add to water
Iron	Dark grey filings	Attracts	Sinks
Sulfur	Yellow powder	Does not attract	Floats
Iron and sulfur mixture	Yellow and grey grains	Iron attracts, sulfur does not attract	Iron sinks, sulfur floats
Reacted product, iron sulfide	Dark grey solid	May attract weakly	Sinks

Iron filings are attracted to the magnet. Covering magnets in plastic (clingfilm or similar) so they can be cleaned easily is a good precaution against the magnets becoming coated in filings. Alternatively, the elements and the mixture can be given to students in sealed containers (see instructions). The mixture separates (roughly) when added to water, although less cleanly than with a magnet. These tests show the iron/sulfur mixture can be separated. Emphasise this separation, as this distinguishes the mixture from the compound.

Next, make the compound: students can do this themselves or watch a demonstration. For students, ignition tubes (10 mm × 75 mm borosilicate test tubes) are advised. For a demonstration, use a borosilicate test tube. Note that initially the mixture will turn red – this is just the iron getting hot. The sulfur turns red as it changes state. Keep heating. When the reaction takes place the mixture *glows* red. At this point stop heating and allow the tube to cool down.

A small amount of sulfur dioxide gas may be produced. Placing mineral wool in the mouth of the test tube should absorb it. Have a damp cloth on hand to place on the mouth of the tube to absorb any sulfur dioxide that escapes. Sulfur dioxide may cause skin burns and eye damage and is toxic if inhaled.

The iron sulfide produced often fuses to the glass tube. The ignition tube may crack, and gentle tapping on the heat-resistant mat may be sufficient to remove the product, noting that eye protection must be worn throughout the whole procedure. If the tube does not crack, tapping may be enough to remove some product from the mouth of the tube once the mineral wool plug has been taken away. If none of this works, the teacher or a technician may need to cover the tube with a cloth then hit it firmly with a hammer.

There is evidence that a reaction has taken place because:

- The separate colours of iron and sulfur are no longer visible.
- A magnet cannot separate the elements. Sometimes a small amount of unreacted iron remains as well as the iron sulfide product and so a slight attraction to the magnet of remaining iron filings may occur. Using a weak magnet reduces the likelihood of this.
- If the product is placed in water, it does not separate into iron and sulfur.

Questions to ask

- Are iron and sulfur present in the product? In what way?
- What was happening when iron and sulfur were heated together?
- What are the differences between iron and sulfur as elements and as a compound?
- What is the difference between an element and a compound?

4.4 Grouping elements based on properties: Groups 1, 7 and 0

The idea that elements are grouped according to their properties is relatively straightforward for students to accept (see sections on Students' prior knowledge and misconceptions and Progression, above). Three groups are discussed: Group 1, the alkali metals; Group 7, the halogens; and Group 0, the noble gases. Studying the groups can follow on from study of atomic structure and electron configurations. The route depends on whether teachers wish students to recognise and know about patterns of properties in groups before explaining them.

Group 1: The alkali metals

Group 1, the alkali metals, is a good starting point. The alkali metals are the first column on the left-hand side of the periodic table (Figure 4.1). They are lithium (Li), sodium (Na), potassium (K), rubidium (Rb), caesium (Cs) and francium (Fr). Lithium, the lightest metal discovered in 1817, is named from the Greek word *lithos* meaning 'stone'. Sodium, first isolated in 1807, is named from the Latin for sodium carbonate, *natrium*. It is the seventh most abundant element on Earth and the fifth most abundant metal. The name of potassium, an element also isolated in 1807, is derived from the English word potash, as one method of extracting certain potassium salts involves putting ashes from burnt wood in a pot. Its symbol, K, is from neo-Latin *kalium*. Spectroscopy (see Chapter 9) enabled identification of rubidium and caesium in 1861, named from Latin *rubidus* for 'deep red' and *caesius* for 'sky blue' after red and blue lines found in their respective spectra. Francium, named after France, was the last naturally occurring element to be discovered, and was isolated by Marguerite Perey in 1939. It is extremely radioactive.

KEY ACTIVITY

Reactivity of the Group 1 metals

This is a teacher demonstration, taking 10–20 minutes. Full instructions are available at https://edu.rsc.org/resources/reactivity-trends-of-the-alkali-metals/731.article Videos showing the reaction of rubidium and caesium with water are available via the RSC interactive periodic table at www.rsc.org/periodic-table/video. The reactivity of the alkali metals increases down the group. This means that lithium is the least reactive and caesium is the most reactive.

Table 4.2 is a suggested results table with outcomes. Flame test colours and information about the solubility of salts of the alkali metals are given in Chapter 9. A visualiser or camera with a screen helps students to see details.

Table 4.2 Group 1: The alkali metals

Element	Symbol	Appearance	Physical properties	Reaction with the atmosphere	Reaction with water
Lithium	Li	Dark grey; shiny grey when cut	Hard to cut with a knife; floats on water	Reacts with oxygen to form lithium oxide	Holds its shape; forms hydrogen gas and lithium hydroxide
Sodium	Na	Light grey; shiny surface	Cuts quite easily with a knife; floats on water	Reacts with oxygen to form sodium oxide	Rolls into a ball; forms hydrogen gas and sodium hydroxide
Potassium	K	Blue-grey; shiny surface	Cuts very easily with a knife; floats on water	Reacts with oxygen to form potassium oxide	Rolls into a ball; forms hydrogen gas which ignites; the heat generates a lilac flame and potassium hydroxide
Rubidium	Rb	Silver-grey	Very soft; prepared in glass ampoules; floats on water	Very reactive; forms rubidium oxide	Rolls into a ball and reacts vigorously; forms hydrogen gas and rubidium hydroxide

4 The periodic table

Element	Symbol	Appearance	Physical properties	Reaction with the atmosphere	Reaction with water
Caesium	Cs	Pale gold; shiny	Very soft; prepared in glass ampoules; sinks in water	Very reactive; forms caesium oxide	Explodes, forming hydrogen gas and caesium hydroxide
Francium	Fr	Unknown, possibly red	Radioactive; probably liquid at room temperature	Cannot be tested but probably similar to caesium	

Note: Elements shaded grey cannot be tested in a school laboratory

Appearance

Show students the metals lithium, sodium and potassium in turn. They generally have a grey, white or blue crust on the outside, which is the metal oxide or, for lithium, a mixture of nitride and oxide. The metals are shiny when freshly cut.

Physical properties

Draw students' attention to alkali metals being stored under oil, due to their high reactivity with the atmosphere. Lithium can be cut using a scalpel, although this is fairly hard work. The metal conducts electricity: test by connecting the sample into a simple circuit (see instructions from link above). Lithium is the lightest metal: put a sample into a petri dish, tape it closed, and pass this around so students can observe the cut surface and sense the density. This can be repeated with the other metals for comparison. Sodium is easier to cut than lithium; likewise, potassium is easier to cut than sodium. The metals become softer down the group. Students can research melting points of the metals and predict how easy it may be to cut rubidium, caesium and francium. Videos about each element may help here. The alkali metals float as they have low density compared to other metals. Students may benefit from seeing that a lump of iron or copper sinks.

Reactions with water

Full instructions are provided via the link above. In addition, consider these points:

- Add one drop of washing up liquid to the water before the alkali metal to reduce the surface tension. This prevents the metals from just sticking to the side of the trough and reduces the likelihood of the metal spitting out.
- Indicator can be added to the water: phenolphthalein is colourless at the start and turns pink where the metal passes. Universal Indicator is green and turns purple. The indicator changes colour as the metal forms the metal hydroxide, which immediately dissolves to form an alkaline solution. Avoid students thinking that the metal causes the colour change: adding indicator at the end of the reaction, or testing the water with litmus/Universal Indicator paper, may be preferable.
- Ensure metal pieces added to water are about rice-grain sized. Replace all other pieces in their bottles and seal the lids. From a safety perspective, slightly more lithium can be used but keeping sizes constant provides a better comparison of properties.

4.4 Grouping elements based on properties: Groups 1, 7 and 0

- Blot off oil, take the sample in tweezers and hold it over the water. Do not flick the metal piece: open the tweezers and tap them firmly on the trough to keep control of where the metal lands. Once the metal is in the water, place a Perspex sheet over the trough.
- Health and safety advice recommends that the water is changed between each metal in case an increase in temperature occurs that contributes to a significant increase in the rate at which the metals react with water.
- Again draw students' attention to the fact that alkali metals float on water due to their low density. Demonstrate that iron and copper sink if not done already.
- Videos of rubidium and caesium reacting with water finish the pattern. Some videos fake the reactions of these metals, so take care to choose genuine ones.
- Francium completes the group but is radioactive, so its reactions are unknown. Challenge students to predict the expected properties given the pattern already seen.

Establish that all the alkali metals react with water to produce the metal hydroxide and a gas, hydrogen. The trend in increasing reactivity is demonstrated by showing that the reaction with water becomes more vigorous as we proceed down the group from lithium to sodium then potassium.

The heat produced may be sufficient to ignite the hydrogen gas, creating a lilac (potassium) or orange (sodium) flame for these metals. An equation for the reaction (see Chapter 2), with lithium as an example, is:

$$\text{lithium} + \text{water} \rightarrow \text{lithium hydroxide} + \text{hydrogen}$$

$$2\text{Li(s)} + 2\text{H}_2\text{O(l)} \rightarrow 2\text{LiOH(aq)} + \text{H}_2\text{(g)}$$

The metals also react with oxygen in the air, for example:

$$\text{sodium} + \text{oxygen} \rightarrow \text{sodium oxide}$$

$$2\text{Na(s)} + \text{O}_2\text{(g)} \rightarrow \text{Na}_2\text{O(s)}$$

Overall, reactivity increases down the group: sodium is more reactive than lithium, and potassium is more reactive than sodium.

Questions to ask

- Find the alkali metals on a periodic table: what trends in patterns occur in the reaction with water and the metals' physical properties?
- Write word (and, if appropriate, symbol) equations for the reactions between:
  - sodium and water
  - potassium and oxygen
  - lithium and water.
- Predict the properties of rubidium and caesium. Complete the table to suggest the elements' appearance, physical properties and chemical reactions.

Videos of these reactions can be shown for students to see whether their predictions were correct.

Group 7: The halogens

Halogen means 'salt-former' from the Greek *hals* for 'salt' and '—gen' meaning 'giving birth to', recognising that a salt forms in reactions involving these elements. The halogens are the non-metals fluorine (F), chlorine (Cl), bromine (Br), iodine (I) and astatine (At). The group is the second column from the right-hand side of the periodic table (Figure 4.1). Fluorine is the most reactive non-metallic element. At room temperature and pressure, fluorine and chlorine are gases, bromine is a liquid and iodine and astatine are solids. The first four elements exist as diatomic molecules, F_2, Cl_2, Br_2 and I_2. The synthetic superheavy element (SHE) tennessine (atomic number 117, symbol Ts) may also be a halogen. Astatine and tennessine are radioactive and their chemistry is unknown.

Table 4.3 Group 7: The halogens

Element	Symbol	Appearance	Physical state at room temperature	Chemical properties
Fluorine	F	Yellow	Gas	Extremely reactive.
Chlorine	Cl	Green-yellow		Weak acid in solution, turns blue litmus paper red then bleaches white
Bromine	Br	Dark orange	Liquid	
Iodine	I	Purple-black	Solid	Partially soluble in water. Stains litmus paper brown
Astatine	At	Unknown		Unknown
Tennessine	Ts			

Fluorine

The extreme reactivity and toxicity of fluorine means it is only available in research and industrial laboratories. A video demonstrating this is available at www.rsc.org/periodic-table/video. Fluorine reacts with all other elements except neon and helium and reacts on contact with most materials: a video of the reaction between fluorine and cotton wool makes this point well.

Chlorine

Chlorine can be generated in a fume cupboard. Show students that it is a yellow-green gas. The gas is highly toxic. Full instructions are available at https://edu.rsc.org/resources/liquefying-chlorine-gas/1740.article Test the procedure with an experienced chemist before attempting it in the classroom. Chlorine dissolves in water forming a clear colourless solution that is weakly acidic. Demonstrate that chlorine has bleaching properties by placing damp litmus paper into a gas jar of chlorine, ensuring the jar remains in a fume cupboard throughout.

Bromine

Bromine is reactive and very toxic, so must be handled only in a fume cupboard with great care. The element is normally purchased in small, fragile glass ampoules. Elemental bromine requires experience and expertise in handling. Bromine is a dark orange liquid that vaporises easily. Bromine is soluble in water, forming a yellow-red liquid, bromine water, which is safer to demonstrate. Note that layers of bromine and bromine water may form if the solution is left standing over a long period.

Iodine

Iodine is a purple-black crystalline solid that may be shown, with care, outside a fume cupboard. The solid stains skin on contact. Iodine sublimes on heating, becoming a gas without first melting to liquid. Iodine is partially soluble in water, but dissolves more readily in ethanol. In schools, iodine solution is a brown liquid comprising iodine dissolved in a solution of potassium iodide. Iodine solution has antiseptic qualities and was used in medical settings.

Astatine and tennessine

Chemists know little about astatine and tennessine, as they are very radioactive. Students can predict the physical and chemical properties of these elements.

KEY ACTIVITY

Reactivity trends in Group 7

Reactivity of the halogens decreases down the group, showing the opposite trend to the alkali metals. Fluorine is the most reactive and tennessine (in theory) is the least reactive. To show trends in halogen reactivity, a combination of laboratory-based and video material is valuable. In practice, only chlorine, bromine and iodine are tested in school laboratories.

Displacement reactions

This can be a demonstration or a class experiment. The experiment involves reacting chlorine water, bromine water and iodine solution with solutions of potassium chloride, potassium bromide and potassium iodide in displacement reactions. A more reactive halogen displaces a less reactive halogen from its compound in aqueous solution. Full instructions are available at www.rsc.org/learn-chemistry/resource/res00000733/reactions-of-halogens-as-aqueous-solutions Table 4.4 shows expected results. Note the names of elemental halogens end in —ine whereas compounds containing halogens end in —ide.

4 The periodic table

Table 4.4 Displacement reactions of chlorine, bromine and iodine

	Reaction with		
	potassium chloride solution	potassium bromide solution	potassium iodide solution
Chlorine water	No reaction	Orange-yellow colour of bromine appears	Dark brown colour of iodine appears
Bromine water	No reaction	No reaction	Colour darkens from orange-yellow to brown
Iodine solution	No reaction	No reaction	No reaction

Source: https://edu.rsc.org/resources/reactions-of-halogens-as-aqueous-solutions/733.article

Solutions of potassium bromide and potassium chloride are colourless. In the reaction between chlorine water and potassium bromide solution, chlorine displaces bromide ions from potassium bromide solution, producing bromine and chloride ions. Bromide ions become bromine atoms, which pair into molecules. The colour changes from pale-yellow to orange because bromine forms an orange solution. Chlorine atoms become chloride ions:

chlorine + potassium bromide → bromine + potassium chloride

$$Cl_2(aq) + 2KBr(aq) \rightarrow Br_2(aq) + 2KCl(aq)$$

Chlorine is more reactive than bromine, displacing it from aqueous solution. Potassium ions are spectators and do not participate. Similar exchanges occur when chlorine water is added to potassium iodide, and when bromine water is added to potassium iodide.

bromine + potassium iodide → potassium bromide + iodine

$$Br_2(aq) + 2KI(aq) \rightarrow 2KBr(aq) + I_2(aq)$$

The colour change in the reaction above is from orange to brown as iodine forms a brown solution. Solutions of potassium bromide and potassium iodide are colourless. Thus, chlorine displaces both bromine and iodine, bromine displaces only iodine, and iodine does not displace either chlorine or bromine.

Questions to ask
- Which halogen is the most reactive? How do we know?
- Which halogen is the least reactive? How do we know?
- Write word/symbol equations for the reactions between:
  - bromine and potassium iodide
  - chlorine and potassium iodide.
- How does the trend in reactivity relate to the positions of the halogens in Group 7?
- Predict how astatine and tennessine would react with chlorine water or bromine water.

4.4 Grouping elements based on properties: Groups 1, 7 and 0

Group 0: The noble gases

The noble gases are helium (He), neon (Ne), argon (Ar), krypton (Kr), xenon (Xe) and radon (Rn). The noble gas group is located on the right-hand side of the periodic table, numbered 0 (see Figure 4.1). These gases are relatively unreactive (but see Enrichment). Helium gas is available relatively easily from suppliers, but the heavier gases are very expensive: at the time of writing, for example, a balloon of xenon gas costs over £100 in the UK. They are best explored using video material, such as the lecture available at www.rigb.org/christmas-lectures/watch/2012/the-modern-alchemist/air. Segments of this lecture are referred to below. Short videos providing information about individual noble gases can be found via the RSC's interactive periodic table at www.rsc.org/periodic-table/video.

The elements are all colourless, odourless gases at room temperature and pressure. They tend not to form compounds easily. There are no known compounds of helium and neon: these two elements do not even react with fluorine. The SHE oganesson (number 118, symbol Og) is classified as a noble gas, but with a boiling point of 110 °C is solid at room temperature and pressure.

Density

The extreme lightness of helium can be demonstrated relatively easily using helium-filled balloons. Density of the noble gases increases down the group. Several very good videos demonstrate this, particularly the segment from 22:58–25:18 (minutes: seconds) in the video lecture *The Modern Alchemist* referenced above (see also Chapter 3). Allow students to see how balloons filled with the first four noble gases, helium to krypton, float or sink in air. Stop the video, then provide the short diagnostic task on xenon (number CSU4.1) available at www.stem.org.uk/best/chemistry-earth-science/big-idea-substances-and-properties. In this task students predict the density for xenon. Once they have completed the diagnostic, show the last part of the video, then discuss their answers. Relate this to increasing sizes of atoms of the elements (see below). Compare these with sizes of atoms of elements in the air (nitrogen, oxygen).

Electrical conductivity

Noble gases glow brightly when a high voltage passes through them. This property creates coloured lighting, particularly used for advertising signs. Segment 25:23–26:07 in *The Modern Alchemist* shows colours created by the first five noble gases. Videos about each element also illustrate this point.

Other uses of the noble gases

The lack of reactivity of the noble gases has led to uses, particularly in medicine. 26:15–28:14 in the lecture referenced above illustrates the use of helium in medicine to enhance the quality of images of human lungs. The medical uses of xenon are explored in the segment from 28:14–34:42.

Questions to ask

→ What is the trend in density of the noble gases?
→ Why are the noble gases also called the inert gases?
→ Where do we see the noble gases in everyday life?

> **Enrichment**
>
> Neil Bartlett (1932–2008), an English chemist from Newcastle-upon-Tyne, was the first person to make compounds from xenon, a noble gas element in Group 0. Chemists believed for many years that noble gases were completely unreactive, calling them inert. In March 1962, Bartlett was working alone one evening in his laboratory. He prepared a glass apparatus containing a very reactive red gas, platinum hexafluoride (PtF_6), on one side, with xenon (Xe), a colourless gas, on the other side of a glass seal. When Bartlett broke the seal, the gases reacted immediately at room temperature, producing an orange-yellow solid. Bartlett realised what had happened, saying 'At once I tried to find someone with whom to share the exciting finding, but it appeared everyone had left for dinner!' Convincing others of his discovery that the noble gases are not inert was the next challenge. Eventually, the orange-yellow solid was identified as xenon hexafluoroplatinate ($XePtF_6$), the world's first noble gas compound. Bartlett's experiment destroyed the theory that the noble gases are completely unreactive.

4.5 Atomic structure and electron arrangements

Students often ask why they need to learn about atomic structure. One answer is to explain and understand patterns in the properties of elements in the periodic table. Relating the properties of the elements to their electronic structures and position in the periodic table allows prediction of properties of other elements and their likely reactions. This predictive potential makes the periodic table an extremely useful tool in chemistry.

4.5 Atomic structure and electron arrangements

Atomic structure

Chapter 3 describes the size of atoms and development of atomic theory. This section discusses sub-atomic particles and their arrangement in atoms. Apart from hydrogen atoms, which comprise one proton and one electron only, all atoms comprise protons, neutrons and electrons. Table 4.5 shows who discovered each of these (see also Chapter 6 in de Winter and Hardman, 2021).

Table 4.5 Discovery, charge and mass of sub-atomic particles

	Date discovered	Scientist	Charge	Mass in atomic mass units
Electron	1897	J.J. Thomson	−1	5×10^{-4}
Atomic nucleus	1911	Ernest Rutherford	+1 to +118	1–294
Proton	1919		+1	1
Neutron	1932	James Chadwick	0	1

The diameter of an atomic nucleus is about $10\text{--}15 \times 10^{-15}$ m, which is about 100 000 times smaller than the diameter of an atom. To put this in perspective, if the nucleus were the size of a football, the atom would be the size of a football stadium (not the pitch). The nucleus contains protons and neutrons, while electrons are in a cloud around the nucleus. Atomic number (Z) is the number of protons present in an atomic nucleus. For example, carbon, atomic number 6, has six protons. Atoms are electrically neutral, so the number of electrons (negative) and protons (positive) are equal. Carbon therefore has six electrons in each atom. The mass number, symbol A (from German *Atomgewicht*, meaning atomic weight), is the number of protons plus the number of neutrons. The number of neutrons may vary, leading to isotopes: that is, atoms with different masses although they are the same element. Scientists calculate relative atomic mass (A_r) as a weighted average of masses of atoms of an element (see Chapter 9).

Hydrogen has the lightest atoms, comprising one proton and one electron only. Oganesson has the heaviest atoms, comprising 118 protons and 176 neutrons. The periodic table places elements in atomic number order. Students can carry out independent research using the periodic table to answer questions about chemical elements and atomic structure (see above).

Questions to ask

→ Which element has the lightest atoms? Write down the name, symbol, atomic and mass numbers of this element. What particles are present in an atom of this element? Where is this element found in nature? What properties does it have?

4 The periodic table

→ Which element has the heaviest atoms? Write down the name, atomic and mass symbol of this element. What particles are present in an atom of this element? What properties might this element have?

→ Students could be given a selection of elements and asked to write down the mass number, atomic number and from that the numbers of protons, neutrons and elements in the atoms of that element.

→ Are there any connections between the atomic number, mass number and position of the elements in the periodic table?

KEY ACTIVITY

Finding out about what we cannot see

This activity imitates scientists' investigations of materials, substances and ultimately the atoms comprising them. As atoms are invisible to the human eye, students are often curious (or even disbelieving) that scientists know their structure.

Using sealed boxes or tins each containing a small object can help them understand the principle of investigating something that is unknown/invisible. A simple example is a matchbox containing a paperclip or something similar. Alternatively, prepare a set of small, identical metal tins with lids, placing a different object in each. Seal the lids on the tins in place with tape. Challenge students to explore the tins *without opening them* and discuss what they can and cannot deduce about the objects inside. Moving and shaking the tins, listening to the sounds made, feeling how weight shifts inside the tins and making comparisons between them gives information about the contents. Ask students to make their best guess of the object(s) in each container. Ask for tests that would confirm these (other than opening the containers). Some may suggest weighing the tins to give information about the relative mass. The activity mimics experiments carried out on atomic structure, where indirect evidence contributed to understanding, although questions remained and it was not possible to see what was being investigated.

Electron configurations

Chemical properties of elements rely on electrons. A commonly held view is that electrons orbit atomic nuclei like planets around a sun. This is a useful model but is not scientifically correct. Danish physicist Neils Bohr used quantum mechanics (outside the scope of this book) to explain that electrons are found in areas of three-dimensional space that are determined by the energy they possess. Bohr's model proposes that discrete energy levels hold specific numbers of electrons, and that electrons generally stay in those energy levels. The Bohr model of the atom explains the structure of the periodic table. His atomic model has been superseded (see Chapter 3). Nevertheless, the convenient shell structure taught today arises from Bohr's ideas.

4.5 Atomic structure and electron arrangements

Each electron 'shell' (or space where electrons can be found) has an energy level, the value of which is based on complicated calculations. As a result, shells fill in order, from lowest to increasingly higher energy levels. The principle, known as the 'building up' or 'Aufbau' principle, is based on achieving the configuration with the lowest total energy level for an atom. Energetic factors mean that from level 4 onwards, shells with higher numbers partially fill before lower numbered ones are complete. This is because the energy corresponding to placing the first two electrons in the higher numbered shell is lower that needed to fill the remaining spaces in the lower numbered shell. To illustrate this, we see that the first two energy levels, corresponding to periods 1 (hydrogen - helium) and two (elements lithium to neon) take two and eight electrons each respectively. However, the third energy level takes up to 18 electrons and is filled in two sections. Period 3 (sodium to argon) corresponds to the first eight electrons of level three. Next two electrons are added to the fourth energy level, leading to the configurations for potassium and calcium. After this, the ten remaining spaces in level 3 fill, corresponding to elements scandium to zinc.

An electron configuration states the arrangement of electrons in a specific atom. For a sodium atom with eleven electrons, the configuration is 2.8.1. Reading from left to right, this represents two electrons in the lowest energy level, eight in the next energy level, and one in the third energy level. By convention, when writing an electron configuration a full stop is placed between numbers.

The last number in a configuration corresponds to the element group. Sodium is a Group 1 element. All Group 1 elements have 1 as the last number in their configurations; 2.1 represents lithium and 2.8.8.1 represents potassium. Similarly, Group 2 elements have 2 as the last number in their configuration, and so on to Group 7. The noble gases, Group 0, have 2 (helium) or 8 (neon, argon, krypton, xenon) as their last numbers.

The *number* of numbers in electron configurations corresponds to an element's period. For sodium, the configuration 2.8.1 has three numbers, corresponding to Period 3. Hydrogen has an electron configuration of 1 and is in Period 1. Elements in Period 2, lithium to neon, have electrons in two energy levels, so two numbers appear in their configurations; Period 3 elements, sodium to argon, have three numbers, and so on. Across a period, each element has one more proton in the nucleus and one more electron in the energy levels surrounding the nucleus. This leads to trends in physical and chemical properties across the periods. These trends become more marked in periods further down the periodic table than 1–3.

4 The periodic table

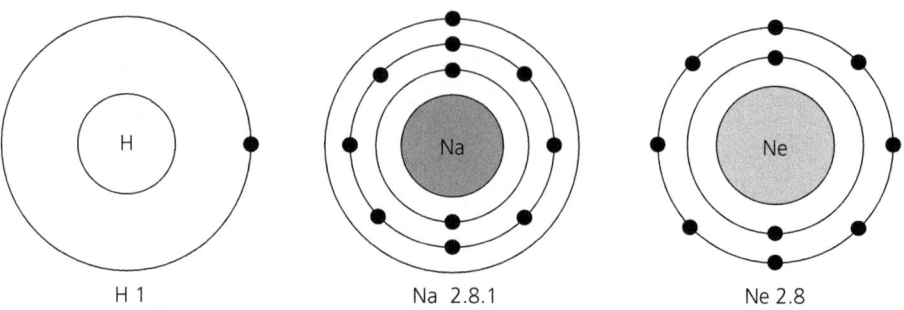

Figure 4.4 Electronic structures of hydrogen, sodium and neon

KEY ACTIVITY

Writing electron configurations

Students need to write electron configurations for elements from hydrogen to calcium. Challenge students to find the number of electrons for a given element (the same as the atomic number) and to draw and write their configurations. Providing templates with pre-drawn concentric circles and marked spaces for electrons (not all of which will be filled) will help. Students can place counters (or similar small items) to represent electrons for specific atoms, given their atomic numbers.

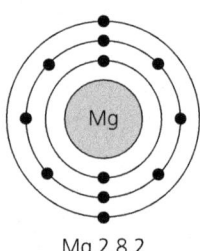

Figure 4.5 Magnesium atom showing twelve electrons

Note that this model of the atom has limitations (see chapter 5). Careful discussion is required to ensure students understand this is just an image representing the number of electrons in an atom in the first twenty elements in line with its position in the periodic table. The drawings do not represent how electrons are arranged in three-dimensional space.

Some students may appreciate drawing electron configurations for some transition metals. These metallic elements, known as 'transition' metals are found in the centre of the periodic table. Their electron configurations follow a different pattern based on energy levels of electron shells (see p 117). Table 4.6 shows the expected answers.

Table 4.6 4.5 Electronic configuration of some transition metals

Element	Symbol	Atomic number	Electron configuration
Scandium	Sc	21	2.8.9.2
Titanium	Ti	22	2.8.10.2
Iron	Fe	26	2.8.14.2
Copper	Cu	29	2.8.18.1
Zinc	Zn	30	2.8.18.2

Questions to ask
- What do you notice about the configurations of the elements in the same group?
- What do you notice about the configurations of the elements in the same period?

Relating properties of chemical elements to periodic table position

Relating properties of chemical elements to their electronic structures and positions in the periodic table allows prediction of reactions and properties. This predictive power makes the periodic table a useful tool. Mendeleev compiled his version of the periodic table not knowing *why* it was structured that way. This makes the accuracy of his table even more surprising.

The reaction of the alkali (group 1) metals with water reinforces the predictive power of the periodic table (see Table 4.2). Students should be able explain why all Group 1 elements react similarly. These elements all have electron configurations ending with '1', which leads to similar reaction patterns. The reactivity of Group 1 metals arises from the low energy level associated with the last electron in their electron configurations. This means that, in solution, little energy investment is needed to produce singly charged positive ions (cations). It is tempting to describe the reactions as metal atoms 'needing' or 'wanting' to lose an electron. Atoms cannot make decisions. Reactions occur because of energetic considerations. See also Chapter 5 and references at the end of the chapter (Kind, 2004; Taber 2002 and 2003).

Reactivity increases down the alkali metal group because an extra shell of electrons is present in each atom as the group is descended. This is reflected in the electron configurations as the number of numbers in a configuration increases by one for every period. The effect of the additional electron shell is to 'shield' outer electrons from the positively charged atomic nuclei. Increased shielding reduces the amount of energy required to form a singly charged positive ion (cation) from an atom of an alkali metal.

The reactivity pattern is reversed for atoms of the halogens (group 7). Prompting students to understand that reactivity trends are reversed in groups 1 and 7 indicates they understand how the periodic table relates atomic structure and element reactivity. The halogens are closely related elements that share common chemical properties which vary smoothly down the group. Fluorine is the most reactive due to the small size of its atom and its electron configuration. Chlorine is the next most reactive, followed by bromine, iodine and other halogens. In chemical reactions, the most energetically favoured outcome is that halogen atoms gain an additional electron, creating a singly negative ion (anion). This creates a full shell configuration. The ease with which this occurs is greatest at the top of the group and least at the bottom.

Noble gas atoms have either two or eight electrons in their outermost electron orbitals. These numbers result in energetic stability. A lot of energy and/or very favourable reaction conditions are required to make noble gas atoms participate in chemical reactions. Accordingly, they form few compounds, and none are currently known for helium and neon. Xenon forms several compounds including di-, tetra- and hexafluorides (XeF_2, XeF_4 and XeF_6). Gases lower in Group 0 are more reactive than those towards the top, following the same pattern as Group 1 metals. This pattern is due to increased shielding, meaning that electrons in outer shells are less attracted to the atomic nuclei so less energy is required to react.

KEY ACTIVITY

Writing a script for demonstrating the reaction between potassium and water

To assess students' understanding of electron configurations and chemical reactions, they can write a script describing the demonstration of the reaction between potassium metal and water (or a similar reaction). Ask students to draw electronic configurations of lithium, sodium and potassium. Refer to these while demonstrating lithium and sodium reacting with water (see earlier activity) and explaining the trend in reactivity. Ask students to write a script for the reaction between potassium and water. Show the reaction without saying anything. What should the demonstrator say to describe and explain the reaction? Provide a list of key words, such as electron, cation, more/fewer electron orbitals, more/less reactive.

4.5 Atomic structure and electron arrangements

Questions to ask
- Using electron configurations, explain why the alkali metals react increasingly vigorously with water.
- Predict the reactions of rubidium and caesium with water.
- How do sodium and chlorine react together? Show a video without the commentary or demonstrate the reaction. Ask students to explain what is happening.

Encourage students to avoid anthropomorphic language such as atoms needing or wanting to lose/gain electrons, finding partners or holding hands. Focus on energetics (see Chapter 6). Energy invested in making reactions occur is recouped when compounds form: the vigorous reactions observed between the halogens and many metals (see above and Chapter 1) produce highly stable compounds. The references listed in Section 4.6 may help develop this.

Enrichment

In periodic tables used in schools, groups are numbered 1–7 and 0. Groups headed by lithium and beryllium are numbered 1 and 2; those headed by boron, carbon, nitrogen, oxygen and fluorine are Groups 3–7 and the noble gases, headed by helium, are Group 0. For Groups 1–7 the numbers correspond to the number of electrons present in the outermost shells of the element atoms. For the noble gases, outermost shell electron numbers make it energetically very difficult for them to react, reflected by the 0. Older periodic tables number noble gases as Group 8, as all except helium have eight electrons in their outermost shells. IUPAC numbers all columns of the periodic table – including those in the centre, headed by scandium, titanium and vanadium and so on – so the halogens and noble gases are groups 17 and 18 respectively. This helps advanced chemists develop precise group chemistry.

Careers

Understanding the periodic table is a pre-requisite for professional chemists. Detailed understanding of reactions of chemical elements leads to a career as an inorganic chemist. This might include being a laboratory technician, health and safety officer, public analyst, forensic scientist, microbiologist, biochemist, materials scientist, particle physicist or engineer. Understanding that different materials have different properties and using appropriate materials for their properties will be important in roles which involve design such as architecture, aircraft design, jewellery design, and civil or structural engineering. Further details including profiles of individuals' careers are available at www.rsc.org/careers/future/.

4.6 Resources

References and further reading

Aldersley-Williams, H. (2011) *Periodic tales: The curious lives of the elements*, Viking.

Emsley, J. (2011) *Nature's building blocks*, OUP.

Petrusevski, V.M. and Cvetkovic, J. (2017) On the "true position" of hydrogen in the periodic table. *Section of Natural Mathematical and Biotechnical Sciences, MASA* 38 (1): 83–90.

Kind, V. (2004) *Beyond appearances: Students' misconceptions about basic chemical ideas*. Available at https://edu.rsc.org/resources/beyond-appearances/2202.article

Taber, K. (2002) *Chemical misconceptions: prevention, diagnosis and cure*, Royal Society of Chemistry.

Taber, K. (2003) The atom in the chemistry curriculum: fundamental concept, teaching model or epistemological obstacle? *Foundations of Chemistry*, 5(1): 43–84.

de Winter, J. and Hardman, M. (2021) *Teaching Secondary Physics* (3rd edition), Hodder Education.

Websites

The periodic table

Dimitri Mendeleev and the periodic pattern: www.stem.org.uk/resources/elibrary/resource/131977/dimitri-mendeleev-and-periodic-pattern

This interactive periodic table provides information about each individual element, with links to videos, class experiments and careers information: www.rsc.org/periodic-table

The Royal Institution Christmas lecture series The Modern Alchemist comprises three one-hour lectures about the periodic table: www.rigb.org/christmas-lectures/watch/2012/the-modern-alchemist

Tracing the history of organising the chemical elements: https://en.wikipedia.org/wiki/Timeline_of_chemical_element_discoveries#cite_note-10

Additional activities relating to the periodic table are available via this link: www.stem.org.uk/resources/community/collection/12562/periodic-table

The chemical elements

This classic song names at high speed the elements of the periodic table song known at the time: www.youtube.com/watch?v=rz4Dd1I_fX0

Chemical element 'top trumps' and the history of the periodic table: https://edu.rsc.org/resources/getting-to-grips-with-the-periodic-tables-structure/4010545.article

Discovering new elements: www.stem.org.uk/resources/elibrary/resource/78146/discovering-new-elements

Marie Curie: www.stem.org.uk/elibrary/resource/30086

Reacting elements to form compounds

The reaction between aluminium and iodine: www.rsc.org/learn-chemistry/resource/res00000715/reaction-between-aluminium-and-iodine

The reaction between sodium and chlorine: www.rsc.org/learn-chemistry/resource/res00000732/heating-group-1-metals-in-air-and-in-chlorine

Group 1 – the alkali metals

Details for the demonstration of the alkali metals in water: https://edu.rsc.org/resources/reactivity-trends-of-the-alkali-metals/731.article

The reaction between fluorine and caesium: www.youtube.com/watch?v=TLOFaWdPxB0

The reaction between sodium and water (large scale) and caesium and air: www.rsc.org/learn-chemistry/resource/res00001124/ri-christmas-lectures-2012-the-alkali-metals#!cmpid=CMP00002105

Explaining why elements become more reactive down Group 1 and more reactive towards the top of Group 7: https://www.bbc.co.uk/bitesize/clips/z7j34wx

Group 7 – the halogens

The properties and reactivity of the halogens, including displacement reactions: www.rsc.org/learn-chemistry/resource/res00000733/reactions-of-halogens-as-aqueous-solutions

Video of the halogen displacement reactions practical: https://www.youtube.com/watch?v=rBhMWpyO7Ts

Reactions of fluorine with iron wool and cotton wool: www.periodicvideos.com/videos/009.htm

Group 0 – the noble gases

Floating gold on xenon: www.youtube.com/watch?v=kdmsVa0CosQ

Video of noble gases, their properties and uses: www.youtube.com/watch?v=qNaBMvJXdJ4

Neil Bartlett and the reactive noble gases: www.acs.org/content/acs/en/education/whatischemistry/landmarks/bartlettnoblegases.html

The chemistry of the noble gases: www.stem.org.uk/elibrary/resource/27439

Other resources

Theodore Gray's element cards: www.amazon.co.uk/Photographic-Card-Deck-Elements/dp/1603761985/ref=asc_df_1603761985/?tag=googshopuk-21&linkCode=df0&hvadid=310867999190&hvpos=1o1&hvnetw=g&hvrand=12736179088024480598&hvpone=&hvptwo=&hvqmt=&hvdev=c&hvdvcmdl=&hvlocint=&hvlocphy=1006688&hvtargid=pla-406163960153&psc=1&th=1&psc=1

Chemical bonding

Hannah Sevian and Edenia Amaral

Introduction

The question addressed in this chapter is: How can teachers develop students' thinking about chemical bonds in the chemistry classroom? Discussion begins with consideration of students' thinking about chemical bonding to aid teaching in a manner that connects students' experiences and prior knowledge to the scientific view. The focus is on teaching that adopts scientifically appropriate and accurate language to represent chemical bonding. This ensures that students learn about chemical bonding in ways that support their progression to more complex ideas, or, alternatively, complete their studies of chemistry with sound, scientifically based understanding.

The chapter presents chemical bonding concepts in a teaching sequence that begins with metallic bonding, continues to ionic bonding and concludes with covalent bonding and intermolecular bonds. This may seem unusual, as this reverses the more traditional approach. However, this strategy permits students to gradually assimilate chemical bonding concepts by starting with the structures and properties of materials. Understanding is built by introducing chemical bonding as 'holding power' the strength of which determines the structure of a material, for which solids provide an initial focus. The approach enables students to connect chemical bonding to particle theory (Chapter 3), as holding power offers an explanation for state changes. Also, the chapter takes electrostatic attractions between oppositely charged particles as the lead into understanding all types of chemical bond. This affords a linguistic and conceptual clarity for students that facilitates understanding.

The chapter adopts a different style to others in the book. Although some activities are proposed, the main focus is on presenting information to aid specialist and non-specialist teachers seeking support for developing students' conceptual understanding of this challenging topic.

Teaching sequence overview

The content follows a route that offers background information, resources, and justification for the chosen approach to teaching chemical bonding. First, the chapter summarises key ideas about students' thinking about chemical bonding from research literature.

Knowing about difficulties that students face as they learn chemical bonding is helpful for planning instruction. The most frequent student misconceptions about chemical bonds are discussed and aligned with the scientific views (Table 5.1). This provides information to create effective teaching strategies that illuminate thinking about chemical bonding. Readers wishing to follow references up in detail will find further information at the end of the chapter. Students are familiar with the world at a macroscopic scale. So, we start from questions about macroscopic phenomena, leading to explanations for these based on chemical bonding models.

In summary, to organise issues for teaching and learning about chemical bonding, the chapter brings together:

→ Awareness of students' previous knowledge and potential barriers to learning this topic.
→ Contextualisation of non-scientific ideas and modelling of scientific ideas of chemical bonding.
→ Ways students think about and use chemical bonding concepts to reason about the world around them.

 Understanding how students think about chemical bonding

Children bring ideas from primary school and life experiences to their learning in secondary school science. Many are emerging ideas directed toward developing understanding of scientific ideas. Some are misconceptions that inhibit development of scientific understanding. Others are alternative conceptions, which are views affecting interpretation of evidence. All should be discussed directly as they arise. Researchers have identified students' ideas about chemical bonding. The chemical language and deeper understanding of chemical processes students use are often dismissed as trivial. The temptation is to ignore them (often for considerations of time) and teach the 'correct' view without discussion. However, studies show repeatedly that students need support to develop accurate understanding about chemical bonding. Time spent on finding out what students already know and working to establish basic understanding will be repaid later on.

Innovative proposals for teaching chemical bonding have been developed and tested (see References). Researchers in Israel who asked how often strategies and pedagogy mislead students claim that many students' alternative conceptions are 'pedagogical learning impediments' which actually originate from science teaching. That is, these researchers think students' misconceptions arise from failures in teaching. Teaching often centres on models, linguistics and phenomena explained through a sub-microscopic dimension. This contrasts with students' starting point, which is the macro-scale (see Chapter 1). The authors propose a four-step assessment-driven 'scientific and effective teaching approach' based on problem-solving. The steps are: identify key-learning goals; use big ideas for all phases of chemical bonding; design activities; and constantly assess formatively.

This chapter takes students' previous knowledge on chemical bonding as a crucial starting point for planning effective strategies for teaching. When introducing this topic, knowing students' thinking in depth is helpful, so we begin by summarising students' ways of thinking about chemical bonding. These enable the planning of teaching strategies taking students from simple to more complex ideas. For example, to develop students' understanding of how particles are held together in materials, start with macro-scale properties that students recognise and/or measure. Once this is secure, progress to scientific models that characterise types of chemical bonding, the sub-microscale. Show students how the macro-scale properties relate to the sub-microscale bond types. Finally, consider how to represent chemical bonds in symbols.

Researchers who have studied ways students think about bonds have identified misconceptions that include anthropomorphism (giving inanimate objects human characteristics), confusion between ionic and covalent bonds, poor understanding of energy in chemical bonding and incorrect representations of bonding. Table 5.1 shows students' ideas about bonding reported in research literature, together with scientific views.

Table 5.1 Students' conceptions about chemical bonding from science education literature

Students' understanding	Categories	Scientific view
Students present anthropomorphic explanations for chemical bonding, suggesting that bonding happens because atoms 'want' or 'need' a particular electron arrangement or number of bonds.	Anthropomorphism	Atoms are inert objects that do not share human characteristics like wanting and needing. Bonding is driven by energetics.
Students think of bonds as something material – glue, sticks, springs or other physical connectors rather than an electrical force.	General ideas	Bonds are non-material. Take care to emphasise a bond is not a physical entity in space. This is a limitation of molecular models.
Students may think the bonding in the atom as being like that due to magnets.		Bonds arise from electrical forces, rather than magnetic forces. This is important as electrical charges are monopoles, while magnets are dipoles (see p 134).

5 Chemical bonding

Students' understanding	Categories	Scientific view
Students visualise metallic bonding as a sea of electrons among cations without thinking about the balance of charges.	Ionic and metallic bonds	A metallic lattice is electrically neutral, the total charge of positive cations is balanced by the total charge of delocalised negative electrons.
Substances with metallic and ionic bonds naturally form molecules.		Substances with metallic and ionic bonds form lattices in the solid state.
Ionic bonds are the electron transfer, rather than electrostatic attractions between the ions that result from the electron transfer. Bonds form only between the atoms that have donated and accepted electrons.		Ionic bonds are electrostatic attractions between oppositely charged ions. Electron transfer forms ions from atoms. Bonding is not limited to the atoms which donated or accepted electrons. Some ionic bonds exist in nature without electron transfer. Some ionic bonds form, for example, in precipitation reactions without electron transfer.
Ionic bonds are unidirectional under the same rules as covalent bonds.		Ionic bonds are multi-directional.
An ionic lattice is constructed from discrete molecules. For example, a sodium chloride 'molecule', formula NaCl, is a discrete entity.		There are no discrete molecules within an ionic lattice. For example, each sodium cation is attracted to each chloride anion. NaCl (and other ionic formulae) represent the ratio of ions in the lattice, not an individual molecule.
Electronic configurations determine the number of ionic bonds formed.		Ionic bonds are electrostatic attractions between any oppositely charged ions. Bonding does not depend on electronic configuration.
Ions interact with oppositely charged ions around them but for those not involved in donating and accepting electrons, these interactions are only forces.	Bond or force?	The electrostatic interaction between oppositely charged ions is an ionic bond.
Students think that all atoms involved in covalent bonds form macromolecules. They confuse intramolecular bonds and intermolecular bonds.	Covalent bonds and intermolecular forces	Many covalently bonded substances form small molecules.
Covalent bonds are weak, as covalent compounds have low boiling points.		Covalent bonds are not affected when a substance changes state. State changes in covalent compounds only affect intermolecular bonds.
Covalent bonds are disrupted when a substance changes state.		
In covalent bonds, a single electron is shared between two atoms.		In covalent bonds, an electron pair is shared between the nuclei of two atoms.
Covalent bonds involve total transfer of electrons.		
Pairs of electrons are equally shared in all covalent bonds.		Sharing of electrons within a covalent bond depends on the atoms involved. Equal sharing is most likely if the two atoms are identical.
Covalent bonds are higher in status than ionic and metallic bonds.		Covalent, ionic and metallic bonds are all strong bonds.
In a covalent bond, the atoms are kept together because they share electrons.		In a covalent bond, atoms are kept together by electrostatic attraction between the electron pair and each atomic nucleus.

Students' understanding	Categories	Scientific view
Sodium reacts with chlorine because the octet rule causes chemical reactions to occur.	Octet rule	Reactions occur due to energetics. The octet rule is a simplification that helps initial understanding of the formation of a small number of compounds.
Each atom in a stable molecule must obey the octet rule, particularly for the atomic centre of molecules.		Many substances form stable molecules without obeying the octet rule. Examples include nitrogen oxides (see text).
Students think a chemical bond is like a spring, which releases energy when broken.	Energy	Bonds are electrostatic attractions, so energy is released when bonds are formed. Breaking a bond requires overcoming the electrostatic attraction, this requires energy.
Students think bonds are physical entities, so making bonds requires energy and breaking bonds releases energy.		
Representations for formation of NaCl(s) often used in textbooks lead students to think that a single sodium atom reacts with a single chlorine atom forming a single ionic pair that is NaCl, not taking into account the crystalline network formed.	Representations	The reaction between sodium and chlorine involves forming sodium cations (Na+) and chloride anions (Cl−). Billions of ions form in this reaction. The NaCl formula unit represents the ratio of sodium ions to chloride ions in the lattice. Electrostatic attraction between oppositely charged ions forms ionic lattices. Ionic bonds form without electron transfer in precipitation reactions.
Electron shells are visualised as material shells which enclose and protect atoms. Electron clouds are visualised as material clouds in which electrons are soaked.		Shells and clouds are metaphorical images for the distribution of electrons in three-dimensional space.

Chemical bonding: teaching a scientific view

Chemical bonding is a theoretical idea that is essential to chemists' sub-microscopic scale models of the structure of matter. Students will be aware of macro-scale properties of materials from their experiences within and outside the classroom. However, once the particle model has been introduced (Chapter 3), students should be able to understand that particles in different states have holding power of varying strengths. Holding power causes particles to clump together in solids and, to some extent, in liquids. Particles with sufficient energy can overcome this holding power and the substance moves from the solid state, becoming liquid or gas (adapted from Taber, 2002).

Explaining chemical bonding as holding power

Substances can exist in a solid, liquid or gas state, depending upon how the constituent particles are held together. Plasmas, which are

5 Chemical bonding

considered to be a fourth state of matter, form in special conditions usually involving high-voltage electric fields. Most substances do not exist as plasmas under normal conditions on Earth. Before introducing the term bonding, ask students how they think particles hold together in ice, liquid water and water vapour.

> **KEY ACTIVITY**
>
> ## Introducing chemical bonding as holding power
>
> Illustrating this discussion with images or a physical demonstration of change of state would be helpful. To demonstrate change of state of water, have ready some ice cubes to represent solid water, liquid water from a tap or bottle, and prepare to boil water using standard laboratory equipment (beaker, heatproof mat, Bunsen burner, thermometer).
>
> 1. Challenge students to think about what happens to water particles as water changes physical state (see Chapter 3).
> 2. Invite students to discuss how holding power changes according to particle movement: at higher temperatures, particles move faster so the holding power is less effective, so particles move apart. Changing holding power can be connected to change of state. Take care to use 'particles' rather than 'molecules' as this is a general term applicable to all substances.
> 3. Discuss other substances. For example, introduce the physical properties of liquids ethanol and/or a vegetable oil such as sunflower oil. These have boiling points of 78 °C and 230 °C respectively, while water boils at 100 °C. Sunflower oil freezes at −17 °C, so it may be possible to make oil cubes in a domestic freezer. Pure ethanol freezes at −114 °C. An ethanol–water mixture freezes at temperatures closer to 0 °C. A 40% ethanol–water mixture freezes at −23 °C, which may also be possible to achieve in a domestic freezer.
> 4. Discuss with students which particles have stronger holding power, requiring more energy to disrupt. Of these, sunflower oil particles have the strongest holding power, as the higher boiling point shows more energy is required to separate the particles compared to ethanol and water. The composition and structure of particles of ethanol and sunflower oil mean that the holding power between them is different and both are different to the holding power between water particles.
>
> ### Questions to ask
> - What happens to water particles as water changes state from solid to liquid to gas? What does this tell us about the holding power between particles?
> - Draw an image to represent the holding power between water particles in ice, liquid water and water vapour.
> - Do you think the holding power is the same in every solid/liquid/gas? Why/why not? The correct answer is that the particles of different substances have holding powers of varying strengths. It is important that students understand this. Details about the factors that affect the strengths of holding powers will be discussed later.

5.2 Chemical bonding: teaching a scientific view

> - Thinking about substances other than water, what do we learn about holding power from their melting points and boiling points? Which substance has the strongest holding power between particles? How would you know (what evidence tells you this)?
> - What do you think holding power is like? This question should permit students to state their ideas about chemical bonds – they may say bonds are like magnets, glue, or possibly joining sticks, depending on previous exposure to scientific concepts and molecular models. Accept all answers and say that they will be learning more about holding power which may change their ideas.

Chemical bonding is the formal term for holding power. Substances do not all have the same type of chemical bonding. Some substances have particles organised in closely structured crystalline arrangements: for example, sodium chloride, diamond and granite. In other substances, holding power leads to the formation of weakly bound, haphazard clumps: for example, soil, clay, cement. Chemists seek explanations for the types of holding power and the effects these have on structures. The bond type depends on the nature of the particles forming the bond. The basic principle of chemical bonding is that positive and negative electrical charges attract to form a bond. All bonds, whether intermolecular or intramolecular, form due to electrostatic attractions. Describing a bond as a force or attraction adds unnecessary complications for students who are learning these ideas for the first time, so should be avoided.

Chemical bonding models suggest what happens at sub-microscopic scales when chemical reactions occur. In reactions, combinations of particles that start by being bonded together in reactants break apart. The particles rearrange, forming new bonds between them and so creating the products (see Chapter 2). A useful notion is that rearrangements of particles made in reactions are more stable than those in the starting reactants. That is, product particles hold together more tightly, or are more strongly bonded than particles in reactants. Chemical reactions occur so that electrical charges rearrange into more stable patterns. Often, these involve attractions between positively charged atomic nuclei and negatively charged outer shell electrons. Later, we learn this is not the whole story. This is a starting point for a scientific explanation based on energetics that is understandable by secondary students. This explanation avoids the temptation to use language such as particles 'wanting' or 'needing' to form bonds or 'find partners', which is unhelpful and incorrect. To summarise: teaching this topic using scientifically sound models helps students realise chemistry explains natural phenomena.

5 Chemical bonding

The nature of the chemical bond

In secondary science, chemical bonding is regarded as arising from electrostatic attractions. Atoms contain electrically charged particles: they comprise positive nuclei (containing positively charged protons) surrounded by negatively charged electrons. Like charges repel one another, so positive atomic nuclei repel each other and negative electrons repel each other. Particle physics explains why positively charged particles can exist extremely close together in atomic nuclei without flying apart due to electrostatic repulsion. Opposite charges attract each other: atomic nuclei attract electrons and electrons attract atomic nuclei. In a chemical bond, atomic nuclei and electrons are at a distance where similarly charged components are close enough so repulsions and attractions balance. The arrangement is stable because electrical attractions between nuclei and electrons are sufficiently strong to resist small disturbances. This principle applies to atoms, ions, molecules, lattices and other structures. Chemical change involves reconfigurations of negative electrons surrounding positive atomic nuclei.

Chemical bonding is taught using models that represent arrangements of nuclei and electrons. The models have limitations, so make explicit that (a) the models are representations that are not necessarily fully accurate in real life; and (b) using models to explain sub-microscopic events that cannot be directly observed is normal in chemistry. Models are thinking tools that represent limited descriptions of nature. Discourage students from thinking that models are true accounts of the molecular world. Using this approach avoids the need to say that previous accounts were 'wrong' when more sophisticated models are introduced to develop understanding further as students progress.

Encourage students to realise that abstract ideas in science are often represented by models. Also, theories are partial, limited accounts of nature, based on our best available understanding. Focus on teaching ideas of intellectual merit, rather than using language and ideas that lead students towards misconceptions that hamper future progress.

Quantum theory

The electrical charge model for bonding does not explain why stable arrangements produced by electrostatic attractions often result in particles with regular electronic configurations (see Chapter 4), for example, two electrons in the first shell and, often, eight in an outermost shell of an atom or ion. Detailed explanations that build on knowledge and understanding of the periodic table and atomic structure are developed in post-16 chemistry and beyond. In principle, the explanation relates to how, at the tiniest scale, everything is quantised: quantities such as energy and angular momentum exist in 'packets' of a minimum size. Matter and charge are also quantised, giving basic units such as electrons that cannot be sub-divided. Particle physics explores this in greater depth. The idea that charge exists only as multiple of a fundamental, minimum amount seems counterintuitive but is one that students accept if they are given opportunities to explore and discuss particle models of matter.

The observation that electrons are often found in pairs in atomic and molecular structures is a clue that the electrostatic attraction model cannot be the whole story. If electrostatic attractions were the only factor, electrons would not pair up because, as both are negatively charged, they naturally repel each other. Nonetheless, electron pairs form covalent bonds between atomic cores, while non-bonding pairs of electrons (that is, those not involved in covalent bonds) sometimes contribute to intermolecular bonds. Students rarely raise electron–electron repulsion as an issue. Students raising this objection are thinking about matter at the sub-microscopic scale in the way chemists do. To explain why electrons are found in pairs despite both having negative charges, note that the electrons in a pair spin in opposite directions, reducing the repulsion effect. Paired electrons can be considered (modelled) as tiny magnets arranged anti-parallel (N–S and S–N), so magnetic attraction counters electrical repulsion.

5.3 Modelling scientific ideas for chemical bonding

Matter exists because particles are held together in various ways: electrical interactions occur between oppositely charged ions, between atomic nuclei and electrons. Electrostatic attractions between neutral molecules balance the pull between opposite charges and repulsion between similar charges. Allow time for students to voice any misconceptions they may have as ideas are introduced. Information from Table 5.1 may help with this.

Teachers often think they have taught something, only to find later that students have not actually learned it. For chemical bonding, using correct language aids students' learning. If language is incorrect, likely outcomes are that students will use scientific terms without understanding their meanings or/and rote-memorise 'explanations'. They will not learn the concepts.

Chemical bonding does not involve human feelings

Table 5.1 shows students use anthropomorphism to explain ways particles hold together stating atoms want or need to be linked. Teachers are often responsible for encouraging this attractive but faulty notion. The association has power as an analogy: for example, feeling = attraction, movement toward stability = togetherness. Expert chemists may talk about chemical reactions using the same analogies. Students, however, cannot map the human analogy to the scientific view, as they do not understand the context correctly. This reinforces the anthropomorphic model, so students believe it is true. This model implies atoms are alive, have brains and can choose how to act. Particles are not alive. Never describe atoms as wanting or needing to form bonds or gain, lose or share electrons. It is very important not to use this language, and to avoid models and role plays that require anthropomorphic language to describe them. Focus on using correct scientific terminology, introducing this carefully and reinforcing understanding as students progress through the topic.

Chemical bonds are not material objects

Students may intuitively think chemical bonds are material objects, that is, solid items that can be picked up and moved from place to place. This originates from the idea that electrons, ions, atoms and molecules are objects, which helps understand how particles move in space (kinetic particle theory). Students may consider that particles connect through some type of glue, or elastic bands, stating that components 'stick together'. Images in textbooks and molecular models reinforce these ideas. Modelling kits represent bonds using wooden rods or plastic sticks between 'atoms' that are wood or plastic spheres, coloured to distinguish between chemical elements. However, bonds are regions of space where electrically charged particles may be found, not physical objects. Models represent our best approximations at representing something which is very difficult to imagine.

Chemical bonds can be represented in molecular models

Support students' understanding by presenting a range of molecular models for well-known substances. Discuss the limitations of these macroscopic models, explaining that chemists don't have sub-microscopic toolkits to link electrons, ions, atoms and molecules with tiny sticks. Making comparisons can be effective, such as stating the bond acts *like* glue or *like* an elastic band joining two things together. Then introduce abstract ideas as scientific models.

For example, Figure 5.1 shows four paper-based representations of a methane (CH_4) molecule. Each of these illustrates a different property of the molecule, so all are important. The first image uses a dot-and-cross representation of covalent bonds, showing electrons in the outermost orbitals of carbon and hydrogen atoms. The second shows bonds as lines connecting the atoms. The third shows how the atoms are distributed in three-dimensional space. The fourth represents the bonds as clouds of space in which electrons can found. Each of these has specific qualities that tell us something about a methane molecule. Chemists may choose to use one representation over another depending on what they need to know about methane.

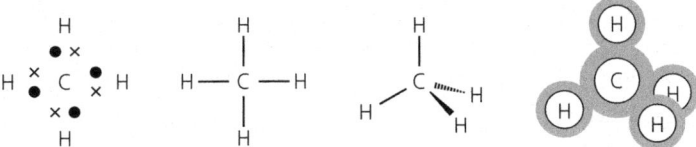

Figure 5.1 Several representations of the methane (CH_4) molecule

KEY ACTIVITY

Introducing molecular models

1. Prepare a set of images or real-life models of everyday substances such as water, methane, sodium chloride, oxygen, diamond, copper.
   The models could be: a Molymod® model; a ball-and-stick model made from, say, table-tennis balls and lollipop sticks, or wooden modelling balls stuck together with Velcro®; a commercial model in which bonds are represented as metal springs; images of the molecules/lattices of the substances drawn from the internet.
   Using different strengths of Velcro/other adhesive material introduces the notion that chemical bonds have different characteristics.
2. Taking one substance at a time, discuss with students:
   a. The nature of the substance – what physical properties does it have? (For example, melting point, boiling point, colour, state at room temperature, smell.)
      What is the substance used for, and why?
   b. What do these properties tell us about the holding power (or chemical bonding) between the particles of the substance?
      What evidence do we have for these ideas?
      Is it possible to 'see' the particles?
   These steps reprise the previous activity.

5 Chemical bonding

> 3. Next, introduce the models as ways of representing the particles in these substances. Taking each substance in turn, examine the models.
>   What do they tell us about the holding power (chemical bonding) between the particles?
>   Is this really what the chemical bonds are like?
>   What can we explain about the substance using these models?
> 4. In discussion, allow students to express their views. They may think that bonds are rigid and one directional (stick, plastic link) until they see a spring as a connector. The spring illustrates that chemical bonds are flexible. Also, models are limited in that they show all bonds as being the same, but differences in physical properties suggest this cannot be true. If students can accept this, a good platform is formed for discussing different types of chemical bond.

Teachers confident in their understanding of magnetism may choose to use magnets as an analogy of chemical bonding. The analogy is powerful because electricity and magnetism are related mathematically. However, the magnets analogy has shortcomings, including that each magnet has two dipoles (two charges on one object) while electrically charged particles are monopoles (one charge on one object). Nonetheless, associating chemical bonding with magnets requires the notion that electrostatic forces act over a distance of empty space, which is essentially correct. Differentiate between magnetic poles and electrical charges in a particle. Emphasise electrical charges to introduce chemical bonding based on electrostatic attractions. Chemical bonds hold together because attraction and repulsion forces between charged particles balance.

 5.4 Metallic lattices

Many metals are chemical elements (see Chapter 4). Their relatively simple structures provide a good starting point for modelling bonding. The key points are that cations form a regular three-dimensional structure and that metals are bonded by electrostatic attraction between positive cations and negative electrons. Models of metals show a regular arrangement of atomic cores, that is, metallic cations, which – despite all having a positive charge – are held in a lattice while electrons move around and between them.

Note that diagrams emphasise certain points rather than being realistic images of what metals are actually like. Metallic bonding involves interactions between positively charged atomic cores (cations) and delocalised electrons moving about the structure. Positively charged atomic cations are packed into a tight regular

5.4 Metallic lattices

lattice, surrounded by negatively charged electrons, often described as a 'sea of delocalised electrons'. This term implies free electrons are available in excess; however, it does not indicate that electrical charges between the atomic cores and electrons must balance (Table 5.1). The stability of a metallic lattice depends upon overall neutrality. When teaching metallic lattices, stress that the number of electrons *per cation* equals the positive atomic core charge. For example, in sodium (Na) metal, each atomic core, or cation, has a charge of +1. This means one electron per cation is delocalised (that is, present in the 'sea'), and shared between atomic cores. In magnesium (Mg) metal, each cation has a +2 charge, meaning two electrons per cation are delocalised. (Figure 5.2). Explain the term delocalised before moving on to properties of metals.

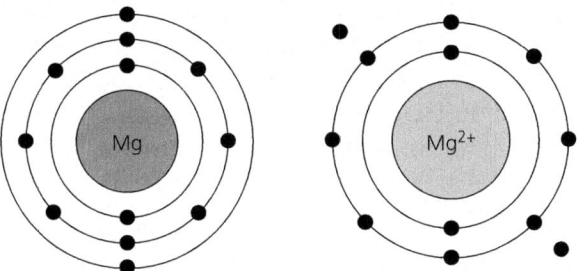

Figure 5.2 Magnesium atom as an atomic core (Mg^{2+} cation) plus two valence electrons

A magnesium metal lattice has twice as many electrons as a sodium lattice with the same number of atomic cores. Both metallic lattices are neutral overall. Neutrality is important for the stability of the lattice. Students may consider the sea of electrons to be a glue holding cations together, leading to a static image. In fact, electrons in metals are mobile and move in random directions. When a voltage is applied, net movement is in one direction, creating an electrical current. In copper, as in magnesium, two electrons are delocalised and each copper cation has a +2 charge. Recall that copper is used extensively in applications where good electrical conductance is required. The delocalised electrons contribute to copper being a very good conductor. Make clear that the electrons moving between cations are negatively charged particles with almost no mass. Chemists argue that metallic bonds between atomic cores do not exist, but instead that a giant metallic bonding orbital (the sea of delocalised electrons) holds the whole lattice together. Figure 5.3 represents an electrically neutral metal lattice of magnesium. Note that in the two images, the electrons (shown as dots) are in different places, but their number is unchanged. This represents the movement of electrons within the cation lattice.

5 Chemical bonding

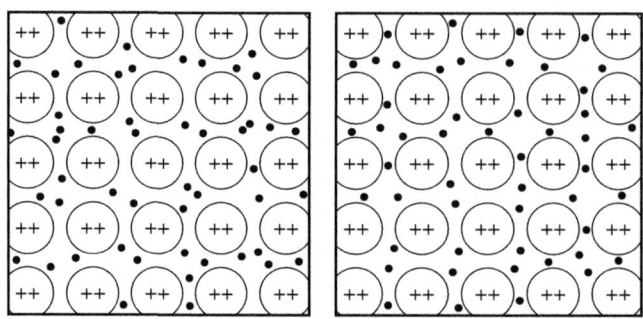

Figure 5.3 Metallic bonding holds together the lattice in metals such as magnesium. The bond comprises the mutual attraction between the metallic atomic cores and the delocalised electrons able to move around the structure.

Metal cooking pans and metallic devices for conducting current are helpful for exploring how metal structures are related to their properties. Good thermal and electrical conductivity can be explained by the model of metallic bonding described above. Students' most frequent non-scientific ideas about metallic bonding are in Table 5.1. Questions to ask to elicit students' thinking are suggested in the next activity.

KEY ACTIVITY

Metallic bonding

Have ready a series of everyday objects made from metals, such as cooking pots and pans, wires, devices that rely on thermal or electrical conductivity. Try to include a range of metals.

The purpose of the activity is to apply the model of metallic bonding (cation lattice, electron sea) to explain the structure and properties of metals. Students can examine one or more items, depending on the time available, and invited to share their explanations. Formative assessment via peer feedback using application of previously agreed criteria that relates to the metallic bonding model may be useful here. If misconceptions are apparent, use the model above and Table 5.1 to correct students' thinking before moving on to the next type of bonding.

Questions to ask

- Draw how you imagine the arrangement of metal particles in the object.
  How many electrons does each atom contribute to the delocalised sea?
  What is the charge on each metal cation?
- Why does electrical charge pass across a piece of metal?
- Why does thermal energy pass across a piece of metal?
- Why can some metals be shaped into many different objects?
- Why are some metals harder to shape or/and scratch than others?
- Are all metals electrical conductors? If so, why?
- In what ways are all metals the same? In what ways are metals different from each other?

5.5 Ionic lattices

Be aware that students commonly misunderstand ionic bonding. Their ideas include that metallic and ionic bonds form molecular compounds (so, for example, sodium chloride exists as NaCl molecules); ionic compounds obey the same rules as covalent bonds; electronic configurations determine the number of ionic bonds in a compound; ionic bonds only form between atoms that donate and accept electrons; ions interact with other ions; and 'only forces' link units that are not connected ionically. Some misunderstandings derive from students misapplying ideas about covalent bonding to ionic bonding.

When introducing ionic bonding, emphasise that bond type (metallic, ionic, covalent) depends on the nature of the particles. Particles in ionic bonds are generally metal ions (cations, positively charged, for example, Na^+, Mg^{2+}, Cu^{2+}) and charged non-metal atoms/molecules (for example, oxide ion, O^{2-}, chloride Cl^-, carbonate ion CO_3^{2-}). Telling the difference between molecular structures and ionic lattices depends on identifying the particles. Referring to ions *in comparison to* discrete atoms losing or gaining electrons, showing these side by side, is helpful to emphasise formation of ionic lattices in contrast to metallic lattices. Focus attention on electrostatic attractions holding ions together. Next, think of each interaction between adjacent oppositely charged ions as a discrete ionic bond. Note that, in an ionic lattice, an ion interacts with all ions in its immediate vicinity, so electrostatic attractions that comprise ionic bonds are multi-directional.

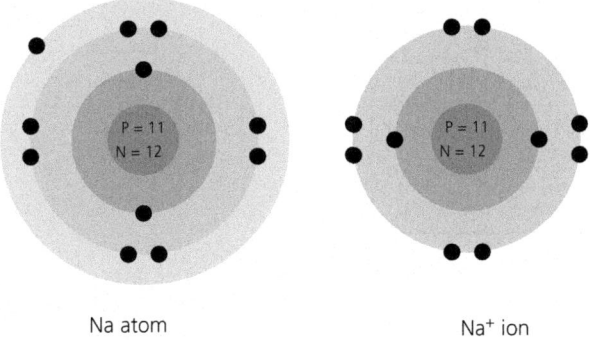

Figure 5.4 Sodium: a discrete atom and an ion

Ionic bonding is often taught by considering one metal atom and one non-metal atom interacting to form ions, which bond together. This is commonly represented as shown in Figure 5.5.

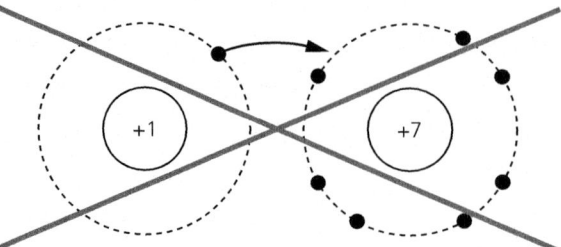

Figure 5.5 A common student misconception of the ionic bond

This approach may lead students to learn incorrectly that electron transfer is the ionic bond. Electron transfer is not involved in formation of ionic lattices: it is a thought experiment rather than an actual chemical process. An ionic bond is the electrostatic attraction between ions of opposite charge. Drawing atoms with electrons in concentric circles shown as dots or crosses is a powerful tool for representing stable configurations (see Chapter 4). However, using such diagrams to represent ionic bonding leads to incorrect and incomplete learning of the actual process. For example, this approach assumes that ionic bonds only form from atoms. In fact, ions exist naturally in many substances. Chemical reactions in which new ionic bonds form, occur between ionic substances. Precipitation reactions illustrate this point (see Chapter 9).

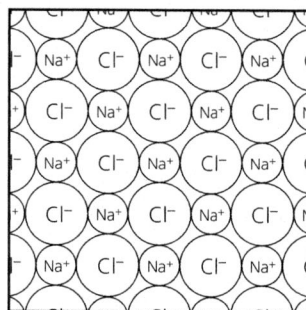

Figure 5.6 Representation of the ionic sodium chloride lattice in two dimensions

To avoid drawing electrons as dots or crosses hopping between atoms, focus on molecular models (see above). Use models of ionic compounds to show ionic bonds are present between oppositely charged ions arranged in regular crystalline lattices. This helps students realise that ionic bonds form in multiple directions and that ionic compounds do not form discrete molecules but regular lattices of billions of ions. Show ionic compounds reacting with each other, for example, acid–base reactions (see Chapter 7). Salts formed in these reactions can be obtained from solution and found to be crystalline solids. Figure 5.6 represents the sodium chloride lattice in two dimensions.

> **KEY ACTIVITY**
>
> ## Ionic bonding
>
>
>
> Repeating the activity suggested above for metallic bonding would be helpful, substituting a group of ionic substances for the metallic objects. Images and chemical representations can be examined.
>
> The substances may include samples of sodium chloride as rock salt (halite) and in a pure form, such as that obtained by evaporation of a salt solution; silicon dioxide as sand or/ and quartz; sodium hydrogencarbonate (bicarbonate), that is, baking soda and aluminium oxide (alumina, also ruby and sapphire).
>
> Students can examine one or more ionic substances and prepare answers to questions for peer assessment as before. Microscopes and hand lenses will help students to examine small crystals.
>
> Invite students to make molecular models that represent an ionic substance. Plasticine® and cocktail sticks could be used to make ball-and-stick models, or Molymod kits could be used.
>
> ### Questions to ask
> - Draw how you imagine the ions are arranged in this substance.
>   Where are the ionic bonds?
>   How many bonds are formed by each ion?
> - What properties does the substance have?
>   How are these properties explained by ionic bonding?
> - What does the model show about the arrangement of ions in the substance?
>   What does the model not tell you about the substance?
> - Where does the substance come from?
>   How was it formed?
> - In what ways are ionic substances similar? How do they differ from each other?

Ionic bonds are electrostatic attractions between oppositely charged ions. Each ion can form more than one ionic bond, as ionic charge is distributed across the whole ion. Ions pack together in a regular arrangement known as a lattice. The effect of many ionic bonds in a regular lattice forms a rigid crystalline structure. These lattices are often solid at room temperature and pressure. Ionic substances often have high melting and boiling points as a lot of energy is required to disrupt ionic lattices.

Most material on Earth originates from nuclear processes that took place in stars (see Chapter 1) where temperatures are too high for individual atoms to exist. In these conditions, matter exists as plasma (see above), a kind of gas comprising separate nuclei and electrons. At the end of the star's life, the material cooled, forming stable combinations of nuclei and electrons: atoms, ions or molecules. Most substances, for example, silicon dioxide (SiO_2, sand) and aluminium oxide (AlO_3) are formed as ionic lattices. Very little of the matter that became the Earth exists as discrete atoms. The noble (inert) gases (Group 0, see Chapter 4) are an example.

Cross-disciplinary

We often think the space between stars is an empty vacuum, but this is not true: matter exists in space. The interstellar medium (ISM) is the material that fills space in between stars. Scientists estimate that about 15% of the visible matter in the Milky Way is interstellar dust and gas. Molecules, ions and atoms are present in the ISM. This material is ejected from stars or was left over after these stars formed. The ISM contains simple structures such as molecular hydrogen, carbon monoxide, ammonia and sulfur dioxide, as well as quite complex molecules such as ethanol, ethanoic acid and benzene. In the densest regions of the ISM the density of particles may be 100 to 1000 billion per cubic metre, which is in turn is 100 to 1000 billion times less than the density of breathable air. The temperature of the ISM is around 10–30 K (−263 °C to −243 °C). The ISM is very important to astronomers developing our knowledge and understanding of the Universe as it traces the chemical evolution of atoms as they are fused in the cores of stars and then ejected in winds or supernova explosions.

5.6 Covalent bonding

A single covalent bond occurs when the outermost electron shells of two atoms overlap, allowing one (or more) electron pair(s) to be present within the outermost electron orbits of both atoms. This leads to sharing electrons (as opposed to transferring electrons for ionic bonding) in one electron orbit, balanced by the positively charged atomic nuclei. Atoms are often drawn simply with overlapping outer electron shells using the Bohr model (Chapter 4). Covalent bonding occurs usually along the axis between two atomic centres, so is directional and comprises discrete bonds (compared to multi-directional ionic bonds).

Covalent bonding tends to occur when the particles involved are uncharged *and* the substances are single elements. Examples include gases such as oxygen, chlorine, hydrogen, fluorine and allotropes of carbon, sulfur and phosphorus. Additionally, covalent bonding explains how atoms combine in ions such as carbonate (CO_3^{2-}), sulfate (SO_4^{2-}), ammonium (NH_4^+) and nitrate (NO_3^-). Plastics such as poly(ethene) - (CH_{2n}) and biological molecules such as carbohydrates, fats, proteins and nucleic acids have covalent bonds in their structures.

In some substances a double or triple covalent bond is present. A double covalent bond involves two electron pairs being shared across two atomic nuclei. In the alkenes, a homologous series of hydrocarbons (see Chapter 10), each molecule has a double bond between two carbon atoms, shown as C=C. Double bonds are also present between oxygen atoms in O_2 molecules and between carbon and oxygen atoms in carbon dioxide molecules (CO_2). Triple bonds occur where three pairs of electrons are shared between two atomic nuclei. Nitrogen gas molecules (N_2) are bonded with a triple bond N≡N.

Table 5.1 shows examples of students' thinking about covalent bonds and the scientific views. Some are intuitive ideas, others arise from teaching. Misconceptions include that a single electron is shared between two atoms; pairs of electrons are shared equally; covalent bonds involve total transfer of electrons; covalent bonds break when a substance changes state; covalent bonds are weak, as covalent compounds have low boiling points; covalent bonds have higher status than ionic and metallic bonds; and all atoms in covalent bonds form macromolecules. There is also confusion between *intra*molecular bonds and *inter*molecular bonds (discussed below). To raise and address these points, students may appreciate a version of the activity introduced above.

KEY ACTIVITY

Covalent bonding

Prepare a set of everyday substances with covalent bonds in their structures. Examples include sugar (sucrose), starch, cellulose (paper or wood), graphite (carbon), diamond (perhaps not an 'everyday' substance, but nonetheless a well-known form of carbon), graphene (see Enrichment, below), carbon dioxide gas, oxygen (a gas jar of air labelled 'oxygen' will suffice), sulfur, polyethene, aspirin (salicylic acid), ethanol, sunflower oil, margarine or butter.

As with previous activities in this chapter, the purpose is to apply ideas about covalent bonding to explain the structure and properties of these substances. Including colourless gases may seem strange, but oxygen and carbon dioxide are both covalent substances familiar to students, so it is worth discussing these. There is always a temptation in drawing gases to show particles much closer together than they actually are. Students can be invited to observe one or more items, depending on the time available, and to share their explanations. Using peer feedback to offer formative assessment by applying previously agreed criteria that relate to scientific ideas and language about covalent bonding would enable analysis of students' thinking. If misconceptions are apparent (see Table 5.1), discuss and correct these.

5 Chemical bonding

> **Questions to ask**
> - What particles are present in the substance?
> - Draw how you imagine the arrangement of particles in the substance. Label the covalent bonds.
> - Describe the physical properties of the substance, such as the melting and boiling points, colour, conductivity of heat and electricity, physical state at room temperature. Include in your description what happens to the covalent bonds when the substance changes state.
> - How is the substance used in everyday life?
>   Connect the physical properties with the uses.
> - In what ways are all substances containing covalent bonding the same?
> - In what ways are covalently bonded substances different from each other? Explain the differences as best you can.

Challenges involved in teaching covalent bonding

There is a strong temptation to apply the octet rule as an explanation for covalent (and ionic) bonding. The octet rule states that elements bond such that each atom in the bond has a full outermost electron orbit, most frequently with eight electrons. The octet rule applies only to a small number of substances, so is a simplified statement of facts. If it is over-used, students may reason falsely, stating for example, that sodium metal atoms react with chloride ions *because* the octet rule *causes* chemical reactions to occur (see Table 5.1); and/or that each atom in a stable molecule *must obey* the octet rule. Many substances form stable molecules without obeying the octet rule. For example, under the octet rule, elements boron and aluminium (both in Group 3) would make five bonds, so that their outermost electron orbital contains eight electrons. These elements never form five bonds. Nitrogen atoms with five outermost electrons form compounds such as nitric oxide (NO) and nitrogen dioxide (NO_2) that do not obey the octet rule. Ensure students do not state the octet rule as a definitive explanation for the formation of a molecule and understand that it does not apply in every case.

Students' ideas about chemical bonding include thinking of bonds as springs that release energy when broken; that bonds hold atoms together and release energy when ruptured and that the formation of a bond requires energy and breaking it releases energy (an idea related to the misconception that chemical bonds are physical entities). These are persuasive, but incorrect. Energy is released when chemical bonds *form*. This can be illustrated via demonstrating chemical reactions (see Chapter 2). Energy is required to disrupt a bond. This can be illustrated by inviting students to break apart molecular models – energy has to be exerted to do this. Further discussion of these points is provided in Chapter 6.

Separating covalent and ionic bonding prompts the assumption that most bonds are one or the other. In fact, very few bonds are 100% covalent, and no bonds are 100% ionic. The ionic bond represented in many chemistry texts is non-existent: compounds considered ionic do not have full ionic bonding. Fully covalent bonds usually only exist between atoms of the same elements, such as between hydrogen atoms in hydrogen molecules (H_2). Carbon–carbon (C–C) bonds in diamond are fully covalent, as carbon atoms are only bonded to other carbon atoms. In contrast, the C–C bond in ethanol (C_2H_5OH) is not fully covalent, as one carbon atom is bonded to an oxygen atom: the carbon–oxygen bond distorts the electron pair in the carbon–carbon bond, so electrons are not shared 50:50. Once students have grasped the principles of the two bond types, introduce the notion that most substances have bonds that are not purely one or the other. Ionic and covalent bonds are extremes and most bonds do not fit exactly into one or other category. Discussing the continuum between ionic and covalent bonds is helpful to students progressing to post-16.

> **Enrichment**
>
> Carbon atoms are able to form up to four covalent bonds with other atoms. This property is seen in multiple structures that elemental carbon can adopt. In diamond, each carbon atom is bonded to four other atoms in the familiar tetrahedral arrangement. In graphite, carbon atoms each form three covalent bonds to other carbon atoms to create a flat hexagonal sheet. Every carbon atom shares a fourth electron across to another sheet in the structure, allowing graphite to conduct electricity. In 2004, scientists at Manchester University isolated graphene, a one-atom thick sheet of carbon atoms arranged in an interlinked honeycomb of hexagons. Graphene can be made using sticky tape, a piece of white card and one flake of graphite or a 6B pencil. Take 10 cm sticky tape, fold the ends up so that they can be used as tabs that allow the tape to be repeatedly closed with the sticky sides together and re-opened. Open the tape. Place a single flake of graphite on the sticky side, or make a dot or mark with the pencil. Fold the tape over the flake or mark. Rub for a few seconds, then peel apart. Repeat this process, each time placing a clean piece of tape over the graphite. The graphite should spread thinly over the tape. Stick the tape to the card. Faint smudges of graphite should be visible on the tape. The folding and unsticking separates layers of graphite. Some may be one atom thick – but these are too small to be visible!

5 Chemical bonding

 ## 5.7 Intermolecular bonds: hydrogen bonding and van der Waals forces

Some textbooks and chemists refer to intermolecular bonds as intermolecular forces or attractions. Intermolecular bond is a preferable term because the main explanation for chemical bonding is electrostatic attractions, a holding power keeping particles (electrons, ions, atoms, or molecules) together. Students need to understand that intermolecular forces conceptually belong with chemical bonding as they also rely on electrostatic attractions between opposite charges. The energy involved in individual intermolecular interactions is much less than in most ionic, covalent or metallic bonds. The term force (or in some texts attraction) is used to make this distinction. However, this can create confusion for students, who are novice chemists. We strongly recommend teaching this as an aspect of chemical bonding to ensure consistency. Once students have understood the nature of intermolecular bonds, terminology can be revised to draw distinctions between these and other bond types.

Intermolecular bonds are essential to understanding the structure and properties of many substances. Materials with discrete covalently bonded molecules are solid at room temperature because of interactions *between* molecules. For example, wax and poly(ethene) (polythene) contain molecules that have covalent intramolecular bonding, but their structures are held together due to bonds between molecules – intermolecular bonding. Each intermolecular bond in wax and poly(ethene) is significantly weaker than a single covalent intramolecular bond in either molecule (each one requires a tiny amount of energy to disrupt). But the molecules of wax and poly(ethene) comprise many atoms, each of which is able to make an intermolecular bond with an atom on one or more other molecules. The cumulative effect of millions of intermolecular bonds is significant. The presence of intermolecular bonds means that both substances are solid at room temperature but have relatively low melting points for their high relative molecular mass values.

These intermolecular bonds form as molecules move close to each other. Electron orbitals on each molecule become distorted due to repulsion of negative charges. The effect is to create, or induce, a minute temporary polarisation or temporary dipole as electrons move between and around atomic cores. A temporary positive charge around an atom on one molecule attracts a temporary negative charge on another. This creates a short-lived bond called a van de

5.7 Intermolecular bonds: hydrogen bonding and van der Waals forces

Waals' force, traditionally named after the Dutch chemist Johannes Diderik van der Waals (1837–1923). A longer, but more accurate, modern name is a temporary dipole–temporary dipole induced bond. These intermolecular bonds hold molecules in irregular solid structures. This information is useful when teaching chemical bonding, as it avoids common misconceptions that molecular solids have covalent bonding throughout, and that covalent bonds are weaker than ionic bonds because covalent solids have lower melting points than ionic solids (Table 5.1).

Students studying biology and those progressing to post-16 science will meet hydrogen bonds. These are the strongest type of intermolecular bonds. They occur between two molecules in which hydrogen is bonded to either an oxygen, fluorine or nitrogen atom. Hydrogen bonds explain some properties of water, proteins and nucleic acids. They arise because hydrogen atoms on one molecule have a partial positive charge (symbol: delta plus, $\delta+$) while the oxygen, nitrogen or fluorine atom has a delta minus charge ($\delta-$). The partial charges arise because electrons are not shared perfectly equally in covalent bonds in each molecule (see above). The bond that forms between a $\delta+$ hydrogen atom on one molecule and a $\delta-$ oxygen, fluorine or nitrogen bond on another molecule is a hydrogen bond.

Cumulatively, hydrogen bonds require significant amounts of energy to disrupt. Consequently, they contribute to the stable structures of large molecules such as DNA and proteins. Water (H_2O), a molecule with a low relative molecular mass (M_r value) of only 18, has a much higher boiling point (100 °C) than similarly sized molecules such as methane (CH_4, $M_r = 16$), which do not have hydrogen bonds between their molecules. Hydrogen bonds also explain why ice floats on liquid water: in ice, molecules adopt a regular open-lattice structure in which hydrogen bonds hold them in position. This structure is less dense than liquid water. This can be demonstrated using molecular models.

When teaching, ensure that students consider all of the electrostatic interactions that hold structures together. Explain that when we say some chemical bonds are much stronger than others, we mean more energy has to be transferred to disrupt them.

Students may use the words bond, bonding, force and attraction interchangeably. Precision in language used in teaching helps avoid this. Resources to support development of correct ideas are listed at the end of the chapter.

5 Chemical bonding

> **Enrichment**
>
> We grow up assuming that no two snowflakes are alike. Wilson Bentley, affectionately known as Snowflake Bentley, a farmer who lived in Vermont, north-east USA from 1865–1931, first made this claim. He took the first photograph of a single snow crystal and photographed over 5000 in his lifetime, never finding two the same. He said, 'Under the microscope, I found that snowflakes were miracles of beauty, and it seemed a shame this should not be seen and appreciated by others. Every crystal was a masterpiece of design, and no one design was ever repeated. When a snowflake melted, that design was forever lost.'
>
> Snowflakes form in clouds when the temperature falls below 0 °C. Snowflakes land on the ground if the air temperature at ground level is also below 0 °C. The shape of snowflakes arises because hydrogen bonds between water molecules form a hexagon-based lattice. Depending on the temperature and conditions, the basic hexagon shape can grow into flat star shapes, long columns, or needles. Classic star shapes form at −15 °C. Consider if it can possibly be true that no two snowflakes can be alike. Use a molecular model kit to make water molecules and investigate snowflake shapes.

5.8 Summarising metallic, ionic, covalent and intermolecular bonding

At this point, students have met several types of bonding. Create a comparison table to summarise similarities and differences between them. Table 5.2 provides an example highlighting that bonding (a) refers to electrostatic attractions between opposite charges that hold particles together; and (b) results in different types of chemical bonding.

5.8 Summarising metallic, ionic, covalent and intermolecular bonding

Table 5.2 Comparing metallic, ionic, covalent and intermolecular bonding

Features	Metallic bonds	Ionic bonds	Single covalent bonds	Double or triple covalent bonds	Intermolecular bonds
Nature of the bond	Electrostatic attractions between cations and electrons; electrons free to move.	Multi-directional attractive electrostatic forces between oppositely charged ions.	Unidirectional electrostatic attractions between an electron pair and two atomic nuclei.	Unidirectional electrostatic attractions between multiple electron pairs and two atomic nuclei.	Electrostatic attractions between small temporary positive and negative charges on different molecules.
3-dimensional arrangements of units	Regular lattices when solid; cations surrounded by free electrons.	Crystalline lattices when solid; ions with opposite charges connected.	Molecules in lattices when solid, in looser but organised arrangements when liquid, and separated with no formal arrangement when gaseous; positively charged atomic centres connected by pairs of electrons.		Responsible for the arrangements of molecules in solids and liquids; strong intermolecular bonds increase the melting and boiling points of molecular substances.
Examples of substances	Sodium (Na) – a metallic lattice of Na+ cations surrounded by free electrons; solid at room temperature. Magnesium (Mg) – a metallic lattice of Mg^{2+} cations surrounded by free electrons. Solid at room temperature.	Sodium chloride (NaCl) – an ionic lattice of Na+ cations and Cl– anions; solid at room temperature. Calcium carbonate (CaCO$_3$) – an ionic lattice of Ca^{2+} cations and CO$_3^{2-}$ anions; solid at room temperature.	Chlorine (Cl$_2$) – a molecule comprising a single covalent bond between two chlorine atoms (Cl–Cl); a gas at room temperature. Ammonia (NH$_3$) – a covalent molecule; a gas at room temperature. Diamond and graphite (C) each have single C–C bonds arranged in different lattices.	Oxygen (O$_2$) – a molecule comprising a double covalent bond between two oxygen atoms (O=O); a gas at room temperature. Nitrogen (N$_2$) – a molecule comprising a triple covalent bond between two nitrogen atoms (N≡N); a gas at room temperature.	Water (H$_2$O) – a covalently bonded molecule; has a higher boiling point than most small molecules due to electrostatic attractions between the molecules so liquid at room temperature. Poly(ethene) – covalently bonded molecules held together by intermolecular bonding; this makes poly(ethene) a solid at room temperature.

Representational challenges for chemical bonding

Students are liable to interpret representations of chemical bonds and formulae to make meanings that are different from those intended. For example, they incorrectly interpret NaCl(s) to mean that a single sodium atom reacts with a single chlorine atom to form a single ionic molecule, formula NaCl, and believe that these molecules are arranged regularly in crystals. Similarly, they incorrectly think that HCl(aq) means that molecules of hydrogen chloride are present in hydrochloric acid, so when magnesium metal is added, magnesium 'swaps partners' with the chloride ions, making magnesium chloride molecules, $MgCl_2$(aq). These faulty ideas arise from teaching. Avoid them by ensuring that discussion of ionic bonds does not emphasise the electron transfer model but illustrates that ionic bonds are present in an ionic lattice as interactions between ions. Following the suggestion above to discuss ionic bond formation between existing ions during reactions will help avoid using the swapping partners phrase.

In ice, H_2O(s), students may not realise water molecules are bonded in a regular crystalline arrangement but believe the organisation is haphazard, as in liquid water. Without the explanation offered by hydrogen bonds, students may think incorrectly that covalent bonds between hydrogen and oxygen atoms break when water evaporates and reform when water condenses on a cold surface. Introducing hydrogen bonds to explain the properties of water avoids this.

Using the Bohr atomic model leads some students to visualise electrons in shells with hard structures, like eggshells, which encapsulate and protect the atomic core, or in layers, like those of an onion, which get smaller closer to the centre. For other students, electron orbitals are cloud-like structures in which electrons are embedded. Drawing electron shells may give the impression that they are solid structural elements of atoms, not tiny three-dimensional regions of space where electrons may be found. Representations of molecules and chemical bonds are static, a feature often taken literally by students. As a result, they think that electrons in a bond cannot move. Reinforcing that a chemical bond results from electrostatic attractions between particles of opposite charge is helpful. Also, ensure students realise that one limitation of molecular models is the way they make chemical bonds appear fixed in one place.

Students may find the variation in representations of chemical bonding confusing, as the same bond type is pictured differently in textbooks, online images and in models, with no explanation. Emphasise the limitations of representations, as these are often a compromise between what can easily be drawn and specific points that need to be taught. Avoid discussion of which representation is 'best'. Make students aware that molecules are not easily drawn, and that scientists often model them through representations that stress relevant features.

> ### Careers
>
> Studying molecular structures and knowledge and understanding of materials are essential to a number of careers. X-ray diffraction and crystallography techniques that investigate structures are commonplace in biomedical sciences and materials science. In medicine, an X-ray crystallographer may work in a pharmaceutical team growing and crystallising new molecules that may be drugs, structural proteins or new materials suitable for use within the body.
>
> Materials scientists work in a variety of industries developing and testing new substances from raw materials to meet customer specifications or new applications. They apply their knowledge and understanding of the structure of molecules to predict how new molecules may form in chemical reactions. Materials scientists may use a wide range of techniques to gain information about the substances they make.
>
> Geologists (see Chapter 11) study the matter that constitutes the Earth and other planets, gaining an understanding of the processes that shaped and formed solids, liquids and gases. Geologists detailed knowledge is essential in the energy and mining industries. They learn a variety of techniques in a wide range of fields including mineralogy, hydrogeology, petrology and sedimentology.

Resources

References and further reading

Anderson, J.S., Hayes, D.M. and Werner, T.C. (1995) The chemical bond studied by IR spectroscopy in introductory chemistry: An exercise in cooperative learning. *Journal of Chemical Education*, 72(7): 653–655.

Barker, V. and Millar, R. (2000) Students' reasoning about basic chemical thermodynamics and chemical bonding: What changes occur during a context-based post-16 chemistry course?. *International Journal of Science Education*, 22(11): 1171–1200.

Boo, H.K. (1998) Students' understandings of chemical bonds and the energetics of chemical reactions. *Journal of Research in Science Teaching*, 35(5): 569–581.

Coll, R.K. and Treagust, D.F. (2001) Learners' mental models of chemical bonding. *Research in Science Education*, 31: 357–382.

Erman, E. (2017) Factors contributing to students' misconceptions in learning covalent bonds. *Journal of Research in Science Teaching*, 54(4): 520–537.

Hapkiewicz, A. (1991) Clarifying chemical bonding. *The Science Teacher*, 58(3): 24–27.

Levy Nahum, T., Mamlok-Naaman, R. and Hofstein, A. (2013) Teaching and learning of the chemical bonding concept: Problems and some pedagogical issues and recommendations in *Concepts of Matter in Science Education* Tsaparlis, G. and Sevian, H. (editors) pages 373–390, Springer.

Taber, K. (2002). *Chemical misconceptions – prevention, diagnosis and cure* Volumes I and II, Royal Society of Chemistry.

Taber, K. (editor) (2012) *Teaching Secondary Chemistry*, 2nd edition, Hodder Education.

Taber, K.S. (1998) The sharing-out of nuclear attraction: or "I can't think about physics in chemistry". *International Journal of Science Education*, 20:1001–1014.

Tan, D.K.C. and Treagust, D.F. (1999) Evaluating students' understanding of chemical bonding. *School Science Review*, 81(294): 75–84.

Websites

Kind, V. (2004) *Beyond appearances: Students' misconceptions about basic chemical ideas.* Available at https://edu.rsc.org/resources/beyond-appearances/2202.article

BEST evidence in science teaching project at the University of York Science Education Group: www.york.ac.uk/education/research/uyseg/research-projects/bestevidencescienceteaching/

5.9 Resources

http://molview.org/ is a free website that allows common molecules to be displayed and rotated in a range of models

RSC Teach Chemistry, https://edu.rsc.org/ has a wide range of resources for teaching and learning chemistry, including worksheets, simulations, practicals and demonstrations.

Other resources

Molymod® – molymods.com – available from school equipment suppliers such as Better Equipped, Philip Harris, SciChem and TimStar.

Energetics, rate and extent of chemical change

Ann Childs and Neil Dixon

Introduction

This chapter offers information that helps to explain how far, how fast and why chemical reactions occur. Enabling students' understanding of these questions supports a secure base knowledge of chemistry. The chapter begins by discussing energetics, which explains how energy is transferred in bond-breaking and bond-making, and how energy changes can be measured. This underpins many concepts needed to understand rates of reaction as well as many other aspects of chemistry. The study of rates of reaction provides a rich variety of opportunities for learners to develop their practical and investigative skills, and their confidence in analysing and interpreting data. The final section examines reversible reactions, leading to an understanding of the conceptually demanding topic of chemical equilibria.

Students' previous knowledge and experience

Students' experience of chemical reactions in primary education is likely to be characterised as changes resulting in the formation of new materials, such as combustion, rusting, and cooking. They may be aware these changes are not usually reversible. Students may know that changes such as dissolving, mixing, melting and boiling may be reversed. In addition, they will have met many chemical reactions, although they may not recognise these changes as chemical reactions without some prompting.

Progression

Energetics of chemical change

Students investigate and classify exothermic and endothermic reactions. They learn that exothermic reactions release energy, increasing the temperature of the surroundings. This links to physics by discussing that energy is never destroyed but can be transferred. Students then learn about bond-breaking, bond-making, activation energy and reaction profiles. These concepts explain why reactions are exothermic or endothermic. Further study at higher levels applies these concepts to understanding energy cycles, Hess's law, entropy and Gibb's free energy.

Rates of reaction

To begin with, students start from familiar examples of reactions, recognising which are faster or slower than others. Next they learn ways to measure and calculate rate of reaction. Collision theory is introduced to explain how changes to conditions affect the rate of a reaction.

Extent of chemical reactions

Understanding whether or not a chemical reaction goes to completion is essential for chemists. Primary and lower-secondary school students assume chemical changes are always complete. They will not be familiar with the fact that some reactions may be incomplete or reversible. They may realise changes of state can be reversed by heating and cooling as necessary.

Next, teachers establish that chemical reactions, in principle, can be reversed, and that reversible reactions in closed systems achieve dynamic equilibrium. Finally, the effects of changing conditions like temperature, concentration and pressure on equilibrium are investigated. The effects of changing conditions on equilibrium positions will be used to explain the reaction conditions employed in some industrial processes, such as the Haber process, which rely on equilibrium reactions.

Teaching sequence overview

Energetics of chemical change

→ Energy changes during exothermic and endothermic reactions.
→ Measuring energy changes in chemical reactions.
→ Reaction profiles linked to bond formation, bond-breaking and activation energy.

Rates of reaction

→ What is rate of reaction? How is rate of reaction measured?
→ Collision theory (including activation energy) as a framework to understand rates of reaction.
→ Factors influencing rate of reaction:
  - temperature
  - concentration

6 Energetics, rate and extent of chemical change

- pressure of a gas
- surface area
- presence of a catalyst.

Extent of chemical change

→ An introduction to reversible reactions and dynamic equilibrium.
→ Factors affecting equilibrium position and the yield of the desired product.

6.1 Energetics: how is energy involved in chemical reactions?

Reactions are classified as exothermic or endothermic by measuring the temperature change occurring in their surroundings. Exothermic reactions emit heat energy to the surroundings, so temperature increases. Endothermic reactions absorb heat energy from the surroundings, so temperature decreases. This is explained further below.

The Greek letter delta, written Δ, is used in sciences to represent change. Thus, ΔT means temperature change. Enthalpy (H) is the formal term from physics that describes the total internal energy of a system, plus the product obtained by multiplying the pressure and volume of the system. This is represented by the equation below:

$$H \text{ (enthalpy)} = E \text{ (internal energy)} + PV \text{ (pressure} \times \text{volume)}$$

This sounds complicated, but this is the term often applied when discussing energy changes in chemical reactions. The words energy and enthalpy are often used interchangeably but are not the same. Chemists measure enthalpy change, symbol ΔH. An exothermic reaction is 'ΔH negative' because the reaction, that is, the reactants and products together, transfer energy to the surroundings. Correspondingly, the temperature of the surroundings increases. The internal energy of the system decreases, so a negative sign is required. An endothermic reaction is 'ΔH positive' because energy from the surroundings transfers to the reaction. The internal energy of the system increases, so ΔH is positive, while the temperature of the surroundings has decreased.

Many reactions in school chemistry (for example, neutralisation) take place in solution so inserting a thermometer into the liquid conveniently measures any temperature change occurring. For exothermic reactions such as combustion, the temperature change can be measured using a simple calorimeter comprising a metal beaker of water with a thermometer immersed in the water. The beaker and thermometer are set slightly above the reaction vessel so heat energy that is released transfers through the metal wall of the beaker into the water and is measured as a temperature change by the thermometer.

Students' misconceptions about energy changes

Energy is a challenging concept to understand, so inevitably students may harbour misconceptions and misunderstandings about chemical reactions and energetics. Some student ideas that may need addressing when teaching this topic, each followed by a way of addressing the problem, are described below.

Students confuse an increase in temperature with an increase in chemical stored energy

Consider placing a thermometer into a boiling tube containing a dilute acid reacting with magnesium ribbon. Students may think the increase in temperature means that the chemicals *gain* energy, classifying the reaction as endothermic. They are measuring temperature change of the *solvent* (water) which is the *surroundings*. The internal energy *of the reaction* decreases.

Students think energy is stored in a system ready to be released

The first law of thermodynamics states that energy cannot be created or destroyed. Students may interpret this to mean that energy in chemicals is released when they react. Language such as 'fuels are energy stores' or 'fuels contain energy' does not help. The notion sidesteps what actually happens, creating the perception that burning (and by extrapolation, other chemical reactions) is a destructive event in which fuel is 'used up'. This contravenes the law of conservation of mass (Chapter 2). Instead, students need to realise chemical reactions involve reactants and products, and some reactions release energy to the environment.

Students think energy is used up when we use fuels

Fuels are used to 'do' something, such as move a motor vehicle, cook a meal or heat a house. There is a strong link between fuel use and fuel as an 'energy source' for the task. Students may think energy is consumed as the task is done. Calculations of fuel consumption in a car engine or of gas consumed by a boiler contribute to this. Showing students that we transfer energy from usable to non-usable forms is helpful. The overall amount of energy in the Universe remains constant.

Students think burning is a state change, not a chemical reaction

Students may forget that oxygen from the atmosphere is involved in burning. They perceive changes only to the fuel. For example, they think using petrol in a car engine means turning the petrol to exhaust gas, without oxygen being involved; similarly, burning a candle involves wax melting, not reacting with oxygen.

Students think that bond-breaking releases energy

A consequence of teaching that fuels are energy stores is that students think fuel molecules release energy when they burn. Rather like an egg, which releases the contents when the shell is broken, students' reason that when a bond is broken, energy is released. Even students who know bond-breaking requires energy may think this, on the grounds that a small amount of energy may be needed to start off the break, but the actual breaking releases far more. This misconception makes it very difficult to understand energetics properly.

KEY ACTIVITY

Investigating exothermic and endothermic reactions

A good starter activity to key students in to the ideas behind energy change in reactions is the reaction of sodium hydrogencarbonate with citric acid (see Chapter 7). This reaction occurs when sherbet is placed in the mouth. Sherbet powder contains a mixture of solid citric acid and solid sodium hydrogencarbonate (baking soda). The water in saliva dissolves the reactants and allows them to react, producing carbon dioxide, giving the sherbet its fizz. The slightly endothermic reaction means the tongue is cooled as energy transfers to the reacting chemicals, adding to the sensation when eating sherbet.

Students can investigate exothermic or endothermic reactions by measuring temperature changes produced using the following pairs of reactants:

- sodium hydroxide solution + dilute hydrochloric acid or dilute sulfuric acid
- copper(II) sulfate solution + magnesium powder
- dilute sulfuric acid + magnesium ribbon

Full instructions can be found at https://edu.rsc.org/resources/exothermic-or-endothermic/406.article

When carrying out these reactions, expanded polystyrene cups are commonly used as a reaction container since these insulate reaction systems reducing heat transfer to and from the air. This enhances accuracy of measurement of the temperature change compared to using non-insulating glassware. Placing the cups into 250 cm³ beakers to prevent them falling over is helpful. If polystyrene cups are unavailable, boiling tubes can be used. Use just enough reactants to allow the thermometer bulb to be fully immersed when making measurements.

Start by recording the temperature of the first reactant added to the cup/boiling tube. Take the next reading after adding the second reactant, when the temperature reaches the highest or lowest point. Students may find it difficult to know when the temperature is at this point. The best way is to keep taking readings until there is no further change. Students should write down temperature changes with positive or negative signs. Record temperature changes as positive if the reaction temperature is higher than the starting value and negative if the reaction temperature is lower than the starting value. Remind students they are measuring the temperature of the water in which one or more reactants and products are dissolved.

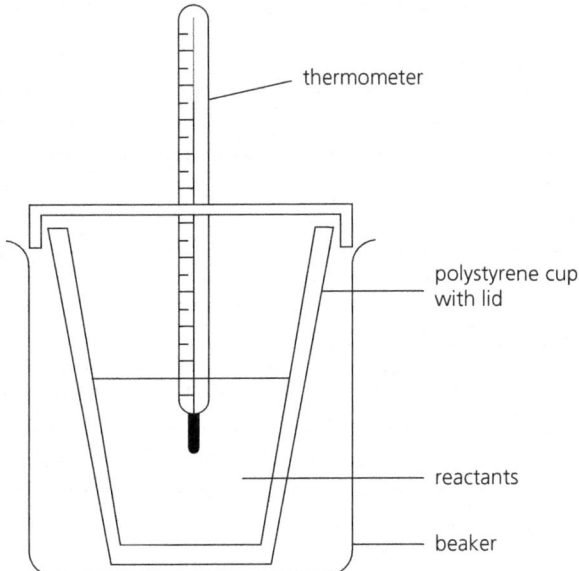

Figure 6.1 The apparatus used to investigate the enthalpy change of the reaction between sodium hydrogencarbonate and citric acid.

The familiar reaction of magnesium with dilute hydrochloric or sulfuric acid has links with investigations of the reactivity series, production of and testing for hydrogen, and reactions of acids to make salts.

The reaction of a dilute acid with any metal above hydrogen in the reactivity series (Chapter 8) will be exothermic. The non-metal ion from the acid (chloride or sulfate in this example) is a spectator that does not take part in the reaction. Highly reactive metals generate larger temperature changes. The higher metal in the series, the more reactive it is; that is, the faster, more vigorous and more exothermic the reaction. It

6 Energetics, rate and extent of chemical change

is not recommended to react very reactive metals (for example, calcium, sodium, potassium, lithium) with dilute acids for safety reasons.

When a reactive metal is placed into an aqueous solution containing ions of a less reactive metal, a displacement reaction occurs (see Chapter 8). In the displacement reaction between copper(II) sulfate solution and magnesium, the less reactive metal ion, copper(II), is reduced to elemental copper. Magnesium atoms are oxidised and replace the less reactive metal in the aqueous solution. Displacement reactions are always exothermic. For this reaction, the equation is:

magnesium + copper(II) sulfate → magnesium sulfate + copper
$Mg(s)$ + $CuSO_4(aq)$ → $MgSO_4(aq)$ + $Cu(s)$

Larger differences between two metals in the reactivity series generate more energy than small differences (assuming other variables are controlled).

Questions to ask
- Write down the temperature changes that were measured for each reaction in order from lowest to highest.
- Write word equations for each reaction. Identify any ions that did not take part in the reactions (these are 'spectator' ions).
- What do you notice about the reactions?

Teacher demonstrations

These teacher demonstration reactions help learners engage with the topic and develop their theoretical understanding of energy change and reactions. Teachers are strongly advised to try them with an experienced technician or colleague prior to demonstrating them in front of students for the first time.

Dissolving ammonium nitrate in water

Endothermic reactions that are easy to investigate are rare. Although dissolving a solid in water can be considered a physical change (see Chapter 2), ammonium nitrate dissolves in water generating a large negative temperature change. This is a convenient way of demonstrating an endothermic enthalpy change. If done in a boiling tube, the temperature decrease can be sensed by hand. Full details can be found at https://edu.rsc.org/resources/energy-in-or-out-classifying-reactions/679.article This demonstration helps students understand an application of a useful endothermic reaction (see Enrichment). The experiment could be a class experiment for post-16 students only.

> **Enrichment**
>
> A variation of the cooling ammonium nitrate reaction occurs inside single-use chemical ice packs. Physiotherapists working pitch-side treat sports injuries with these. When the ammonium nitrate dissolves in water in the icepack, the endothermic reaction cools damaged tissue, reducing inflammation. Ice packs can be purchased from suppliers of first-aid equipment.

The reaction of ammonium chloride with barium hydroxide

This is a captivating spontaneous endothermic reaction, due to the very low temperature achieved. A standard laboratory thermometer ranging from −10 to 100 °C will not record the lowest temperature for this reaction as it is likely to be −15 °C. This represents a drop from room temperature of around 35 °C. The steady temperature decrease over time is exciting to witness. The reaction can take place in a glass beaker. Start with the beaker on a wet tile or watch glass onto which a few drops of water have been placed: the cooling effect is sufficient to freeze the water, sticking the beaker to the wet tile or watch glass.

The equation for the reaction is:

hydrated barium hydroxide + ammonium chloride → barium chloride + ammonia + water

$$Ba(OH)_2 \cdot 8H_2O(s) + 2NH_4Cl(s) \rightarrow BaCl_2(s) + 2NH_3(g) + 10H_2O(l)$$

Full details of how to carry out this experiment can be found at https://edu.rsc.org/resources/endothermic-solid-solid-reactions/739.article

This reaction is driven by the large positive entropy change (see the experimental details for an explanation and calculation of the entropy change). This endothermic reaction has one of the largest positive entropy changes of any common reaction.

The thermite reaction

This highly exothermic reaction emits a lot of heat and light (see Chapter 8). The reaction has a large activation energy, which means energy input is required to initiate it. Like the reaction between magnesium and copper(II) sulfate solution above, this is a displacement reaction (see also Chapter 8) but between two solids, aluminium powder and iron(III) oxide powder. A promoter, usually barium peroxide, and a magnesium ribbon fuse provide the activation energy. The reaction produces molten iron at a temperature exceeding 1538 °C, the melting point of iron. The equation for the reaction is:

iron oxide + aluminium → iron + aluminium oxide

$$Fe_2O_3(s) + 2Al(s) \rightarrow Fe(l) + Al_2O_3(s)$$

6 Energetics, rate and extent of chemical change

Full details of how to demonstrate this reaction safely can be found at https://edu.rsc.org/resources/the-thermite-reaction/724.article

Safety

This reaction is hazardous. Teachers intending to carry out this demonstration must consult their employers' risk assessment (usually from CLEAPSS or SSERC), comply with all control measures and are strongly advised to practice before showing the reaction to a class.

The demonstrator must wear eye protection. The students must be at least 4 metres away. A safety screen must be placed between the reaction and the students. Protect the surrounding bench with heatproof mats in case of sparks. The demonstrator must have room to move to a safe distance once the fuse has been ignited. If the thermite mixture fails to react, pour the powder into a large beaker containing 2 mol dm^{-3} sulfuric acid. Leave overnight, then decant the liquid down the sink, double bag the solid waste and dispose of in normal refuse. This reaction can be unpredictable so may not react. Guidance given on the website indicates how to safely dispose of the reaction mixture in the event of non-ignition and it is important to be prepared for this to happen. Ensuring the reactants are dry helps the reaction occur.

Questions to ask

- Why is a promoter needed for this reaction?
- Why does aluminium displace iron from iron(III) oxide?
- Why is iron formed in a molten state?

Reaction between aluminium and iodine

In this spectacular demonstration, which must be done in a fume cupboard, the reaction between aluminium and iodine is initiated by water. Spectacular clouds of purple iodine vapour are emitted but this is unreacted iodine, not a product. The reaction is exothermic: energy is transferred to the iodine solid, some of which sublimes to form the iodine vapour. The remaining iodine reacts with aluminium to from the white solid aluminium iodide (equation below). The reaction can be temperamental: it may take from a few seconds to several minutes to start and, exceptionally, no reaction may occur. Advice for getting this reaction to proceed includes preparing reagents by grinding the iodine to a fine powder, mixing this very carefully with dry aluminium powder, and adding small amounts of warm water to avoid flooding the reaction.

Full details of how to carry out this experiment safely can be found at https://edu.rsc.org/resources/reaction-between-aluminium-and-iodine/715.article which also gives guidance on what to do should no reaction occur and how to dispose of unreacted chemicals. An alternative method in which the reaction is carried out in a large flask is listed in the references.

Equation for the reaction:

$$\text{aluminium} + \text{iodine} \rightarrow \text{aluminium iodide}$$
$$2\text{Al(s)} + 3\text{I}_2\text{(s)} \rightarrow \text{Al}_2\text{I}_6\text{(s)}$$

6.1 Energetics: how is energy involved in chemical reactions?

> Students may suggest that water provides activation energy. In fact, the water allows the aluminium and iodine particles to collide effectively by dissolving some iodine. Excess iodine is liberated as purple vapour. The reaction proceeds with a bright white flash, and a white solid is produced.
>
> **Questions to ask**
> - What is the purple vapour formed?
> - Why is the purple vapour emitted from the reaction mixture?
> - What observations of this reaction indicate that it is an exothermic reaction?
> - What is the name of the white solid formed?

Using bond enthalpies to explain exothermic and endothermic reactions

Chemical reactions involve breaking chemical bonds in molecules of reactants and making new chemical bonds to form products. *Breaking bonds is always endothermic.* Energy is absorbed to break bonds in reactant molecules. Conversely, making bonds releases energy to the surroundings. These are challenging concepts for students to grasp. A theoretical energy level diagram can be used to explain how an exothermic reaction occurs, as in Figure 6.2.

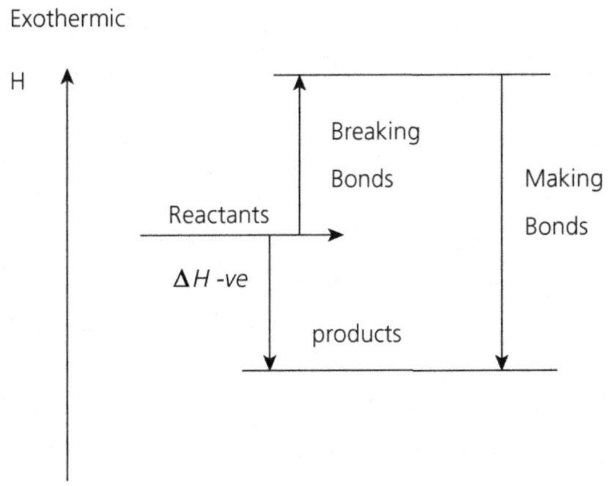

Figure 6.2 A theoretical energy level diagram for an exothermic reaction.

The vertical arrow pointing up from the reactants line represents the energy required to break all bonds in reactant molecules. The highest horizontal line represents the theoretical situation where all atoms are individual. The horizontal downwards arrow on the right represents the amount of energy released to the surroundings when bonds form in product molecules. The bold arrow between the reactant and product lines represents the enthalpy change of the reaction. In Figure 6.2, for an exothermic reaction, less energy is absorbed from surroundings to break bonds in reactant molecules than is released when new bonds are formed in product molecules. This results in a temperature increase in the surroundings.

In reality, the energy level diagram (or energy profile) of an exothermic reaction is quite different. In an exothermic reaction, a *small* amount of energy is absorbed from surroundings to break bonds in a few reactant molecules to initiate the reaction. Atoms from reactant molecules reform into product molecules, making new bonds in the process, releasing energy to the surroundings. Some energy breaks bonds in more reactant molecules, creating a chain effect. The reaction continues until the supply of one reactant is exhausted. Chemists describe this type of reaction as 'going to completion'. An energy profile for a typical exothermic reaction is shown in Figure 6.3.

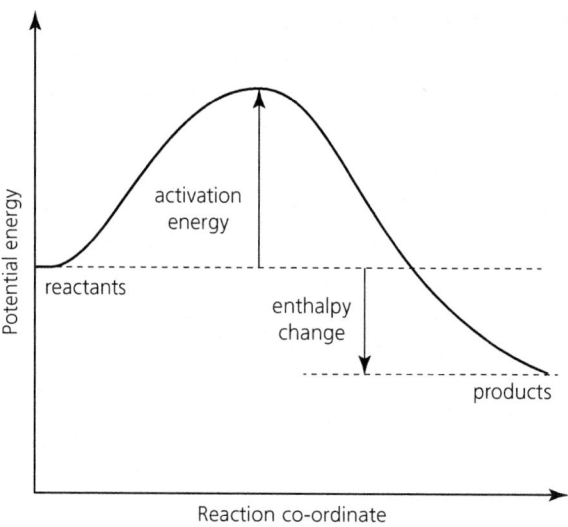

Figure 6.3 An energy profile for an exothermic reaction, including the enthalpy change and activation energy

A familiar exothermic reaction is combustion involving a fuel (wood, gas or petrol, for example) burning in oxygen. Energy from a lighted match, firelighter, burning splint or spark is required to start the reaction. This causes a few molecules of fuel and oxygen to break apart. New substances form, namely, carbon dioxide and water vapour. Energy is released when bonds form in product molecules. Some of this energy breaks bonds in more fuel and oxygen molecules. The rest is emitted as heat and light.

In an endothermic reaction, the amount of energy absorbed from surroundings to break bonds in the reactants is greater than the amount of energy released when new bonds are formed in product molecules. This results in a temperature decrease in the surroundings.

The enthalpy change of a reaction can be calculated from bond energy values. The quantity of energy required to break one mole (see Chapter 2) of a specific bond is called the bond enthalpy or bond energy. Bond enthalpies are measured in units of kilojoules per mole (kJ/mol or kJ mol^{-1}). Some are stated as averages. As explained above, the difference between the quantities of energy absorbed by reactant molecules when bonds break and that released when new bonds form in the product molecules determines whether a reaction is exothermic or endothermic.

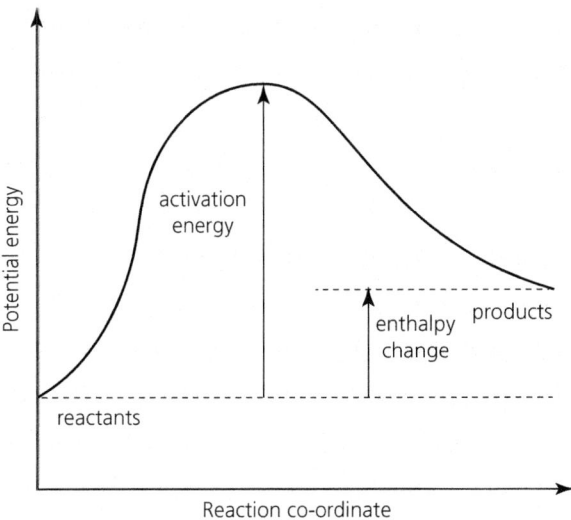

Figure 6.4 An energy profile for an endothermic reaction

Table 6.1 Bond energies in molecules commonly involved in combustion reactions. Note that values will vary from source to source.

Bond	Bond energy/ kJ mol^{-1}
C–H	413
O=O	498
C=O	799
O–H	464
H–H	436
Cl–Cl	243
H–Cl	432
N≡N	945
N–H	388

6 Energetics, rate and extent of chemical change

By convention, absorbing energy from the surroundings is regarded as a positive enthalpy change. This means the enthalpy change of a reaction is calculated by *subtracting* the sum of bond enthalpy values of the product molecules from the sum of bond enthalpy values of the reactant molecules.

Students should understand that energy must be absorbed by reactants to break bonds in reactant molecules. Logically – though not intuitively – energy must be released when new bonds are formed in the products.

Make sure that the balanced symbol equation for the reaction is represented with fully displayed formulae so that the number and type of each bond can be seen and easily counted. Impress that calculations must be set out logically, showing how the answer was achieved.

 Maths

Example 1: Combustion of methane

Methane + oxygen → carbon dioxide + water

$CH_4(g) + 2O_2(g) \rightarrow CO_2(g) + 2H_2O(g)$

Bonds broken in the reactants	Bonds formed in the products
C–H 4 × 413 = 1652	C=O 2 × 799 = 1598
O=O 2 × 498 = 996	O–H 4 × 464 = 1856
Total energy absorbed = 2648 kJ mol⁻¹	Total energy released = 3454 kJ mol⁻¹

Energy change of reaction = 2648 − 3454 = −806 kJ mol⁻¹

The energy change is negative, indicating that the reaction is exothermic.

Ask students to draw an energy level diagram to represent this reaction.

Energy level diagrams could also be drawn for examples 2 and 3.

Maths

Example 2: Reaction of hydrogen with chlorine

$$\text{hydrogen} + \text{chlorine} \rightarrow \text{hydrogen chloride}$$
$$H_2(g) + Cl_2(g) \rightarrow 2HCl(g)$$

H—H + Cl—Cl ⟶ H—Cl
                                          H—Cl

Bonds broken in the reactants   **Bonds formed in the products**

H–H   1 × 436 = 436   H–Cl   2 × 432 = 864

Cl–Cl   1 × 243 = 243

Total energy absorbed = 679 kJ mol^{-1} Total energy released = 864 kJ mol^{-1}

Energy change of reaction = 679 − 864 = −185 kJ mol^{-1}

The energy change is negative, indicating that the reaction is exothermic.

Maths

Example 3: Reaction of nitrogen with hydrogen

$$\text{nitrogen} + \text{hydrogen} \rightarrow \text{ammonia}$$
$$N_2(g) + 3H_2(g) \rightarrow 2NH_3(g)$$

N≡N + H—H, H—H, H—H ⟶ H—N(H)(H) and H—N(H)(H)

Bonds broken in the reactants   **Bonds formed in the products**

N≡N   1 × 945 = 945   N–H   6 × 388 = 2328

H–H   3 × 436 = 1308

Total energy absorbed = 2253 kJ mol^{-1} Total energy released = 2328 kJ mol^{-1}

Energy change of reaction = 2253 − 2328 = −75 kJ mol^{-1}

The energy change is negative, indicating that the reaction is exothermic.

6 Energetics, rate and extent of chemical change

π Maths

Example 4: Decomposition of water

water → hydrogen + oxygen
$2H_2O(l)$ → $2H_2(g)$ + $O_2(g)$

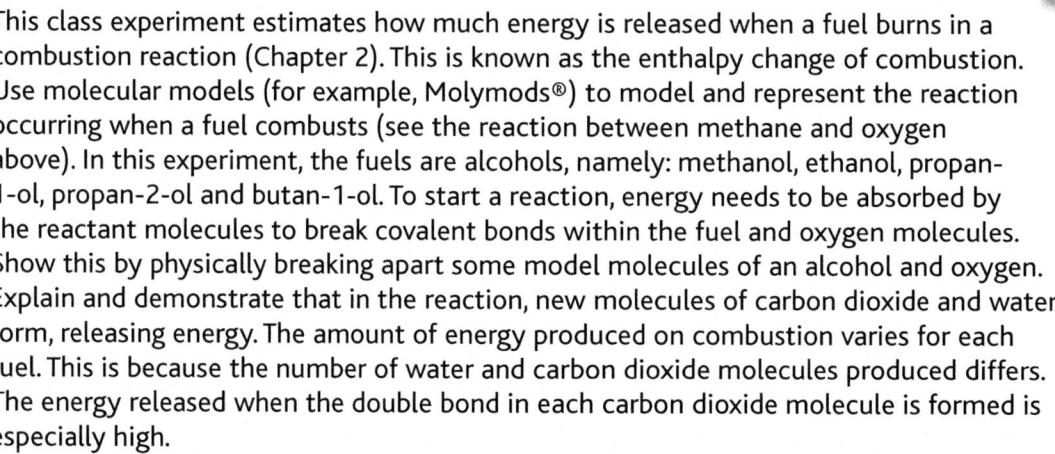

Bonds broken in the reactants

O–H   4 × 464 = 1856

Total energy absorbed = 1856 kJ mol⁻¹

Bonds formed in the products

H–H   2 × 436 = 872

O=O   1 × 498 = 498

Total energy released = 1370 kJ mol⁻¹

Energy change of reaction = 1856 − 1370 = + 486 kJ mol⁻¹

The energy change is positive, indicating that the reaction is endothermic.

KEY ACTIVITY

Burning a fuel

This class experiment estimates how much energy is released when a fuel burns in a combustion reaction (Chapter 2). This is known as the enthalpy change of combustion. Use molecular models (for example, Molymods®) to model and represent the reaction occurring when a fuel combusts (see the reaction between methane and oxygen above). In this experiment, the fuels are alcohols, namely: methanol, ethanol, propan-1-ol, propan-2-ol and butan-1-ol. To start a reaction, energy needs to be absorbed by the reactant molecules to break covalent bonds within the fuel and oxygen molecules. Show this by physically breaking apart some model molecules of an alcohol and oxygen. Explain and demonstrate that in the reaction, new molecules of carbon dioxide and water form, releasing energy. The amount of energy produced on combustion varies for each fuel. This is because the number of water and carbon dioxide molecules produced differs. The energy released when the double bond in each carbon dioxide molecule is formed is especially high.

Details can be found at https://edu.rsc.org/resources/heat-energy-from-alcohols/1733.article

There are three methods of handling the data from this experiment depending on the age and previous experiences of your students. The first method is the fastest and simplest, so would be appropriate for younger or lower attaining classes. The second method is appropriate for students who have not yet met, or will never meet, mole calculations. For

11–14-year-olds learning the mole later, this approach works well. The third method is preferred, as long as pupils have prior learning in place, including knowledge of moles and competence with calculations. A spreadsheet with some simple formulae is a good way to do these calculations.

Method 1

Students calculate temperature change per gram of fuel burned using:

$$\text{temperature change per gram of fuel (°C g}^{-1}) = \frac{\text{temperature change (°C)}}{\text{mass of fuel (g)}}$$

Method 2

Students calculate energy released per gram of fuel burned. This measures 'energy density'. This uses specific heat capacity (SHC, from physics) of water to calculate energy change:

$$\text{energy change of the water (J)} = \text{mass of water (g)} \times \text{SHC (J g}^{-1}\text{°C}^{-1}) \times \text{temperature change (°C)}$$

The SHC of water is 4.18 J g^{-1}°C^{-1}

Energy density is given by the equation:

$$\text{energy density (J g}^{-1}) = \frac{\text{energy change of the water (J)}}{\text{mass of fuel (g)}}$$

Method 3

Students calculate amount of energy released per mole of fuel burned by converting the mass in grams of fuel burned to moles. To do this they need the relative formula mass (M_r) of each fuel. To find the number of moles:

$$\text{amount of fuel (moles)} = \frac{\text{mass (g)}}{M_r}$$

Then divide energy released in kilojoules (see Method 2 above, converting joules to kilojoules by dividing by 1000) by the amount of fuel in moles.

$$\text{energy released per mole (kJ mol}^{-1}) = \frac{\text{energy change of the water (kJ)}}{\text{amount of fuel burned (moles)}}$$

Students use their analysed data to compare the estimated energy released by each fuel. If the experiment has controlled appropriate variables, expected results are that the amount of energy increases as the number of carbon atoms in the alcohol molecule increases (per mole of fuel burned). In all cases, the figures represent estimates of the *enthalpy change of combustion* of the alcohols.

The calculated values are substantially smaller than accepted values for enthalpy of combustion of these alcohols quoted in Table 6.2. This activity helps students focus on identification of errors. There are two main sources of error in this experiment. First is heat lost to the surroundings, rather than being absorbed by the water; and second, incomplete combustion occurs leaving soot (carbon) on the bottom of the copper beaker. Also, the water will boil if the temperature reaches 100 °C, and any additional energy transferred to the water will not increase the temperature further. Additionally, as the water temperature increases, so does the rate at which hot water loses energy to the surroundings.

Students may ask why there is no need to control the amount of fuel burned. The answer is that the amount of fuel (in grams or moles) is accounted for when calculating energy density or energy released per mole.

The results can also be compared with bond energy calculations (see above) for each alcohol. Table 6.2 shows standard, (that is, accurately measured) enthalpy of combustion values for alcohols having a number of carbon atoms ranging from one (methanol) to eight (octan-1-ol). The enthalpy of combustion changes by about 650 kJ mol^{-1} for each alcohol because a CH_2 group is added each time. This results in a corresponding regular change in value.

Table 6.2 Standard enthalpies of combustion for alcohols

Alcohol	Number of carbon atoms	Formula	$\Delta H_{combustion}$/ kJ mol^{-1}	Change in $\Delta H_{combustion}$ from previous alcohol/ kJ mol^{-1}
methanol	1	CH_3OH	−726	
ethanol	2	CH_3CH_2OH	−1367	641
propan-1-ol	3	$CH_3(CH_2)_2OH$	−2021	654
butan-1-ol	4	$CH_3(CH_2)_3OH$	−2676	655
pentan-1-ol	5	$CH_3(CH_2)_4OH$	−3329	653
hexan-1-ol	6	$CH_3(CH_2)_5OH$	−3984	655
heptan-1-ol	7	$CH_3(CH_2)_6OH$	−4638	654
octan-1-ol	8	$CH_3(CH_2)_7OH$	−5294	656

Students can plot the number of carbon atoms (x-axis) against energy released per mole (y-axis). This graph should be a straight line due to the regular addition of CH_2 groups. The line goes through the origin because no carbon and hydrogen atoms means there is nothing to combust. Plotting values obtained in their own experiment and the standard values from Table 6.2 on the same graph will prompt discussion of errors in measurement, as mentioned above.

 The rate of chemical change: how fast do chemical reactions go?

Study of rates of reaction provides many opportunities for practical investigations that support and develop students' understanding of theoretical aspects of the topic. The experiments develop practical skills and afford opportunities for data analysis.

Introduce the concept of rate of reaction by talking about familiar examples of fast and slow reactions. Fast reactions include the formation of silver chloride precipitate; neutralisation of an acid and alkali; combustion of fireworks. Slow reactions include the formation of crude oil; formation of rust (iron(III) oxide) on iron; souring of milk (formation of lactic acid by bacteria in milk, which lowers the pH); the reaction of magnesium with cold water; and the reaction of zinc metal with dilute hydrochloric acid. Measuring and controlling reaction rate is important in both everyday life and the chemical industry. Students can then be introduced to the conceptual model of collision theory and use this to predict and explain observations and conclusions obtained through practical investigations.

Students' misconceptions about rates of reaction

Aspects of collision theory and activation energy are abstract, so students may have misconceptions about rates of reaction; suggestions of how to deal with these are provided after each misconception.

Students may think of reactions as fast or slow and that rate of reaction is constant until the reaction stops

In almost all reactions, rate slows over time until the reaction stops. This is because the reactant particles gradually reduce in number, so fewer particles are available to collide and undergo successful collisions. Reactions stop when no unreacted limiting reactant is available. The limiting reactant is the chemical which is no longer present when the reaction has finished.

Students think fast reactions are always exothermic

Students may link fast rate to energy production. This may arise because exothermic reactions are frequently used to illustrate reactions that have a high rate. Rusting is an exothermic reaction that is slow. Endothermic reactions can also be spontaneous and fast, such as that between baking soda (sodium hydrogencarbonate) and vinegar (ethanoic acid).

Students may struggle to comprehend very slow reactions

The time required for some reactions is so long that humans are unable to sense them happening, for example, formation of fossil fuels. Students may consider charcoal production in which wood (and biomass generally) slowly reacts to form carbon, carbon dioxide, carbon monoxide and water vapour. Extrapolating this idea may help consideration of coal and crude oil formation.

In a reactant at a specific temperature, students may think that all particles have the same amount of energy

This means they may not understand how collision theory (see below) predicts and explains rates of reaction. Animations are useful ways to address this misunderstanding (see Section 6.4).

Students think increased particle movement is responsible for rate increase at higher temperatures

Students often think reaction rate increases with temperature because reactant particles move faster than at lower temperatures. This is partially true. Increased temperature raises the average energy of particles involved, which means that more particles have the minimum energy needed to react. The increased number of successful collisions between reactant particles increases reaction rate.

Students may claim catalysts lower activation energy

In reality, catalysts provide an alternative reaction pathway with a lower activation energy than the uncatalysed route. Careful explanation, discussion and allowing time to revisit this whenever using catalysts may help students build this idea.

Students may say that catalysts work by providing particles with more energy

This is not true. The only way to provide particles with more energy is to transfer energy to them. Again, careful explanation, discussion and revisiting the reasons why and how catalysts work can help build robust understanding here.

Teaching sequence

A suggested teaching sequence is given which supports development of theoretical knowledge and practical skills for rates of reaction. There is a variety of reactions and methods for investigating different factors which affect rates of reaction. For example, the effect of temperature on rate can be investigated using the disappearing cross technique described below. This method works well with both the iodine clock reaction and the reaction between sodium thiosulfate and dilute acid where the opacity of the reaction mixture increases. Both can be used as practical activities within this topic. Alternatively,

6.2 The rate of chemical change: how fast do chemical reactions go?

select one then return to the other later, encouraging students to look for similarities between these two procedures and reactions. This helps students apply knowledge in novel contexts.

- → Introduce fast and slow reactions and the meaning of rate of reaction.
- → Consider the ways in which rate of reaction could be measured and expressed.
- → Introduce the collision theory.
- → Investigate rates of reaction using practical activities.

Introducing fast and slow reactions and the meaning of rate of reaction

Show students images or videos of fast and slow reactions. Prompt them to describe which they perceive as fast or slow, placing them in sequence in order of speed (or rate). Examples include an explosion, rusting, cooking an egg, burning a candle, formation of a fossil fuel, making wine, baking a cake.

Discuss chemists' need to measure and calculate reaction rate. Explain that controlling reaction rate is important in the chemical industry because if a reaction is too slow, the amount of desired product produced each week will be lower, reducing product supply and profit. However, if the rate of an exothermic reaction of reaction increases too much, the amount of energy released might cause temperature inside the reaction vessel to increase to a dangerous level, creating a dangerous runaway situation. Within a familiar context, home refrigerators slow decomposition reactions that occur when food perishes.

Measuring rate of reaction

Qualitatively, students should equate rate with speed. However, quantitatively, they need to learn that rate of reaction is calculated by dividing a quantity of substance produced by a measurement of time. Conventionally, rate of reaction is defined as the amount of product made per unit time. This may be easier to understand if, in physics, they have learned to calculate speed or velocity by dividing distance or displacement by time. Link this to equations for calculating rate, and reinforce the connection between units of a scientific quantity and the equation used to calculate this quantity. For example, in investigations, volume of a gaseous product is measured during a certain time period. In this case, rate is calculated using:

$$\text{rate of reaction} = \frac{\text{volume of gas produced}\,(\text{cm}^3)}{\text{time taken}\,(\text{s})}$$

Units for this rate of reaction are therefore $\frac{cm^3}{s} = cm^3\,s^{-1}$.

In another investigation, reaction rate is calculated by measuring reduction in mass of the reaction mixture when a gas like carbon dioxide is liberated into the surroundings.

$$\text{rate of reaction} = \frac{\text{change in mass (g)}}{\text{time taken (s)}}$$

For this calculation, the units of rate will be $\frac{g}{s} = g\,s^{-1}$.

Volume or mass of a product can be converted into moles before dividing by time taken, to give units of rate as $mol\,s^{-1}$. This is preferable for students who have learned mole calculations (see Chapter 2).

Finally, in some reactions, measuring the amount of product made during the reaction is difficult. Examples include iodine clock reactions, and the reaction between sodium thiosulfate and dilute acid. These reactions produce an opaque substance and the time taken for this to occur is measured. In these situations, an approximation of the reaction rate is calculated using the following equation:

$$\text{rate of reaction} = \frac{1}{\text{time taken (s)}}$$

For this calculation, the units of rate will be $\frac{1}{s} = s^{-1}$.

Introduce collision theory

Collision theory is central to understanding reaction rates. A prerequisite is that students understand the particle model of matter (Chapter 3). Particles in all substances move constantly to a greater or lesser extent, depending on temperature, pressure and bonds that exist between them (Chapter 5). Collision theory makes predictions and explains observations from investigations. To summarise, the central principles are:

→ Reactant particles must collide to react. Faster particles have more kinetic energy, so increasing the chances of collision with another particle. Collisions are more likely between particles of gases than in liquids or solids.
→ Reactant particles must have sufficient kinetic energy to react when they collide. The minimum energy required is the activation energy. If the activation energy is exceeded, a reaction occurs and product molecules form. The collision is 'successful'.

- Most collisions occurring between particles have amounts of energy less than the activation energy, so are unsuccessful. This means products do not form, as these particles bounce off one another rather than react.
- Factors that increase frequency of successful collisions increase rate of reaction.
- Factors that increase the proportion of successful collisions increase the rate of reaction. This is because particles have an amount of energy equal to or greater than the activation energy.

Collision theory can be introduced via computer animations and simulations of moving particles colliding then reacting or not reacting. An example can be found at https://phet.colorado.edu/en/simulation/legacy/reactions-and-rates

To help students understand collision theory, demonstrate the reaction between white solids, lead nitrate and potassium iodide (see Chapter 2). Add one spatula of lead nitrate (toxic) to a glass weighing bottle followed by one spatula of solid potassium iodide. Seal the bottle carefully with the lid. Shake the bottle so the powders combine. A pale-yellow solid forms, which is lead iodide (toxic); the other product is potassium nitrate, which is white. The reaction is slow because collisions between individual lead ions with individual iodide ions are infrequent and/or low in energy. All four ions are fixed in crystalline lattices (see Chapter 5), which require energy to disrupt. Repeat the reaction with aqueous solutions (care: lead nitrate solution is toxic). Bright yellow solid lead iodide precipitates instantaneously. The reaction rate is much faster with dissolved reactants because ions move freely in the solvent, so collisions between ions are more frequent and more are successful.

Questions to ask
- Name the reactant and products in the reaction above.
- Using collision theory, explain the slow rate occurring when two solids react together.
- Using collision theory, explain why reaction rate is faster when two solutions of the solids react together.

Investigating factors which affect rate of reaction

Students should understand collision theory theoretically before undertaking investigative work on variables that affect rates of reaction. This is because collision theory enables them to make

6 Energetics, rate and extent of chemical change

predictions about what to expect and helps explain their conclusions. Factors affecting reaction rate are temperature, concentration, surface area of a solid (particle size) and presence of a catalyst.

The suggested sequence of investigations described below starts with a simple method to investigate the effect of changing temperature and then progresses into increasingly complex practical techniques and quantitative analyses. These illustrate how concentration, surface area and catalyst affect reaction rates. Qualitative investigation of catalysts could be done at any point in the teaching sequence.

KEY ACTIVITY

Investigating the effect of changing temperature

Each small group will require:

- 250 cm³ beaker
- 10 or 25 cm³ measuring cylinder
- boiling tubes
- stirring thermometer (−10–100 °C)
- stopwatch
- 40 cm³ of 1 mol dm⁻³ hydrochloric acid (HCl)
- four 2 cm lengths of magnesium (Mg) ribbon

This experiment utilises the reaction between magnesium and hydrochloric acid. The reaction rate is measured by timing how long it takes for the magnesium ribbon to completely react. At this point, fizzing will stop and no magnesium ribbon is present. This simple method introduces practical skills, specifically, using a water bath to control temperature, and requires access to hot water (from a kettle) and ice cubes.

Students prepare boiling tubes containing dilute acid at different temperatures. Add 10 cm³ of dilute hydrochloric acid to each of four boiling tubes.

Prepare:

- one boiling tube below room temperature (for example, 10 °C)
- one boiling tube at room temperature (about 20 °C)
- two boiling tubes above room temperature (say 30 °C and 40 °C – see safety note below)

To gently warm the acid, place the boiling tube in a water bath of a beaker of warm water. To cool the acid, place the boiling tube in a beaker of ice and water. It is the temperature of the acid which is important, not the temperature of the water bath.

For each boiling tube:

- Check the temperature of the acid using a thermometer.
- Add a piece of magnesium ribbon to the acid in the boiling tube and start the stopwatch. Students can use the thermometer to stir the acid during the reaction.

6.2 The rate of chemical change: how fast do chemical reactions go?

- Observe the reaction until fizzing has stopped and all the magnesium has reacted. Note and record the time taken for the reaction to complete at each temperature.

As 'fizzing' is a subjective judgement there will be some variation in the results, encourage students to discuss this.

The data analysis involves calculation of 1÷ time (see above). This gives an approximate rate of reaction, with the units of s^{-1}. Table 6.3 gives sample results for the reaction of magnesium ribbon with dilute hydrochloric acid.

Table 6.3 Rate of reaction of magnesium ribbon with dilute hydrochloric acid

Temperature of acid/°C	Reaction time/s	Rate (= 1÷ time)/s^{-1}	Rate × 1000/s^{-1} × 1000
10	98	0.0102	10.2
20	47	0.0213	21.3
30	24	0.0417	41.7
40	12	0.0833	83.3

Students need to calculate 1÷ time for each temperature. These values will be very small, much smaller than 1, and in order to make the numbers more manageable for students to plot on their graphs it may be helpful to multiply all of the resultant values for rate by a specific number (for example, by 1000 as in the table above). Figure 6.5 shows the results from Table 6.3 plotted with temperature (x-axis) against the calculated value of 1÷ time multiplied by 1000 (y-axis). The graph shows a clear trend with a curve that starts near to the origin and increases exponentially demonstrating that as the temperature increases the reaction rate increases. This is because when the temperature is increased, the particles have more energy and so move more quickly. If they move more quickly, they collide more often, and with more energy, so a higher proportion of collisions are successful and the reaction rate increases.

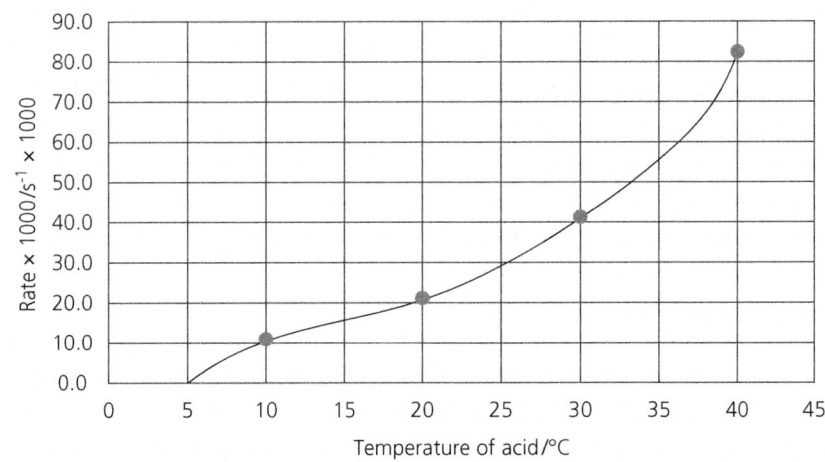

Figure 6.5 Graph showing the effect of changing temperature on reaction rate for the reaction between magnesium and dilute hydrochloric acid

6 Energetics, rate and extent of chemical change

Safety

Hydrochloric acid is an irritant at this concentration. All students should wear eye protection. Magnesium ribbon is flammable. Groups should be handed out four pieces of Mg ribbon by the teacher and not given free access to the tub of pieces or the roll of Mg ribbon. Do not heat dilute acids above 40 °C to prevent the production of gases which are irritants.

Questions to ask

- How do we determine the reaction is complete?
- What happens to the time taken for the reaction as the temperature increases?
- What happens to the reaction rate as the temperature increases?
- Explain why the reaction rate changes as the temperature increases using collision theory.

Discussion points

Increasing temperature of a reactant can be explained using these ideas:

- Increasing temperature gives all particles more kinetic energy. Within a sample of a chemical at a specific temperature, particles have a range of energies. Some particles have low energy and move slowly, while other particles have higher energy levels and move quickly.
- Higher temperatures mean more frequent successful collisions.
- A greater proportion of collisions are successful because more particles have levels of energy above activation energy. This is different from the previous point and should be emphasised.

KEY ACTIVITY

Investigating the effect of concentration by continuous measurements

Details for this experiment can be found at https://edu.rsc.org/resources/the-rate-of-reaction-of-magnesium-with-hydrochloric-acid/1916.article Students aged 11–14 know concentration affects rate of reaction. They may have observed and measured the times for a reaction to go to completion, comparing two or more different concentrations of acid (often the reaction between magnesium and a dilute acid, or a metal carbonate and a dilute acid). Students aged 14–16 will use the same reactions but focus on rate rather than time. This reaction produces hydrogen gas. Measuring gas volume every ten seconds provides students with a data set for one or more concentrations of acid. These can be used to plot a graph of gas volume (y-axis) over time (x-axis). The data for each concentration of acid can be plotted separately on the same graph. The data from this experiment can be analysed to calculate an average rate of reaction or the instantaneous rate of reaction at a given time. To find the instantaneous reaction rate, draw a tangent to the curve at the required time. The gradient of the tangent gives the instantaneous reaction rate (see below).

The reference provides details for 1 mol dm^{-3} hydrochloric acid only, but the experiment can be extended using concentrations of 0.5 mol dm^{-3}, 0.25 mol dm^{-3} and 0.1 mol dm^{-3}.

6.2 The rate of chemical change: how fast do chemical reactions go?

Data can be collected and analysed by:

1. Calculating average rate of reaction for each concentration during the first 30 seconds. This is the simplest quantitative treatment of results obtained, suited to an introductory experiment. Measure the volume of gas produced at 30 seconds. Rate is calculated by dividing the volume of gas produced by the time taken (30 s), with the units $cm^3\ s^{-1}$.

2. Plotting a graph of time (x-axis) against volume of gas produced (y-axis). Measure the volume of gas using the apparatus shown in Figure 6.6. Draw a smooth curve of best fit for the data obtained for each concentration. A typical graph is shown in Figure 6.7.

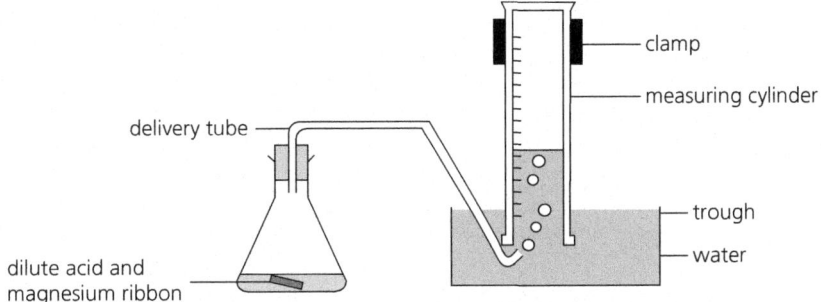

Figure 6.6 Apparatus for continuous measurement of gas produced in the reaction between magnesium with dilute acid

3. Calculating initial rate of reaction for the highest and lowest concentrations of acid. This is done by drawing a tangent to the curve at time = 0 seconds and calculating the gradient of these tangents. The method is shown in Figure 6.8.

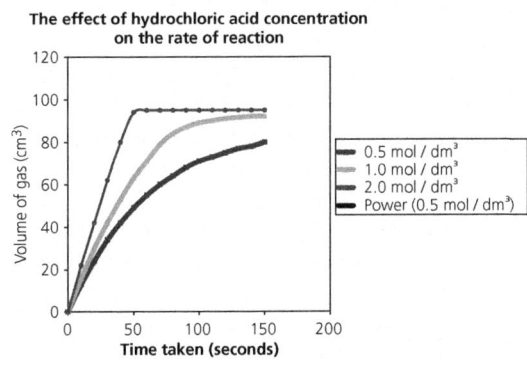

Figure 6.7 Volume of gas evolved from the reaction of magnesium ribbon with hydrochloric acid of different concentrations

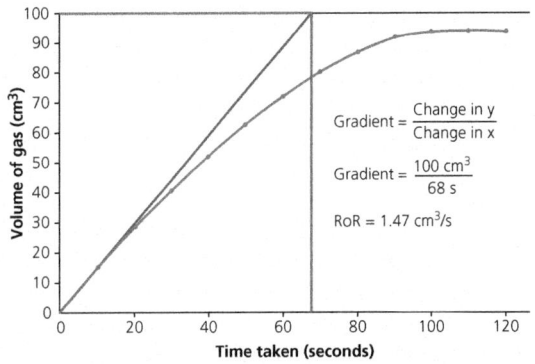

Figure 6.8 Drawing a tangent to the curve at the start of the reaction (t = 0 s) and calculating the gradient to deduce initial rate of reaction

6 Energetics, rate and extent of chemical change

4. Tangents can be drawn at other times – for example, at 10, 20, 30, 60, 90 seconds – and rates of reaction calculated at these points.

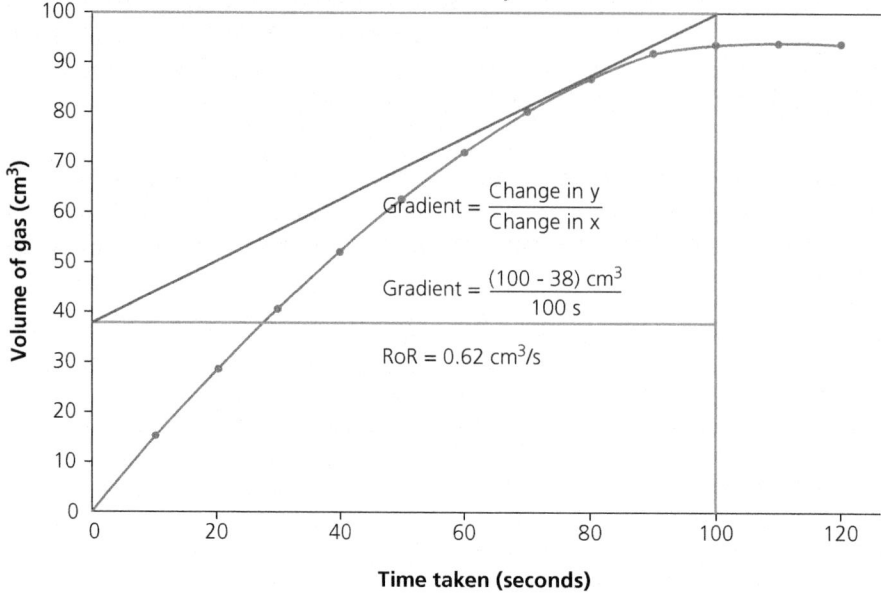

Figure 6.9 As the reaction progresses, the gradient of the curve decreases, and this can be demonstrated by drawing tangents at various points on the curve and then calculating the gradient of each line

Reaction rate decreases as time progresses. The steepest tangent, corresponding to the fastest reaction, is at the start, when magnesium is first added to the acid. When the line for each reaction flattens, the reaction has stopped; the tangent gradient is zero, so the rate of reaction is zero.

Discussion points

Students should discuss their results, explaining these in terms of collision theory. Increasing the concentration of a dissolved reactant increases reaction rate because:

- *There is a greater number of reactant particles per unit volume.* In other words, reactant particles become increasingly crowded at higher concentrations, so the chance of collisions is higher.
- *Successful collisions are more frequent*, although the *proportion* of collisions which are successful does not change.

6.2 The rate of chemical change: how fast do chemical reactions go?

KEY ACTIVITY

Investigating the effect of concentration by change in opacity of the reaction mixture

Student and teachers' notes for this experiment can be downloaded from https://edu.rsc.org/resources/the-effect-of-concentration-on-reaction-rate/743.article

Not all reactions produce gases. Students should be familiar with investigating rate of a reaction that produces a change in opacity from a clear solution. The reaction between sodium thiosulfate solution and dilute hydrochloric acid produces sulfur as a pale-yellow precipitate. The equation for this reaction is:

sodium thiosulfate + hydrochloric acid → sodium chloride + sulfur + sulfur dioxide + water

$Na_2S_2O_3(aq) + 2HCl(aq) \rightarrow 2NaCl(aq) + S(s) + SO_2(g) + H_2O(l)$

The reactants are added to a conical flask. As sulfur is produced, the reaction mixture becomes cloudy. The precipitate will eventually become sufficiently dense to obscure a cross drawn on a piece of paper placed under the flask. Students record the time at the instant the cross becomes invisible when viewed from above the reaction. The volume of sulfur dioxide cannot be measured because the gas is soluble and dissolves easily in water. Small quantities of sulfur dioxide can be smelt during this experiment, so the laboratory should be well ventilated.

Data can be collected for a range of sodium thiosulfate concentrations. For each concentration, an approximation of the reaction rate can be calculated:

$$\text{rate of reaction} \approx \frac{1}{\text{time taken(s)}}$$

This will have units of s^{-1}.

Typical values will be very small, less than 1. Table 6.4 gives results for this reaction and Figure 6.10 is a graph showing the concentration of sodium thiosulfate against rate of reaction.

Table 6.4 Reaction mixtures and rates of reaction data for sodium thiosulfate and dilute hydrochloric acid

Volume of acid/cm³	Volume of Na₂S₂O₃/cm³	Volume of water/cm³	Concentration of Na₂S₂O₃/ g dm⁻³	Reaction time/s	Rate (= 1÷ time)/s⁻¹	Rate × 1000/ s⁻¹× 1000
5	25	0	33	62	0.0161	16.1
5	20	5	27	79	0.0127	12.7
5	15	10	20	110	0.0091	9.1
5	10	15	13	150	0.0067	6.7
5	5	20	7	229	0.0044	4.4

6 Energetics, rate and extent of chemical change

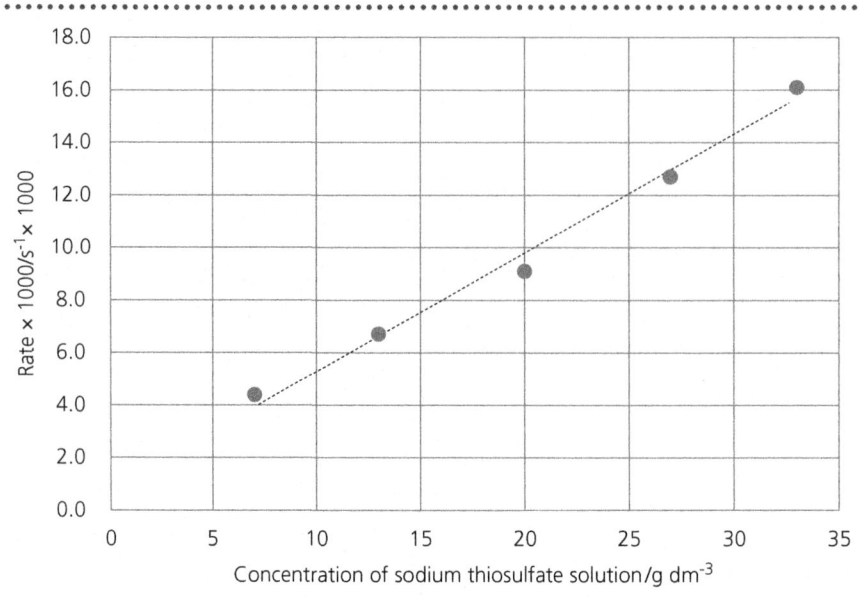

Figure 6.10 Graph of concentration of sodium thiosulfate solution (x-axis) against 1/time (y-axis).

Multiplying the values of 1÷ time by 1000 makes them easier to plot on the y-axis. The graph shows that as the concentration increases so does the reaction rate. Encourage students to explain the trend using collision theory.

KEY ACTIVITY

The effect of concentration on reaction rate: the iodine clock demonstration

Details for carrying out this demonstration can be found at https://edu.rsc.org/resources/iodine-clock-reaction-demonstration-method/744.article

Many 'clock' reactions can be used as demonstrations. In a clock reaction, a sudden, dramatic colour change occurs after a time interval known to the demonstrator. Timing of the colour change can be varied by altering the concentration of one reactant, hence the name clock reactions. Commonly, iodine is produced in the reaction, which in turn reacts with thiosulfate ions in solution. This prevents iodine reacting with starch indicator present in the mixture. The supply of thiosulfate ions is restricted. As soon as all thiosulfate ions have reacted, any further iodine produced forms a soluble complex with starch, which appears immediately in the solution as a dark blue colour.

This demonstration requires preparation of two solutions, labelled A and B. Solution A contains sodium thiosulfate and potassium iodide. Solution B contains ethanoic acid and hydrogen peroxide solution. Halving the concentration of solution B (by diluting it with water) doubles the time required for the dark blue complex to form. Increasing the temperature of both solutions by 10 °C doubles the rate and halves the time required.

6.2 The rate of chemical change: how fast do chemical reactions go?

> **Questions to ask**
> - Why does the reaction rate decrease when the concentration of solution B is decreased?
>   Explain your ideas using collision theory.
> - Why does the reaction rate increase when the temperature of solutions A and B are increased?
>   Explain your ideas again using collision theory.

KEY ACTIVITY

Investigating the effect of a catalyst

Industrial chemical manufacturing processes use catalysts to maintain reaction rate of a desired chemical reaction, so the same amount of product is produced at a lower temperature. This saves fuel and money, ensuring profit. For some reactions, a lower temperature helps to increase yield in a reversible reaction (see Section 6.3). Many catalysed reactions occur in the gas phase, so are impossible to study in the school laboratory. One reaction that can be studied qualitatively is the decomposition of hydrogen peroxide. The equation for the reaction is:

$$\text{hydrogen peroxide} \rightarrow \text{water} + \text{oxygen}$$
$$2H_2O_2(l) \rightarrow 2H_2O(l) + O_2(g)$$

Explain that hydrogen peroxide molecules are unstable, meaning that there is sufficient energy in the surroundings to cause the reaction above to take place slowly. This means that hydrogen peroxide decomposes over time to produce water and oxygen. The slow rate means bubbles of oxygen are unobserved, as the molecules produced dissolve in water, or diffuse gradually into the air. Adding a catalyst on which two hydrogen peroxide molecules can meet increases the rate of decomposition, allowing oxygen production to be observed.

Details for this reaction can be found at https://edu.rsc.org/resources/hydrogen-peroxide-decomposition-using-different-catalysts/831.article

Ensure students identify the catalyst. They sometimes confuse reactants with the catalyst, as both are added to the reaction vessel at the start. Students can compare the efficiency of different catalysts. Older students can examine the energy profile of the reaction with and without the catalyst and discuss what happens at particle level in each case.

A catalyst is a substance that:

- increases rate of a reaction by providing an alternative route with a lower activation energy to produce the products
- does not alter the products of the reaction

- does not take part in the reaction
- is unchanged chemically and in mass at the end of the reaction.

The catalysed, alternative reaction pathway has a lower activation energy than the uncatalysed reaction (see Figure 6.11).

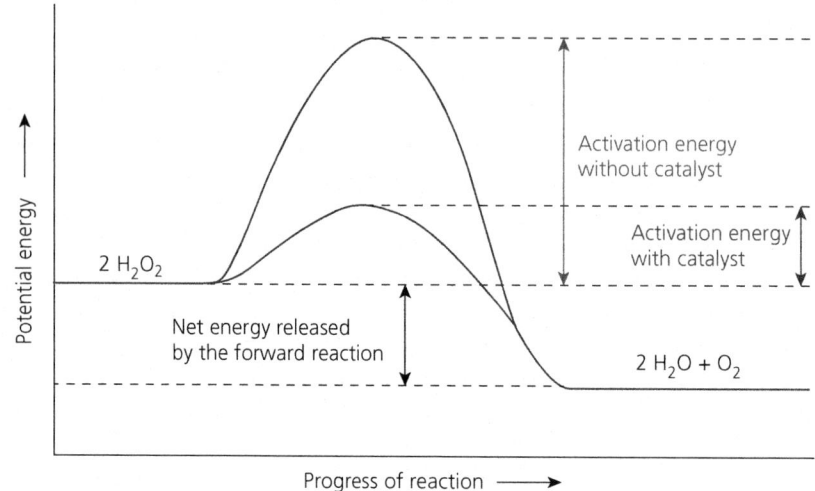

Figure 6.11 Energy profile diagram for the decomposition of hydrogen peroxide, with and without a catalyst

The effect of using a catalyst on reaction rate is explained as follows:

- Catalysts increase reaction rates without being chemically changed.
- Catalysts are usually specific to a reaction.
- Transition metals and their compounds are often used as catalysts, especially in redox reactions. Transition metal ions transfer and accept electrons while existing as more than one stable ion, for example, Fe^{2+} and Fe^{3+}.
- Biological catalysts are enzymes, made from proteins. Like all proteins, they denature at high temperatures and in environments with extreme pH values.
- Catalysts provide an alternative reaction pathway with a lower activation energy than the uncatalysed route. The proportion of particles with energy equal to or greater than activation energy is increased.
- Occasionally, a chemical called an inhibitor is added to slow reaction rate.

Questions to ask

- What metals in the periodic table are often used as catalysts?
  What type of reactions are they used to catalyse? Why?
- How do catalysts speed up the rate of chemical reactions?
  Why are they not used up in the reaction?
- Give an example of a biological catalyst and the reaction that it catalyses.
  Why do biological catalysts only work at low temperatures and within narrow pH ranges?

6.2 The rate of chemical change: how fast do chemical reactions go?

> The 'elephant toothpaste' demonstration illustrates the impact of a catalyst on decomposition of hydrogen peroxide, see Table 2.2 p 38.

KEY ACTIVITY

Demonstrating the effect of surface area on reaction rate

Lego® bricks illustrate the effect of changing surface area on reaction rate (see Figure 6.12). Building a cuboid with 16 Lego bricks allows students to identify the exposed surface area of a single large object. The cuboid is broken in half, and the exposed surface area recounted. This is repeated until 16 single Lego bricks are the result. The counts indicate surface area increases as the cuboid is broken into smaller pieces.

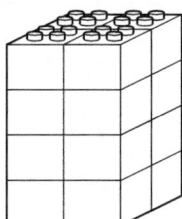 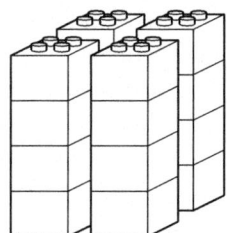 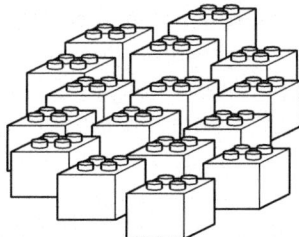

Figure 6.12 To illustrate the effect of changing surface area on reaction rate

There are opportunities for student investigations into the effect of surface area on reaction rates. At the simplest level, this includes adding the same mass of calcium carbonate to boiling tubes containing identical volumes of dilute hydrochloric acid. The surface area of the calcium carbonate (marble) is varied: powdered; small marble chips; and large marble chips. Qualitative observations confirm expected observations that the powder reacts fastest because it has the largest surface area to volume ratio. The equation for the reaction is:

$$\text{hydrochloric acid} + \text{calcium carbonate} \rightarrow \text{calcium chloride} + \text{water} + \text{carbon dioxide}$$

$$HCl(aq) + CaCO_3(s) \rightarrow CaCl_2(aq) + H_2O(l) + CO_2(g)$$

An alternative approach uses a top-pan balance to investigate this reaction rate by measuring mass per unit time as the reaction proceeds. A decrease in mass occurs because carbon dioxide is released into the air. Add 5 g of calcium carbonate in the form of powder, small marble chips or large marble chips and record the mass every 10 seconds. This produces sufficient data for a graph of mass against time. Initial rates of reaction can be found by drawing tangents at time 0 seconds (see above).

6 Energetics, rate and extent of chemical change

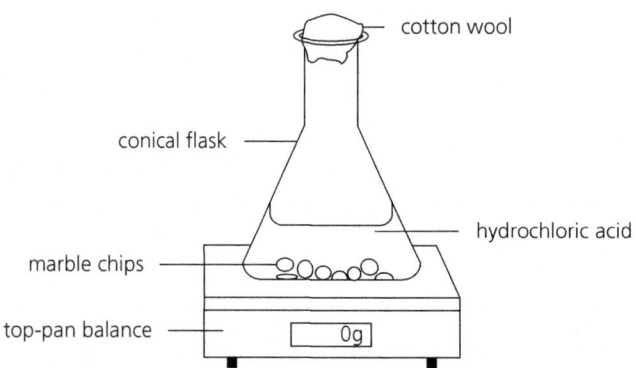

Figure 6.13 Apparatus to investigate how increasing surface area affects the rate of reaction

The effect of using powder can be explained as follows:

- Grinding lumps into a powder increases the surface area to volume ratio of a solid.
- Powders expose a greater number of reactant particles to the other reactant than lumps. This allows more collisions to occur between reactant particles.
- Successful collisions are more frequent.

Safety

Dilute hydrochloric acid is a low hazard, so eye protection should be worn by the demonstrator. Using a loose plug of cotton wool into the neck of the conical flask prevents small splashes of liquid from leaving the flask when bubbles of gas pop at the surface of the reaction vessel.

Questions to ask

- Explain why grinding up a lump of solid into a powder increases its surface area.
- What happens to the rate of reaction if the surface area of a solid in a reaction is increased? Explain your answer using collision theory.

Another interesting demonstration of the effect of surface area on the rate of reaction is the reaction of milk powder in a Bunsen flame. Full details for this experiment can be found at https://edu.rsc.org/resources/burning-milk-powder/830.article The guidance suggests a specific supermarket brand of milk powder but other brands have been found to be effective.

> **Enrichment**
>
> In 1981, a tragedy occurred in a custard powder factory in Banbury, Oxfordshire. The factory used cornflour (starch extracted from maize) to make custard powder. An explosion occurred when the fine cornflour powder spilled from an overfilled hopper. A dust cloud had formed that was ignited accidentally by electrical equipment nearby. The explosion caused the roof of the factory to blow off and nine workers were badly injured. Dust explosions occur when the dust of a combustible material (in this case, cornflour) is mixed with oxygen from the atmosphere and ignited by a spark. The products are carbon dioxide and water.

 The extent of chemical change: how far do chemical reactions go?

Students become familiar with chemical reactions that appear irreversible, that is that they proceed in one direction until all available supplies of a limiting reactant have reacted, and then they stop. Products from these reactions are energetically stable. Reversing these reactions is theoretically possible but requires so much energy this would not be worthwhile. So, it is easier to regard them as irreversible.

In some reactions, the energetics are different: the forward and reverse reactions are both energetically favourable. These are regarded as reversible reactions. When a reversible reaction occurs in a container from which reactants and products cannot enter or leave, the forward and reverse reactions take place spontaneously. They are said to create a *dynamic equilibrium* in a *closed system*. Three characteristics of a dynamic equilibrium are:

→ The rate of the forward reaction equals the rate of the reverse reaction.
→ The amounts of products and reactants remain constant over time.
→ The amount of products does not necessarily equal the amount of reactants.

For example:

$$\text{ammonium chloride} \rightleftharpoons \text{ammonia} + \text{hydrogen chloride}$$
$$NH_4Cl(s) \rightleftharpoons NH_3(g) + HCl(g)$$

The equation shows that ammonium chloride (a white solid) decomposes to ammonia and hydrogen chloride. It also shows that ammonia and hydrogen chloride (colourless gases) can react to form ammonium chloride again. Note that when writing chemical equations for reversible reactions, the usual one-way arrow used in equations is not used. Instead, two arrows are used, each with just half an arrowhead – the top one pointing right, and the bottom one pointing left.

The amounts of reactants and products are described by the term equilibrium position. If the equilibrium position 'lies to the left', the amount of reactants present at equilibrium is larger than the products, so yield is low. Changing conditions may alter equilibrium position to favour the products, that is, the equilibrium 'lies to the right'. In this case, product yield is high while the amount of reactants present in the mixture is low.

→ A chemical reaction regarded as irreversible is represented as
  A + B → C + D
→ A reversible chemical reaction is represented as A + B ⇌ C + D

In organic chemistry, many reactions are reversible. An example is production of an ester and water from an alcohol and a carboxylic acid such as:

$$\text{methanol} + \text{ethanoic acid} \rightleftharpoons \text{methyl ethanoate} + \text{water}$$
$$CH_3OH + CH_3CO_2H \rightleftharpoons CH_3CO_2CH_3 + H_2O$$

Industrial chemical process reactions are often reversible. An example is the Haber process used to make ammonia (NH_3) from nitrogen and hydrogen gases:

$$\text{nitrogen} + \text{hydrogen} \rightleftharpoons \text{ammonia}$$
$$N_2(g) + 3H_2(g) \rightleftharpoons 2NH_3(g)$$

Ammonia is essential in producing fertilisers and thus producing enough food through intensive farming to keep the current population of the world alive.

Enrichment

At one time, scientists believed all chemical reactions were irreversible reactions. In 1803, Claude Louis Berthollet proposed the idea of a reversible reaction after observing the formation of sodium carbonate crystals on the edge of a salt lake in Egypt. Berthollet believed excess salt in the lake formed solid sodium carbonate, which could then react again to form sodium chloride and calcium carbonate:

$$2NaCl(aq) + CaCO_3(aq) \rightleftharpoons Na_2CO_3(s) + CaCl_2(aq)$$

Misconceptions and analogies

The concept of reversible reactions is relatively straightforward, but the concepts of dynamic equilibrium and changing equilibrium position are abstract. Students may have misconceptions about these.

Students may think that a reversible reaction goes first one way and then the other

Such oscillating reactions exist but are rare. In a reversible reaction, the forward and backward reactions occur simultaneously and continue to do so.

Students often think that in a reversible reaction the amount or concentration of the products and reactants are equal

This is rarely true. Chemists use the term yield to describe the amount, percentage or concentration of the desired product. In an industrial process, maximising yield is essential when choosing reaction conditions. Conditions that move the equilibrium position to the right maximise yield of a desired product.

KEY ACTIVITY

Introducing reversible reactions

This experiment introduces students to a reaction that can proceed in a forward and a reverse direction: that is, it is a reversible reaction. Understanding of this concept is aided by showing students one or more familiar reactions generally regarded as irreversible, such as the combustion of methane in a Bunsen burner. Refer to a suitable word or symbol equation and discuss the meaning of the arrow symbol, relating this to the fact that it is impossible that the reaction will reverse spontaneously.

Hydrated copper(II) sulfate ($CuSO_4 \cdot 5H_2O$) is a blue solid at room temperature and pressure. With moderate heat, the solid decomposes into a dirty white solid called anhydrous copper(II) sulfate (Cu_sO_4) and water (H_2O). The equation for the reaction is:

$$CuSO_4 \cdot 5H_2O(s) \quad \rightleftharpoons \quad CuSO_4(s) + 5H_2O(l)$$
$$\text{blue solid} \qquad\qquad \text{dirty white solid}$$

Details for this experiment can be found at https://edu.rsc.org/resources/a-reversible-reaction-of-hydrated-copperii-sulfate/437.article

Questions to ask
- What is the difference between anhydrous and hydrated copper(II) sulfate?
- How is hydrated copper(II) sulfate formed from anhydrous copper(II) sulfate? What do you observe?
- What is a reversible reaction?

6 Energetics, rate and extent of chemical change

Introducing and demonstrating a dynamic equilibrium

When a reversible reaction occurs in a closed system, this eventually results in a dynamic equilibrium being established. Since this is a demanding and abstract concept, it is often helpful to apply analogies to help students understand the key features of a dynamic equilibrium.

→ One way of helping students understand dynamic equilibrium is to show someone staying in the same position on a moving escalator. To stay still, the person must be walking, either taking steps up or down, depending on the direction of the escalator. If the walls of the escalator are opaque, we cannot see this movement but know it must be occurring at the same rate the escalator is moving, or else the person's position would change. This is analogous to a dynamic equilibrium, in which invisible 'movement' is going on. Two reactions occur at the same rate, resulting in no overall change.

Figure 6.14 The idea of a person staying in the same position on a moving escalator can be used to help students understand dynamic equilibrium

→ Imagine an island country where the birth and death rates are equal, and there is no immigration or emigration. The island is a closed system because no people can enter or leave. A dynamic equilibrium is established where the population remains constant over time because the birth and death rates are equal. Ask students if the people alive today will be the same as the people alive next year. Of course they will not. This is analogous to a chemical dynamic equilibrium where the reaction has *not* finished. Chemicals are constantly reacting, in both directions, at the same rate.

6.3 The extent of chemical change: how far do chemical reactions go?

→ Imagine a large department store just before opening time on the first day of a big sale event. There are people queuing outside the revolving door which is the entrance to the store. When the store opens, the rate of shoppers entering the store is high, and initially no shoppers are leaving the store. This is not yet a dynamic equilibrium. After a while, shoppers start to leave the store (the reverse reaction). Eventually the rate of shoppers entering and leaving the store are the same. This is a dynamic equilibrium since the number of shoppers in the store remains constant over time, but of course they are not the same shoppers from one hour to the next. In this analogy, the number of shoppers in the store is analogous to the yield of product. This could be affected by the availability of merchandise, or attractiveness of the prices!

KEY ACTIVITY

Demonstrate a dynamic equilibrium using a carbonated drink. The reaction is between carbon dioxide gas in the space above the liquid and the solution of carbon dioxide, the drink, which contains some carbonic acid (H_2CO_3). Carbon dioxide is pumped into a soft drink under pressure. The gas dissolves in the drink and the bottle is sealed. A small gap, known as headspace, is left above the drink. An equilibrium reaction is established between gaseous carbon dioxide in the headspace and aqueous carbon dioxide in the drink:

$$CO_2(g) \rightleftharpoons CO_2(aq)$$

The forward reaction is carbon dioxide gas molecules going into the drink. The reverse reaction is carbon dioxide molecules diffusing into the headspace. At equilibrium, the rates of these reactions are equal. The sealed bottle, with the top on, is a closed system.

Students can investigate this using a sealed bottle of carbonated drink. Prompt discussion about what is meant by a closed system and what happens to the amount of carbon dioxide in solution and in the space above the drink over time. A full bottle contains far more aqueous carbon dioxide dissolved in the drink than gaseous carbon dioxide in the headspace. If external conditions are unchanged, the bottle does not get bigger, shrink, get harder or softer to press. The amounts of carbon dioxide in the head space and the drink are constant over time. However, some molecules of carbon dioxide are always dissolving into the drink, while other molecules are released from the drink into the air space at the same rate. If the external temperature increases, the drink becomes hotter, disturbing the equilibrium. This releases carbon dioxide from the drink, increasing pressure in the headspace. The bottle becomes harder and anyone opening it will be showered with drink. For the purposes of discussion, leave one bottle unopened.

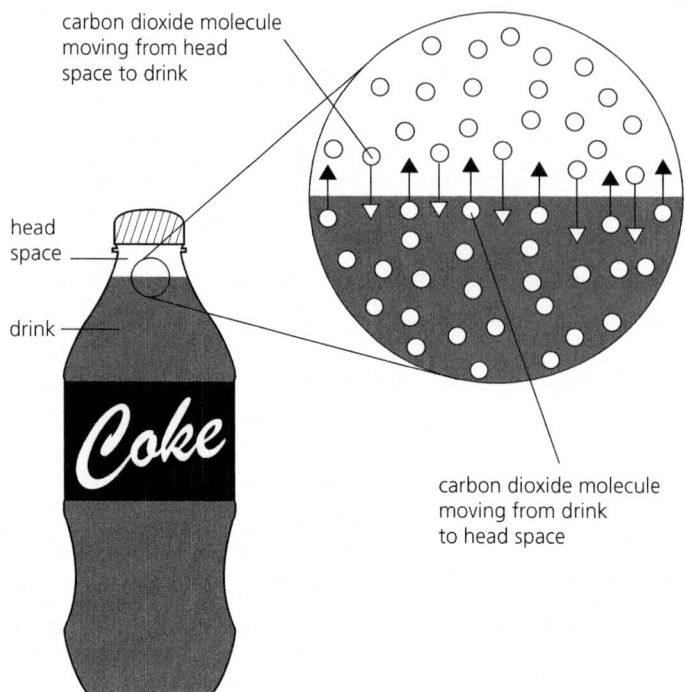

Figure 6.15 Dynamic equilibrium in a stoppered fizzy drink

Have ready a second bottle of carbonated drink. Open the bottle. Draw attention to the fizz of gas escaping when the bottle is opened. Discuss that the fizz occurs as gas under pressure in the headspace escapes into the atmosphere. The bottle becomes an open system and the equilibrium position shifts towards the reverse reaction, releasing carbon dioxide into the headspace by diffusion from the drink. The release of gas can be seen in bubbles in the drink.

Pour some drink from the second bottle into a glass or beaker. Ask students what will happen to this drink and why. The glass and opened bottle are open systems, so carbon dioxide will escape. Prior experience says the drink will go flat as most of the carbon dioxide escapes into the atmosphere.

Pour equal amounts from two new bottles (perhaps down to the level of the label) and replace the caps. Replacing the caps recreates closed systems in each bottle. However, as the bottles are no longer full, the head space above the liquids is enlarged. To restore an equilibrium position, carbon dioxide gas diffuses from the liquid into the headspace. The reverse reaction is favoured. When equilibrium is re-established, concentrations of reactant and product differ from the original, full bottle, situation. The amount of fizz in the drink is lower.

> Place one of these bottles in a beaker of ice and water, and the other into a beaker of hot water. Carbon dioxide is more soluble in water at lower temperatures, and less soluble at higher temperatures. The bottle in the ice/water mixture will become softer and easier to press. Conversely, in the heated bottle, pressure of gaseous carbon dioxide in the gas space rises. The warm bottle becomes harder to press. Of course, in both bottles the liquid is incompressible (see Chapter 3).

Explaining Le Chatelier's principle

Le Chatelier's principle is an established model for predicting and explaining the effect of changing conditions such as pressure, temperature and concentration on the position of equilibrium of a chemical reaction. When chemists talk about the position of equilibrium moving to the right, the amount of product is increased relative to the reactants. This is good if you are trying to produce a product to sell for profit.

Le Chatelier's principle states that when a change in conditions is made, *the position of equilibrium moves to counteract that change*. For example, if pressure is increased in a gaseous reaction at dynamic equilibrium, the equilibrium position moves towards the side with fewer moles of gas, because that reduces the pressure. In the Haber process, the equation shows four moles of gaseous reactants and two moles of gaseous product:

$$N_2(g) + 3H_2(g) \rightleftharpoons 2NH_3(g) \quad \Delta H = -91 \text{ kJ/mol}$$

4 moles of gas $\rightleftharpoons$ 2 moles of gas

When pressure increases, the equilibrium position moves to the right because there are fewer moles of gas on the right (the coefficients are 4 on the left, and 2 on the right, see Chapter 2). The system attempts to counteract the effect of increased pressure. This increases the amount of ammonia, which in practice, is removed as it forms. Pressure within the system reduces and equilibrium is re-established at a different point. In industry, economic factors must be considered. For example, increasing pressure increases energy costs and requires more expensive equipment that is able to withstand higher pressures. For this reaction, a pressure of 200 atmospheres is chosen as a compromise between yield and cost.

The Haber process forward reaction is exothermic, and the backward reaction is endothermic. Heat energy released causes the temperature of the system to increase. Le Chatelier's principle predicts that increasing temperature moves the position of equilibrium to the left, in the endothermic direction, to reduce the temperature.

At higher temperatures, the reaction at equilibrium contains more nitrogen and hydrogen and less ammonia. To maximise yield of ammonia, a low temperature is desirable, favouring the forward reaction. An unwanted consequence is that a low temperature reduces reaction rate. A compromise temperature of 400 °C is selected. This means the forward reaction proceeds at a rate that results in a good yield of ammonia.

A catalyst increases the rate at which equilibrium is reached. The catalyst increases the rates of the forward and reverse reactions equally. Iron is the catalyst for the Haber process, which is cheap and achieves an acceptable yield in an acceptable time. A full explanation of the Haber Process can be found at https://edu.rsc.org/resources/ammonia/17.article A simulation which allows temperature and pressure to be varied to examine the effects on yield is available at www.learner.org/series/chemistry-challenges-and-solutions/control-a-haber-bosch-ammonia-plant/

KEY ACTIVITY

Demonstrating Le Chatelier's principle

This demonstration illustrates the challenging and abstract concept of Le Chatelier's principle. The preparation phase requires the use of concentrated hydrochloric acid, but once the mixture is prepared students can handle the reagents provided these are in stoppered tubes. Details can be found at https://edu.rsc.org/resources/the-equilibrium-between-two-coloured-cobalt-species/1.article As an alternative, copper(II) chloride solution can be used. Instructions are available via this link https://www.sserc.org.uk/subject-areas/chemistry/chemistry-resources/equilibrium-and-le-chatelier-2/

The reaction involves a pink aqueous cobalt complex ion, formula $[Co(H_2O)_6]^{2+}$ and a dark blue aqueous cobalt ion, formula $[CoCl_4]^{2-}$. At the start, a deep purple solution 'halfway' between the two colours is required. Remember that forward and reverse reactions occur at the same rate. The equation for the reaction between these ions is:

$$[CoCl_4]^{2-} (aq) + 6H_2O(l) \rightleftharpoons [Co(H_2O)_6]^{2+}(aq) + 4Cl^- (aq)$$
$$\text{blue complex ion} + \text{water} \rightleftharpoons \text{pink complex ion} + \text{chloride ions}$$

The equation for the reaction includes two complex aqueous ions which might be confusing for some students. The simplified equation underneath the symbol equation may be helpful.

Students should recognise that both the pink and blue substances are present in the warm solution. They can be prompted to explain this using equilibrium reactions shown previously. The answer must be that both reactions occur simultaneously, resulting in the purple colour. The purple solution represents the reaction at equilibrium.

6.3 The extent of chemical change: how far do chemical reactions go?

The purple solution is created when particles of pink complex react with chloride ions at the same time as particles of blue complex react with water molecules. The purple colour arises because both reactions occur at the same rate.

During discussions students may need help to develop the reversible reaction idea. When introducing dynamic equilibrium, they may think an equilibrium is static, see-saw or that a reaction stops entirely at equilibrium. The ongoing nature of the reactions may be difficult to understand. Ask students to explain their thinking or provide evidence for their thinking. For example, if they think the reaction is static, what is their evidence for this? Encourage students to imagine they can see the particles. What would they see?

Questions to ask
- Why is the solution, at equilibrium, purple in colour?
- What colour change would be observed if water is added to the mixture at equilibrium? Explain why there is a colour change using Le Chatelier's principle.
- What colour change would be observed if concentrated hydrochloric acid is added to the mixture at equilibrium? Explain why there is a colour change using Le Chatelier's principle.

Careers

The three topics explored above are crucial in careers in chemical engineering. Chemical engineers are involved in design and development of diverse products. This requires consideration of how to achieve good yields, particularly in equilibrium reactions, at fast rates, with cheaper catalysts, at temperatures as low as possible. These factors are crucial to the success of a product in competitive global markets.

Chemical engineers work in many sectors including water purification and treatment, plastics, food and drink, pharmaceuticals and creating shampoos, soaps and toothpastes. For example, in the food industry chemical reactions occur in foods during processing and storage. Some of these reactions are harmful and need slowing down to avoid food spoiling. Other chemical reactions in food result in enhanced flavour or a desired colour, so need promoting to obtain the best quality food possible. In both cases, this requires good understanding of chemical kinetics.

Chemical engineers working today are involved in cutting-edge industries, making valuable new materials in areas such as nanotechnology and biomedical engineering.

Another career as a chemical technician involves working with chemical engineers testing new products and materials made through chemical reactions. This requires operation of a wide variety of laboratory equipment.

6.4 Resources

References and further reading

Banerjee, A. C. (1991) Misconceptions of students and teachers in chemical equilibrium. *International Journal of Science Education*, 13: 487–494.

Hackling, M.W., and Garnett, P. J. (1984) Misconceptions of chemical equilibrium. *European Journal of Science Education*, 7: 205–214.

Lister, T. (1995) *Classic Chemistry Demonstrations*, Royal Society of Chemistry. Experiments 22 and 23 (pages 48–52) provide details for two clock reactions: the blue-bottle experiment and the Old Nassau clock reaction.

Van Driel, J.H., de Vos, W. and Verloop, N. (1998) Developing secondary students' conceptions of chemical reactions: the introduction of chemical equilibrium. *International Journal of Science Education*, 20(4): 379–392.

Websites

The reaction of sodium hydrogencarbonate with citric acid: https://edu.rsc.org/resources/exothermic-or-endothermic/406.article

The reaction of magnesium with hydrochloric acid or sulfuric acid: https://edu.rsc.org/resources/exothermic-or-endothermic/406.article

Displacement reaction between copper sulfate solution and magnesium: https://edu.rsc.org/resources/exothermic-or-endothermic/406.article

Dissolving ammonium nitrate: https://edu.rsc.org/resources/energy-in-or-out-classifying-reactions/679.article

The reaction of ammonium chloride with barium hydroxide: https://edu.rsc.org/resources/endothermic-solid-solid-reactions/739.article

The thermite reaction: https://edu.rsc.org/resources/the-thermite-reaction/724.article

The reaction between aluminium and iodine: https://edu.rsc.org/resources/reaction-between-aluminium-and-iodine/715.article

An alternative method for the reaction between aluminium and iodine is given here: https://www.sserc.org.uk/subject-areas/chemistry/chemistry-resources/aluminium-iodine-reaction/

Burning a fuel: https://edu.rsc.org/resources/heat-energy-from-alcohols/1733.article

Collision theory: https://phet.colorado.edu/en/simulation/legacy/reactions-and-rates and https://www.bbc.co.uk/bitesize/guides/zsd2bk7/video

Investigating the effect of concentration by continuous measurements: https://edu.rsc.org/resources/the-rate-of-reaction-of-magnesium-with-hydrochloric-acid/1916.article

Investigating the effect of concentration by change in opacity of the reaction mixture: https://edu.rsc.org/resources/the-effect-of-concentration-on-reaction-rate/743.article

The iodine clock demonstration (the effect of concentration): https://edu.rsc.org/resources/iodine-clock-reaction-demonstration-method/744.article

Investigating the effect of a catalyst: https://edu.rsc.org/resources/hydrogen-peroxide-decomposition-using-different-catalysts/831.article and: https://www.youtube.com/watch?v=Ta4DomSDzF8

Introducing reversible reactions: https://edu.rsc.org/resources/a-reversible-reaction-of-hydrated-copperii-sulfate/437.article

Demonstrating Le Chatelier's principle: https://edu.rsc.org/resources/the-equilibrium-between-two-coloured-cobalt-species/1.article

Acids and alkalis

Jane Essex and Despoina O'Flynn

Introduction

The study of acids and alkalis allows exploration of how chemists think about, describe and use materials. It illustrates application of chemistry to everyday contexts, such as food, baking, agriculture, cosmetics and medicines. Overall, the topic provides opportunities for practical activities that illustrate a range of chemical principles and develop manipulative and investigative skills.

Initial encounters with acids and alkalis emphasise the macroscopic level (see Figure 1.1), including appearance, smell and observable effects on other substances. Transition to the curriculum for 14–16-year-olds commonly sees a shift from macroscopic knowledge to sub-microscopic and symbolic accounts of chemical behaviour. In addition, models of acidic and alkaline behaviour are deployed at different levels (see Table 7.3) demonstrating how scientists' understanding changes over time. However, research suggests students may think the models are distinct (or contradictory) models, rather than nested (de Vos and Pilot, 2001). Fragmentation of ideas about acids and alkalis can be exacerbated by the everyday contexts exemplified above.

Students' prior knowledge and experience

Students are likely to have intuitive ideas about acids. These may include that acids taste sour and sharp, as with vinegar and fruit juices; react with living material, so can be hazardous; fizz when added to carbonates and some metals; change the colour of indicators; can be made less acidic by alkalis, a change detected using an indicator; and exist as liquids (in reality, solutions) or solids. Students may have seen the word acid in the full name of DNA (deoxyribonucleic acid).

Similarly, students may be aware of properties of alkalis: for example, that they taste bitter; react with acids; change the colour of indicators; can be made less alkaline by acids; include salts known as carbonates; exist as liquids (in reality, solutions) or solids. Students are less likely to realise that alkalis react with living material, so tend to underestimate their destructive potential. They do not readily recognise household substances, such as soap and toothpaste, as alkaline, in part because they are not named as such.

A common misconception is that acids (and alkalis) comprise a continuous mass of dangerous material and are not made of particles. Practical and conceptual difficulties arise because acids and alkalis are usually encountered in aqueous solution, so appear to be 'just water'. Students often think of alkalis as substances that remove or 'cancel' acidic properties. This leads students to think neutralisation arises due to an acid- (or alkali-) inhibiting substance. Neutralisation is also often believed to involve the acid or alkali 'breaking down' and being permanently destroyed. Students may believe that neutralisation occurs when acids and alkalis are mixed, not reacting. Many students expect neutralisation to always produce a neutral solution. Misunderstandings about neutralisation are heightened by the addition of an indicator.

As representation of acids and alkalis moves to symbolic and sub-microscopic in the curriculum for 14–16-year-olds, further misconceptions arise. Students may accept that hydrogen ions (H^+) cause acidity, but simultaneously believe that when acids react, the acid is present as molecules, for example as HCl, rather than being dissociated into hydrogen ions and an anion (such as the chloride ion, Cl^-). They do not understand that release of hydrogen ions is essential to acidic properties. Students may not grasp that pH arises from a combination of two factors, namely the amount of acid or alkali added *and* its level of dissociation.

Progression

Understanding the properties of acids and alkalis leads to consideration of indicators as substances that detect acidity or alkalinity as two distinct categories. The pH scale is a conceptual step up, involving the idea of a spectrum of acidity and alkalinity. Neutralisation, and especially the products of neutralisation, represents a further development of understanding as students consider the chemical composition of specific acids and alkalis. Precise quantification of exact amounts of an acid and alkali needed to achieve neutralisation involves significant demand on manipulative and numeracy skills and requires secure understanding of the process.

Post-16 chemistry courses feature increasing reliance on mathematical concepts to support understanding of the properties of acids and alkalis in terms of chemical equilibria. Students consider pH and the strength of acids and alkalis quantitatively. They learn the behaviour of complex acid–base systems such as buffer solutions. They consider how scientists have modelled acids and bases, focusing on the Lewis model, which utilises donation and acceptance of electron pairs, rather than donation and acceptance of hydrogen ions.

Teaching sequence overview

The sequence follows curricula for 11–14s and 14–16s. This chapter suggests students start by considering acids and alkalis as two distinct groups of substances. For students aged 11–14, the chemical basis of acidic behaviour may be omitted or covered partially, but this is required for older students. Properties of indicators can be integrated into teaching properties of acids and alkalis, presenting them as a method for detecting levels of acidity or alkalinity. When considering neutralisation, students will develop an understanding of indicators to track chemical reactions between acids and alkalis. Younger students may consider only Universal Indicator, while 14–16-year-olds need to be aware of a range of indicators. Acid and alkali strength can be considered within discussion of general properties, and explicitly taught as these arise in context, for example, in organic chemistry (see Chapter 10).

Students should learn general reaction patterns of alkalis with acids. Products of neutralisation reactions should be discussed as examples of chemical synthesis of useful products (de Vos and Pilot, 2001). Reactions of acids to make products expands the general properties of acids and alkalis, qualifying these for specific contexts.

Neutralisation is a concept that requires explicit teaching to challenge misconceptions and avoid misunderstandings. The neutralisation reaction taught at each curriculum stage derives from the model of acidity being used (Table 7.3). Students may need to quantify amounts of acid and alkali that exactly neutralise each other, typically by titration. Some will need to recognise distinctive patterns of behaviour (including neutralisation), in reactions between strong and weak acids and alkalis.

7.1 Properties of acids and alkalis and the pH scale

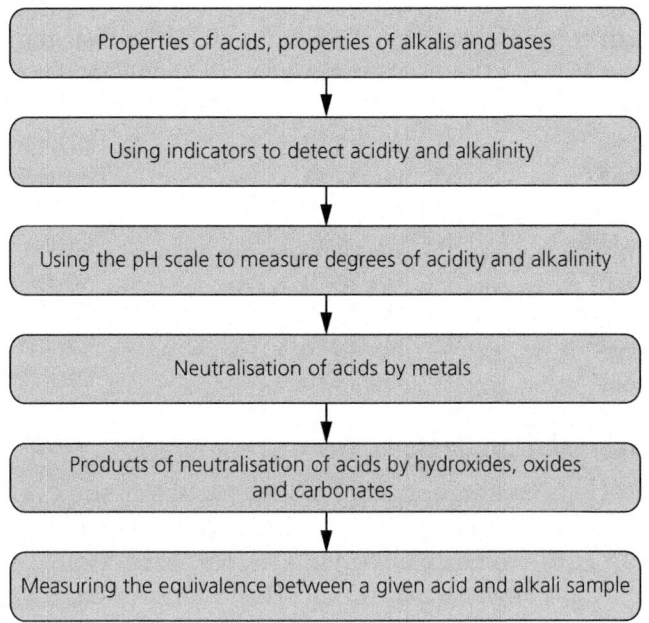

Figure 7.1 Acids and alkalis: teaching sequence overview

 # 7.1 Properties of acids and alkalis and the pH scale

Macroscopic properties of acids and alkalis listed above can be shown systematically via practical exercises. An example is making sherbet.

> **KEY ACTIVITY**
>
> ## Making sherbet
>
> This activity gives a memorable experience of the properties of acids and alkalis and an introduction to neutralisation changing the reacting alkali and acid, not just 'getting rid of' the acid. The fizz (effervescence) and slightly salty taste establish that new products are made during this reaction.
>
> This practical requires a non-laboratory space in which food grade materials can be used. Each student (or pair of students) requires:
>
> - powdered citric acid/tartaric acid – about 10 g
> - powdered sodium hydrogencarbonate (baking soda or baking powder) – about 10 g
> - one clean coffee stirrer (each)
>
> Students put one powder on one side of their tongue and the other powder on the opposite side, using opposite ends of the same stirrer (to avoid pre-mixing). The acid will taste sharp and the sodium hydrogen carbonate will taste bitter. Allowing the powders

199

> to mix in the mouth starts the reaction. Students will notice that the two tastes are lost and they experience a fizz as the reaction between the powders releases carbon dioxide. They will experience a warm feeling because the reaction is exothermic.
>
> **Questions to ask**
> - What do you taste before mixing the powders?
> - Describe what happens when the powders are mixed.
> - What do you taste when the powders are mixed?
> - Explain what is happening in your mouth when the powders mix.

The pH scale

The pH scale corresponds to a measure of the concentration of hydrogen ions. The concept of pH was devised by Danish chemists Søren Peder Lauritz Sørensen (Søren is pronounced 'Suren') and Margrethe Høyrup Sørensen. The husband-and-wife team worked at the Carlsberg Laboratory, run by the Carlsberg brewing company, in 1909. At the time, scientists used different ways of measuring acidity and alkalinity. Sørensen wanted to create a universal system. His interest developed from describing how proteins involved in brewing changed as the level of acidity changed. Sørensen calculated the negative logarithm of hydrogen ion concentration to arrive at a pH number. 'H' represents hydrogen ions. Sørensen never fully explained their choice of 'p', but it is generally thought to represent power or potential. Each unit on the pH scale represents a ten-fold change of hydrogen ion concentration. The highest concentrations of hydrogen ions and most acidic solutions correlate to the lowest numbers on the pH scale, which seems counterintuitive. This means that the concentration of hydrogen ions in a pH 1 solution is ten times higher than in a pH 2 solution, and a thousand times higher than in a pH 4 solution. Neutralising an acidic solution of pH 1 to pH 7 reduces the concentration of hydrogen ions to one millionth of the original value. The pH scale describes alkalinity indirectly.

Acidity and alkalinity are caused by an imbalance of hydrogen ions (H^+) and hydroxide ions (OH^-). In water, some water molecules dissociate (break apart) into hydrogen and hydroxide ions:

$$\text{water} \rightleftharpoons \text{hydrogen ion} + \text{hydroxide ion}$$

$$H_2O(l) \rightleftharpoons H^+(aq) + OH^-(aq)$$

The equation shows hydrogen ions and hydroxide ions are produced in a 1:1 ratio. This means their concentrations are equal,

7.1 Properties of acids and alkalis and the pH scale

so the solution is described as having a neutral pH. At 25 °C, the concentration of hydrogen ions is equivalent to pH7, so chemists usually say pH7 is neutral. However, at higher temperatures, more water molecules dissociate. This releases higher numbers of hydrogen and hydroxide ions. As pH measures the concentration of hydrogen ions, and hot water contains more of these, the pH of hot water is lower than that of cool water. But water is still neutral, regardless of the temperature. This is because hydrogen and hydroxide ions are always produced in a 1:1 ratio, so they are present in equal numbers. Remember that acidity and alkalinity arise due to imbalance in the numbers of these ions present in solution. In acidic solutions, more hydrogen ions are present than hydroxide ions. In an alkaline solution, there is a lower concentration of hydrogen ions than in pure water.

In chemistry, we encounter strong acids, such as hydrochloric acid, and weak acids, such as vinegar (ethanoic acid). Ask students to measure the pH of a 0.1 mol dm^{-3} solution of a strong mineral acid, such as hydrochloric acid, and the same concentration of a weak organic acid, such as ethanoic acid. The pH of the strong acid will be lower than the pH of the weak acid. Next compare the pH of a 0.1 mol dm^{-3} solution of a strong base, such as sodium hydroxide, with the pH of the same concentration of a weak base, such as ammonium hydroxide (formed when ammonia gas dissolves in water). The pH of the strong base will be higher than the pH of the weak base. This illustrates that pH depends on concentration *and* strength of the acid or alkali. Students may think that a strong acid/alkali is one with a high concentration. It is possible to have a concentrated solution of a weak acid or weak base and, conversely, a dilute solution of a strong acid or base. Concentration means the number of acid/alkali molecules dissolved in a specific volume. Strength refers to the number of hydrogen or hydroxide ions that the acid or alkali molecules contribute to the solution. The concentration of hydrogen ions in a solution depends on both the concentration and strength of the acid or alkali dissolved; the latter depends on how easily an individual substance loses or binds to hydrogen ions.

Enrichment

Human activities may cause changes to the pH of water in the environment. For example, burning fossil fuels containing sulfur emits sulfur dioxide, SO_2. Sulfur dioxide is soluble in rainwater, forming sulfurous acid, pH 5.8. This eventually causes river water to become acidic. The pH of water determines the species that live in it. For example, fish living in acidic water are prone to potentially lethal fungal infections. In water of pH below 5, fish cannot reproduce, and if the pH drops below 4, they die. To minimise sulfur dioxide emissions, engineers designed flue gas desulfurization systems to absorb the gas. Similarly, catalytic convertors fitted to cars eliminate toxic gases from exhaust fumes. Water run-off from coal and metal ore mines is also acidic. In the UK, authorities may add lime, calcium oxide, to raise the pH of water in a mining area, preventing heavy metals such as lead from leaching into the soil.

An alkali leakage, such as discharge of domestic waste water containing detergents, may kill fish, damage plant chromosomes and slow root growth. These can be removed by using activated charcoal to adsorb detergents.

Indicators and the pH scale

Chemical indicators monitor acidity or alkalinity of a solution, including during neutralisation reactions (see Section 7.5). A chemical indicator adopts one of two (or more) molecular structures, each with its own colour, depending on the acidity or alkalinity of the surroundings. Some show no colour in one form, and a colour in the other. The colours arise because the structures absorb visible light of different frequencies. Indicators change colour across a narrow range of pH values. Chemists select an indicator that corresponds to the pH range over which neutralisation is expected to occur. This requires experience.

Chemical indicators include litmus, methyl orange, methyl red, thymol blue, bromothymol blue, and phenolphthalein. Table 7.1 shows the colours and pH ranges over which these indicators change.

7.1 Properties of acids and alkalis and the pH scale

Table 7.1 Colour changes of common indicators

Single indicator	Low pH colour	Transition pH range	High pH colour
Thymol blue (first transition)	Red	1.2–2.8	Yellow
Methyl orange	Red	3.2–4.4	Yellow
Methyl red	Red	4.8–6.0	Yellow
Bromothymol blue	Yellow	6.0–7.6	Blue
Thymol blue (second transition)	Yellow	8.0–9.6	Blue
Phenolphthalein	Colourless	8.3–10.0	Fuchsia pink
Litmus	Red	4.5–8.3	Blue

There is no one fixed Universal Indicator. Various mixtures of indicators are made to give colour changes from pH values 1 to 11 or 14 (Table 7.2).

The mixture of indicators in a Universal Indicator can be explored by creating one: see http://science.cleapss.org.uk/Resource-Info/PP057-Making-a-Universal-Indicator-A-Microscale-Approach.aspx for full instructions. In summary, the resource includes a figure to be printed on paper and then laminated with plastic. Single drops of indicator are placed on the laminated surface under the names of the indicator. These act as controls. Next, seven drops of each indicator are placed across the corresponding rows under each pH number, forming a grid of droplets. Universal Indicator is made by putting one drop of each indicator in each of the seven positions in the bottom row. These drops are mixed with the point of a cocktail stick. The indicator colours are created by adding drops of prepared and labelled solutions corresponding to known pH values on each indicator droplet.

An alternative approach is to separate Universal Indicator into its constituent indicators using paper chromatography.

Table 7.2 Colour changes of Universal Indicator

| Universal Indicator component | pH | | | | | | |
	1	4	5	7	9	11	14
Thymol blue	Red	Yellow	Yellow	Yellow	Yellow	Blue	Blue
Methyl orange	Red	Red	Yellow	Yellow	Yellow	Yellow	Yellow
Methyl red	Red	Red	Yellow	Yellow	Yellow	Yellow	Yellow
Bromothymol blue	Yellow	Yellow	Yellow	Blue	Blue	Blue	Blue
Phenolphthalein	Colourless	Colourless	Colourless	Colourless	Fuchsia pink	Fuchsia pink	Fuchsia pink
Universal Indicator	Red	Orange	Yellow	Light green	Blue	Indigo	Purple

7 Acids and alkalis

KEY ACTIVITY

Making red cabbage indicator

Full instructions are available at https://edu.rsc.org/resources/making-a-ph-indicator/422.article Red cabbage (scientific name: *Brassica oleracea* var. *capitata f. rubra*) extract comprises a mixture of over 30 closely related anthocyanin ('flower blue') pigments. These act as a mixture of indicators producing red, purple, blue, green and yellow colours. Red cabbage extract can be separated into its constituent indicator molecules by paper chromatography. Students can compare Universal Indicator and red cabbage extract chromatograms (see Figures 7.2 and 7.3) and evaluate both indicators against factors such as sensitivity to changes in pH, convenience and ease of use.

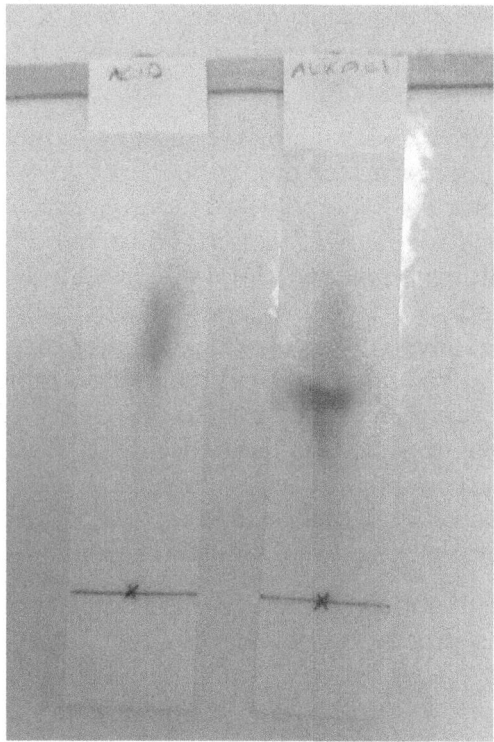

Figure 7.2 Universal Indicator after separating by paper chromatography and treating with dilute acid (left-hand column) and dilute alkali (right-hand column)

7.1 Properties of acids and alkalis and the pH scale

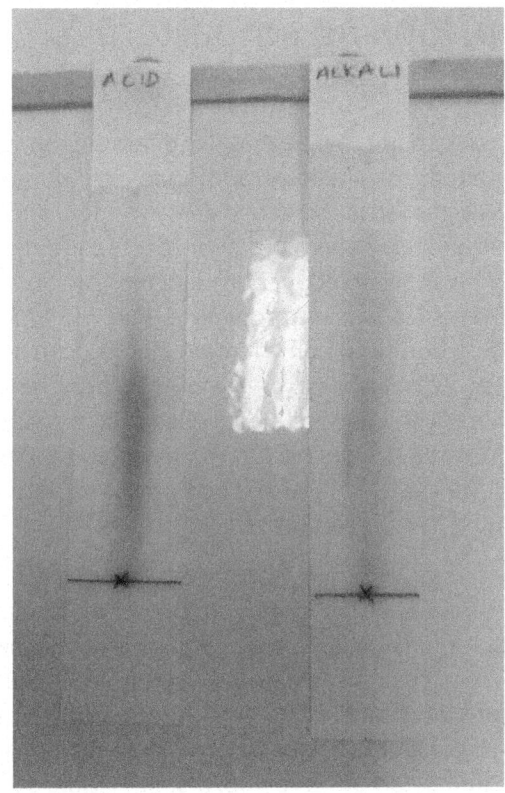

Figure 7.3 Red cabbage extract after separating by paper chromatography and treating with dilute acid (left-hand column) and dilute alkali (right-hand column)

KEY ACTIVITY

The rainbow reaction

The rainbow reaction demonstrates the spectrum of Universal Indicator colours, simultaneously illustrating partial neutralisation of acids and alkalis to create intermediate pH values. Two variations are available, using a burette (http://resources.schoolscience.co.uk/Salters/chemclub2_3.html) or a glass tube (https://edu.rsc.org/resources/universal-indicator-rainbow/700.article). Carbon dioxide gas generated by the reaction provides agitation, while the pH gradient is maintained by an accompanying density gradient. The solutions remain distinctive for long enough to be inspected. Over about 24 hours, diffusion occurs, creating a homogenous solution.

A rainbow can be created in a test tube. Students can be given instructions for the larger scale demonstration and scale this to test-tube size as a problem-solving activity. Perseverance and very gentle inversion of the stoppered test tube are required.

7 Acids and alkalis

KEY ACTIVITY

Burning elements in oxygen, looking for patterns in pH values of oxides

Burning metal and non-metal elements in oxygen illustrates that, on dissolving in water, metal oxides form alkalis while non-metal oxides form acids. Traditionally, elements are burned on a deflagration (burning) spoon in a gas jar of oxygen, water is added, and the pH of the resulting solution tested. Safety issues may arise due to unreacted elements remaining in the gas jar. This method has been superseded by burning elements in a reaction tube. Instructions are available at https://edu.rsc.org/download?ac=12178 Some elements, including magnesium ribbon and carbon (as small charcoal pieces), can be burned by students. Alternatively, a crucible (https://edu.rsc.org/resources/the-change-in-mass-when-magnesium-burns/718.article) or a microscale equivalent (http://science.cleapss.org.uk/Resource-Info/Finding-the-formula-of-magnesium-oxide.aspx) can be used. If technician support or access to a fume cupboard makes the burning of elements in oxygen difficult, videos of elements burning in oxygen can be substituted.

As a second stage, the pH of solutions labelled as 'oxides in water' are tested by students. Stock solutions, such as sulfuric acid, phosphoric acid, sodium hydroxide and potassium hydroxide can be used for the purpose of testing. For 11–14-year-olds, this experiment makes an impressive demonstration that reinforces differences between metals and non-metals. For 14–16-year-olds, the role of water in reacting with oxides and forming an acid or alkali can be explained at a sub-microscopic level. The practical introduces the behaviour of copper(II) oxide with acid. Students can observe and, if safe, handle a range of element samples for classification as metals or non-metals.

 Models of acids and alkalis

Scientists' explanations, or models, of the behaviour of acids and alkalis (Table 7.3) have changed over time. Teaching about the historical development is useful in understanding how we now believe acids and alkalis react. de Vos and Pilot (2001) note this may cause difficulties for students as many models are presented in chemistry without any clear rationale.

7.2 Models of acids and alkalis

Table 7.3 Models of acids and alkalis

Model	Scientist, years of birth and death, nationality	Features of an acid	Features of an alkali	Explanation of neutralisation	Other noteworthy features
Davy	Humphry Davy, 1778–1829, English	Acids always yield hydrogen when reacting with a metal, but do not always contain oxygen, hence all acids contain hydrogen.	Davy used electrolysis to extract elemental metals from alkaline earth compounds. He showed that these alkalis were metal oxides.	None	This model undermines the oxygen theory. Davy could not explain why other substances that contain hydrogen were not acids. Justus von Liebig (1803 to 1873, German) developed Davy's idea, proposing acids had hydrogen that could be displaced by a metal during a reaction.
Arrhenius	Svante Arrhenius, 1859–1927, Swedish	Acids release hydrogen ions (H+) when dissolved in water. The hydrogen ion then bonds to the water to give a hydroxonium ion, H_3O^+.	Alkalis dissolve in water to release hydroxide ions (OH–).	Neutralisation takes place when the concentration of H+ in an acid, or of OH– in an alkali, decreases. This is most commonly caused by the reaction $H^+ + OH^- \rightarrow H_2O$	Arrhenius won the Nobel Prize in Chemistry in 1903 for this work. The work only considers substances dissolved in water, so does not account for insoluble acids or bases.
Brønsted–Lowry	Johannes Nicolaus Brønsted, 1879–1947, Danish. Thomas Martin Lowry, 1874–1936, English	Acids can donate hydrogen ions.	Bases, including alkalis, can accept donated protons (hydrogen ions) by binding to a negative ion, such as OH–, or bonding to a lone pair of electrons, for example on the oxygen atom in a water molecule.	Neutralisation involves transferring protons from acid to base. The reaction is reversible, with the dissociated acid (called a conjugate base) able to accept a proton back from the protonated alkali (called a conjugate acid).	By now hydrogen ions were understood to be a single proton. The model was devised by both scientists simultaneously, but independently of each other. The model extends Arrhenius' model but encompasses the greater knowledge of atomic theory that existed.
Lewis	Gilbert Lewis, 1875–1946, American	Acids can accept a 'lone pair', a pair of electrons in the outer shell of an atom that are not involved in bonding to another atom. Hydrogen ions, which have an empty outer orbital, can accept a lone pair of electrons.	Alkalis can donate a lone pair. The oxygen atom in a hydroxide ion has three lone pairs so can act as a Lewis base.	Neutralisation is transferring lone pairs from acid to base.	Lewis proposed his idea at the same time as Brønsted and Lowry but did not publish until 1938. This model is consistent with the models of Arrhenius and Brønsted and Lowry but widens the chemical identity of acids and alkalis from only those containing or accepting H+. It explains acidic behaviour of substances such as some metal ions.

Alkali is originally an Arabic term meaning '(the) salt' (any ionic compound). Alkalis are defined in Arrhenius' model: they neutralise acids and dissolve in water. The term base includes any substance that neutralises acids, regardless of its solubility in water. The Brønsted–Lowry and Lewis models are required in post-16 chemistry courses. These models broaden the definition to include any materials that neutralise acids, so their definitions incorporate both alkalis and bases. Thus, while base is a more general term, students are more familiar with alkalis in solution than with other bases.

 The reaction of acids with metals to produce salts and hydrogen

The reaction of magnesium with acid illustrates that some metals neutralise acids and displace hydrogen. This recreates Humphry Davy's nineteenth-century experiments, which established that releasing hydrogen makes an acid acidic (Table 7.3).

KEY ACTIVITY

The reaction of magnesium and a dilute acid

This practical demonstrates that hydrogen can be displaced from an acid and, secondly, that higher concentrations of acid displace more hydrogen. Reviewing students' ideas about what happens when a substance dissolves is also worthwhile.

A large-scale quantitative version of the activity can be carried out by measuring the volume of hydrogen produced by collecting it in an inverted measuring cylinder filled with water as described at https://edu.rsc.org/resources/the-rate-of-reaction-of-magnesium-with-hydrochloric-acid/1916.article The reaction can be done simply in test tubes using 10 cm³ of hydrochloric acid at several concentrations from 0.5 mol dm^{-3} to 2 mol dm^{-3}, each with 0.5 cm magnesium ribbon. The presence of hydrogen can be confirmed with a squeaky pop test. Students will observe higher rates of hydrogen production by the increased amount of fizz (effervescence) with more concentrated acid.

Extend the activity by using the reaction to measure acidity. Students test pH of acid samples using Universal Indicator, then add magnesium. Lower pH values correspond to faster reaction rates. Reinforce the connection between hydrogen and acidity by comparing how magnesium reacts with other acids, noting their chemical formulae.

Table 7.4 Names, formulae and reaction of acids with magnesium ribbon

Acid	Chemical formula	Dissociated ions in aqueous solution	Reaction with magnesium
Hydrochloric acid	HCl	$H^+ + Cl^-$	Hydrogen gas and magnesium chloride solution
Nitric acid	HNO_3	$H^+ + HNO_3^-$	Hydrogen gas and magnesium nitrate solution. Note: nitric acid must be cold to avoid nitrogen oxides being produced
Sulfuric acid	H_2SO_4	$2H^+ + SO_4^{2-}$	Hydrogen gas and magnesium sulfate solution
Ethanoic acid	CH_3COOH	Partial dissociation to $H^+ + CH_3COO^-$	Hydrogen gas and magnesium ethanoate solution

The distinction between a strong and a weak acid, such as ethanoic acid, can be shown by including this as a comparator reagent. Ethanoic acid reacts more slowly and produces less hydrogen gas than any of the others (see Chapter 10 for information about organic acids). The metal–acid reaction is one of many that produce soluble salts (see Section 7.4).

Symbolic representation of acid plus metal reactions

The reaction between magnesium and hydrochloric acid can be shown in words as:

magnesium + hydrochloric acid → hydrogen + magnesium chloride

The reaction is a displacement reaction producing hydrogen gas and an aqueous (in solution, or dissolved) salt. The general pattern is summarised as:

metal + acid → hydrogen + salt

For example, Mg(s) + 2HCl(aq) → H2(g) + MgCl2(aq)

Some students will need to understand the reaction as a redox reaction (see Chapter 8). Students may be required to know how metals such as zinc and iron react with hydrochloric and sulfuric acid. These react similarly to the magnesium. Zinc is less reactive than magnesium, while iron is less reactive than magnesium and zinc.

Making salts by neutralisation

The reaction of acids with alkalis and bases produces ionic compounds called salts. Many salts are useful products. Examples include potassium nitrate, which is a fertiliser, rocket propellant and a component of gunpowder; sodium carbonate, used as a water softener and in making glass, paper, soaps and detergents; and calcium sulfate, known as gypsum, which is used in building as plaster and in concrete, as well as for surgical splints, binding clay, conditioning bread dough and, in the form of alabaster, for sculpture. All acid–alkali/base reactions are represented by the general equation:

$$\text{acid} + \text{alkali/base} \rightarrow \text{salt} + \text{water}$$

Neutralisation reactions and their products can be classified according to the acid or the alkali/base. The acid determines the negative ion (anion) in the salt. For example, hydrochloric acid reacts to give chloride compounds, sulfuric acid always yields sulfate compounds and carbonic acid makes carbonates. Changing the negative ions changes the properties of the products. For example, to make a soluble copper salt, consider that copper(II) chloride is moderately soluble, copper(II) sulfate is highly soluble, and copper(II) carbonate is insoluble in water. So either hydrochloric acid or sulfuric acid reacting with copper are the best options for producing a soluble copper salt.

Classifying neutralisations according to the alkali or base, specifically its negative ion, gives information about the reaction characteristics. Bases containing carbonate, hydroxide and oxide ions demonstrate patterns in their reactions with acids. Figures 7.4 to 7.6 show all alkalis react with hydrogen ions, H^+. This property arises from the fact that all three anions, hydroxide, oxide and carbonate contain oxygen, which has lone pairs, and are negative ions. The neutralisation of hydrogen ions is represented in Figures 7.4 to 7.6.

The reaction between hydroxides and acids

Many alkalis are hydroxides. Show that hydroxide ions react with hydrogen ions to make water using ionic jigsaws. Present students with word and symbol formulae for reactants and products of hydroxide and acid reactions on cards. They rearrange the cards and arrive at the general pattern (shown below). For example:

$$\text{sodium hydroxide} + \text{hydrochloric acid} \rightarrow \text{water} + \text{sodium chloride}$$

The reaction between hydrogen ions and hydroxide ions is neutralisation. This is represented as:

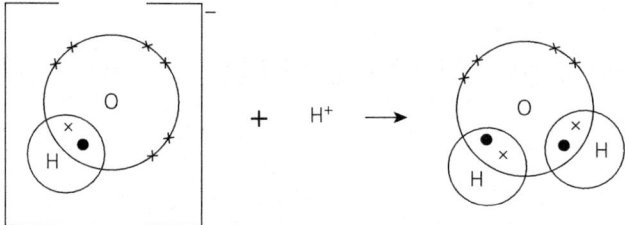

Figure 7.4 The neutralisation of hydroxide ions

$$\text{hydroxide ions} + \text{hydrogen ions (from the acid)} \rightarrow \text{water}$$
$$\text{OH}^-(\text{aq}) + \text{H}^+(\text{aq}) \rightarrow \text{H}_2\text{O}(\text{l})$$

The salt comprises a metal (or other positive) ion from the alkali and the negative ion from the acid. In the example, the sodium ion and chloride ion form sodium chloride in solution. Repetition with different acids and alkalis reinforces these patterns.

Practical activities for neutralising sodium hydroxide solution and hydrochloric acid are carried out as a titration (for example, https://edu.rsc.org/resources/titrating-sodium-hydroxide-with-hydrochloric-acid/697.article). Younger students can use simple glassware for the same reaction. $0.4\,\text{mol}\,\text{dm}^{-3}$ solutions of an acid and an alkali are measured using measuring cylinders. $25\,\text{cm}^3$ of dilute alkali is placed in a conical flask (or beaker). Add acid using dropper pipettes. Using Universal Indicator will reinforce knowledge of its colour changes (see above). Once neutral, a spatula of activated charcoal powder can be added to remove the indicator. The activated charcoal and salt solution are filtered and the colourless solution evaporated to give sodium chloride (table salt). Evaporating to complete dryness is not recommended to avoid possible acidic fumes being produced; and no attempt should be made to taste the salt. Younger students realise that a chemical transformation has occurred in which an acid and an alkali (which they wouldn't think of consuming) react to form the salt added to their food!

> **Enrichment**
>
> Soaps are salts made by neutralisation. Soap-making (saponification) often uses local resources as raw materials. In soap-making, fatty acids (long hydrocarbon chains with an acid group at one end) are obtained from a fat or oil. Mutton (sheep) fat, palm oil and coconut oil are sources of fatty acids. An alkali, usually potassium hydroxide, is obtained from plants, such as burnt seaweed or banana-leaf ash. Fatty acids are heated with concentrated potassium hydroxide solution. A neutralisation reaction occurs forming soap (the salt) and water.

The reaction between oxides and acids

One oxide ion reacts with two hydrogen ions (see Figure 7.5 and the equation below). Students may think of hydroxide ions as a 'half neutralised' oxide ion. Ionic jigsaws (see above) will enable them to explore this.

7 Acids and alkalis

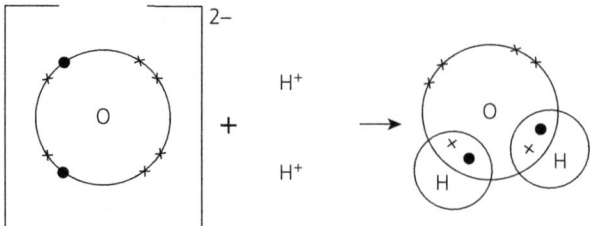

Figure 7.5 The neutralisation of oxide ions

$$\text{oxide ions} + 2 \text{ hydrogen ions} \rightarrow \text{water}$$
$$O^{2-}(aq) \text{ or (s)} + 2H^+(aq) \rightarrow H_2O(l)$$

For example in the reaction:

$$\text{copper(II) oxide} + \text{sulfuric acid} \rightarrow \text{water} + \text{copper sulfate}$$

The oxide and hydrogen ions react to produce water, while copper and sulfate ions form a salt:

$$CuO(s) + H_2SO_4(aq) \rightarrow H_2O(l) + CuSO_4(aq)$$

Oxides enable chemists to change an insoluble base, such as a metal oxide, into a soluble salt. Any unreacted insoluble base can be separated from the salt solution by filtration. The reaction of copper(II) oxide with sulfuric acid to form copper(II) sulfate crystals is an example. There is a visible change as the soluble salt, copper(II) sulfate, is turquoise in colour. Excess copper oxide powder is filtered off and the solution is evaporated to give blue crystals of copper(II) sulfate. Students observe:

$$\text{black solid} + \text{colourless solution} \rightarrow \text{blue solution}$$

Practical instructions can be found at https://edu.rsc.org/resources/preparing-salts-by-neutralisation-of-oxides-and-carbonates/1762.article or for a microscale version at http://science.cleapss.org.uk/Resource/Hydrated-copper-sulfate-preparation-Microscale-method.vid

The reaction between carbonates and acids

Neutralisation reactions of acids and carbonates release carbon dioxide (fizz or effervescence) as well as producing water and an ionic compound:

7.4 Making salts by neutralisation

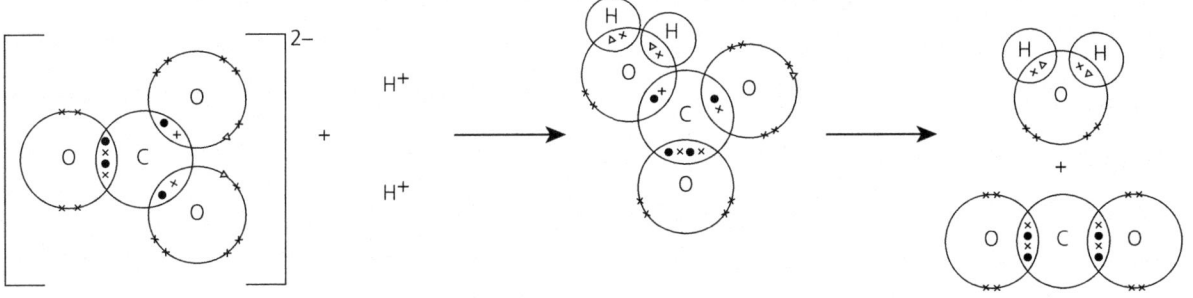

Figure 7.6 The neutralisation of carbonate ions

$$\text{carbonate ions} + \text{hydrogen ions} \rightarrow \text{water} + \text{carbon dioxide gas}$$
$$CO_3^{2-}(aq) \text{ or (s)} + 2H^+(aq) \rightarrow H_2O(l) + CO_2(g)$$

For example,

$$\text{sodium carbonate} + \text{hydrochloric acid} \rightarrow \text{water} + \text{carbon dioxide} + \text{sodium chloride}$$
$$Na_2CO_3(aq) + 2HCl(aq) \rightarrow H_2O(l) + CO_2(g) + 2NaCl(aq)$$

Carbonates are suitable for household products such as baking powder and bath bombs because they are stable in storage and not strongly alkaline in solution. The carbonate ion can be partially neutralised to give hydrogencarbonate ions, HCO_3^-. Hydrogencarbonate ions are less alkaline than carbonate ions, so are preferable for culinary and cosmetic products.

KEY ACTIVITY

Making bath bombs

This activity illustrates the reaction between an acid and a carbonate, and the role of water in acidic and alkaline behaviour. Bath bombs contain a dry powder mixture of citric acid and sodium hydrogencarbonate crystals bound together with a small volume of vegetable oil. When dropped into water, the powders dissolve then react, producing carbon dioxide gas, water and sodium citrate solution. Full instructions for making bath bombs are available at http://resources.schoolscience.co.uk/Salters/chemclub2_3.html

Without water

$$\text{citric acid} + \text{sodium hydrogencarbonate} \quad \text{no reaction}$$
$$C_6H_5O_7 \cdot H_3(s) + 3NaHCO_3(s) \quad \text{no reaction}$$

213

7 Acids and alkalis

> **With water**
>
> citric acid + sodium hydrogen- → water + carbon dioxide + sodium citrate
> carbonate
>
> $C_6H_5O_7 \cdot H_3(aq) + 3NaHCO_3(aq) \rightarrow 3H_2O(l) + 3CO_2(g) + Na_3C_6H_5O_7(aq)$
>
> Citric acid has three groups that can release a hydrogen ion and the formula is written to represent this.
>
> Alternatively, ask students to find the 'best recipe' for a bath bomb, using small quantities of solid citric acid and sodium hydrogencarbonate. Consider criteria for the best bath bomb, such as fizz-ability, time taken to dissolve and water temperature. Students are often surprised by how much alkali is needed in comparison to acid. Students can add a scented oil and spray the bomb 'case' (see instructions) with food colouring.

7.5 Neutralisation and titration

Neutralisation is widely understood as 'removal of acidity'. However, this does not mean fundamental destruction of the acid. Explanations of neutralisation have evolved (Table 7.3). In 11–16 chemistry, the Brønsted–Lowry model is sufficient, so neutralisation can be treated as the reaction of hydrogen ions from an acid with an alkali. The products of neutralisation always include water. The decrease in acidity as the number of hydrogen ions falls causes a corresponding rise in pH value. However, neutralisation does not always result in a pH-neutral solution. The point at which an amount of acid is exactly neutralised by an alkali, but before the alkali is in excess, is the equivalence point. The pH at this point will be closer to neutral than either unreacted acid or alkali. The precise pH value depends on the ionic compound (salt) formed during the reaction. This can be predicted by knowing whether the acid and alkali involved were strong or weak.

> ### Enrichment
>
> Hydrochloric acid (HCl(aq)) has been known since about the sixteenth century. Industrially, it is a by-product of the Solvay (ammonia–soda) process that produces sodium carbonate. Hydrochloric acid is used in production of steel, batteries, fireworks, leather processing and in the manufacture of plastic polyvinylchloride (PVC) among other items. In nature, hydrochloric acid is secreted into gastric juice produced by the

parietal cells in the stomach wall. It helps to establish the correct pH for the formation of pepsin, an enzyme that catalyses breakdown of protein molecules. Citric acid (general formula $C_6H_8O_7$) is a weak acid, forming a solution with pH around 4.8. It was first isolated from lemon juice by Swedish chemist Carl Wilhelm Scheele in 1784. About 2 million tonnes are produced annually worldwide for use as a flavouring agent in food and in the pharmaceutical industry. In nature, citric acid is found in the citric acid cycle which is central to metabolism of many animals and plants.

Strong and weak acids; strong and weak bases

Weak and strong acids release different numbers of hydrogen ions when dissolved in water. Strong acids, including hydrochloric acid, nitric acid and sulfuric acid (mineral acids), completely dissociate in water. All their hydrogen ions are available to react. Weak acids may be around 5% dissociated, with the remaining 95% of hydrogen ions remaining bonded in acid molecules. As a result, fewer hydrogen ions are available for reaction. Weak acids include ethanoic acid, citric acid and other organic acids.

Acid strength determines the concentration of hydrogen ions in solution. Alkalis and bases are similarly classified as weak or strong, depending on the extent to which they release hydroxide or oxide ions in solution. The strengths of the acid and alkali/base determine the pH of the ionic compound that forms in neutralisation. For example, when a strong acid reacts with a weak base the salt is acidic. If a weak acid reacts with a strong base, the salt will be alkaline. A neutral, or near neutral, salt forms when a strong acid is neutralised by a strong base, and when a weak acid is neutralised by a weak base. This pattern is illustrated by pH curves (Figure 7.7). These show how pH changes as acid is added to a solution of base.

7 Acids and alkalis

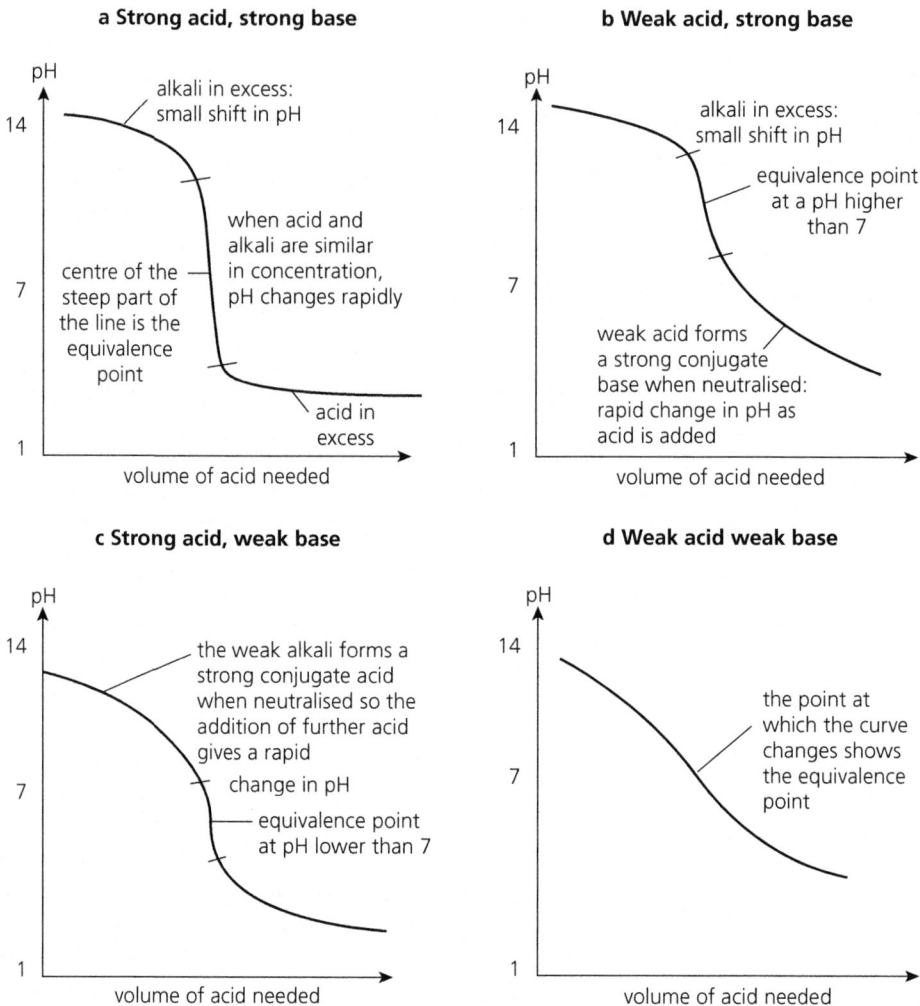

Figure 7.7 pH curves for titrations of strong or weak acids with strong or weak bases

Choosing an indicator to track a neutralisation reaction

A suitable indicator is required to follow change in pH during a neutralisation reaction and to determine end (equivalence) points. Single colour change indicators are commonly used because they give an unambiguous colour change at, or very close to, the equivalence point.

Choosing a chemical indicator depends on the strength of the acid and alkali. A strong acid, such as hydrochloric acid, neutralising a strong alkali, such as sodium hydroxide, gives a pH of near 7 at the equivalence point, so an indicator that changes colour at or around pH 7 would be ideal. Table 7.1 suggests the best indicator is bromothymol blue, as this changes colour in the pH range 6.0–7.6. However, sodium hydroxide solution absorbs carbon dioxide from the atmosphere, which lowers its pH over time. Hence, in practice, the best indicator is methyl orange, or methyl red, which change colour over pH ranges 3.2–4.4 and 4.8–6.0 respectively.

A problem-solving practical, such as evaluation of two antacids (alkalis used to counteract gastric acid), introduces measurement of just enough acid to achieve neutralisation. This prepares students for accurate measurement of reacting quantities at equivalence point.

KEY ACTIVITY

Comparing antacid powders

This activity enables students to explore the notion of equivalence between acid and alkali at the point of neutralisation. It gives them experience of measuring quantities accurately and using indicators other than Universal Indicator. Students discover how little alkali is required near the equivalence point to cause a dramatic change in pH.

Students are given two powders labelled as antacids, sodium hydrogencarbonate and magnesium hydroxide; these can be given names such as Burpo and Tumeeze. Students are told these cure indigestion caused by excessive stomach acid. They investigate which powder has the best acid-neutralising (antacid) properties. Restrictions are that students should use no more than 25 cm^3 of 1.0 mol dm^{-3} hydrochloric acid to react with one sample of antacid and the maximum mass of powder for one reaction should be 3 g. Magnesium hydroxide has greater neutralising capacity for a given mass. The investigation allows for consideration of factors such as whether it makes you burp (produces a lot of gas) and whether it dissolves easily so may be easier to swallow.

Equipment needed includes: a balance, bromothymol blue indicator, hydrochloric acid solution, eye protection, beakers, measuring cylinders and stirring rods. Students are told that bromothymol blue is yellow below pH 6 and turns blue when all the acid is neutralised.

Titration

Students may learn titration as a technique for measuring exact quantities of acid and alkali that neutralise each other. The principle is that a very precisely measured volume of one solution of a known concentration is reacted with a second solution of unknown concentration. By knowing the reacting volumes of both solutions, and the concentration of one, the unknown concentration is calculated.

Titration was established in the nineteenth century as a state-of-the-art technique and remains useful because of its high precision, rapid delivery and low cost. For example, assessing the composition of a water sample by titration is still common. There are several very helpful screen-based simulations that can be used to prepare students for carrying out titrations, for example www.rsc.org/learn-chemistry/resource/res00002077/titration-screen-experiment

Very accurate measurement of the solutions is made possible by use of a burette, a graduated glass tube with a tap at the lower end that measures variable volumes to the nearest 0.05 cm^3; and a pipette, a small glass tube with a marked line, which measures a fixed volume to high accuracy. Usually, the acid is placed in the burette, as alkalis can damage the glass and may crystallise around the tap. The alkali is measured using a pipette and placed in a conical flask.

An indicator is chosen that changes in the pH range at which the end point occurs (see Table 7.1 above). This is added to the alkali in the flask. The volume of acid added to cause the colour change is known as the end point and, if a suitable indicator is chosen, this will be the equivalence point. Often, an operator's reaction time to seeing the colour change and stopping at the end point means that slightly more acid is added than necessary. Overshooting by a few drops of acid can cause the pH to change by many units at or near equivalence point, as even one drop of acid greatly increases the number of hydrogen ions in the reaction mixture.

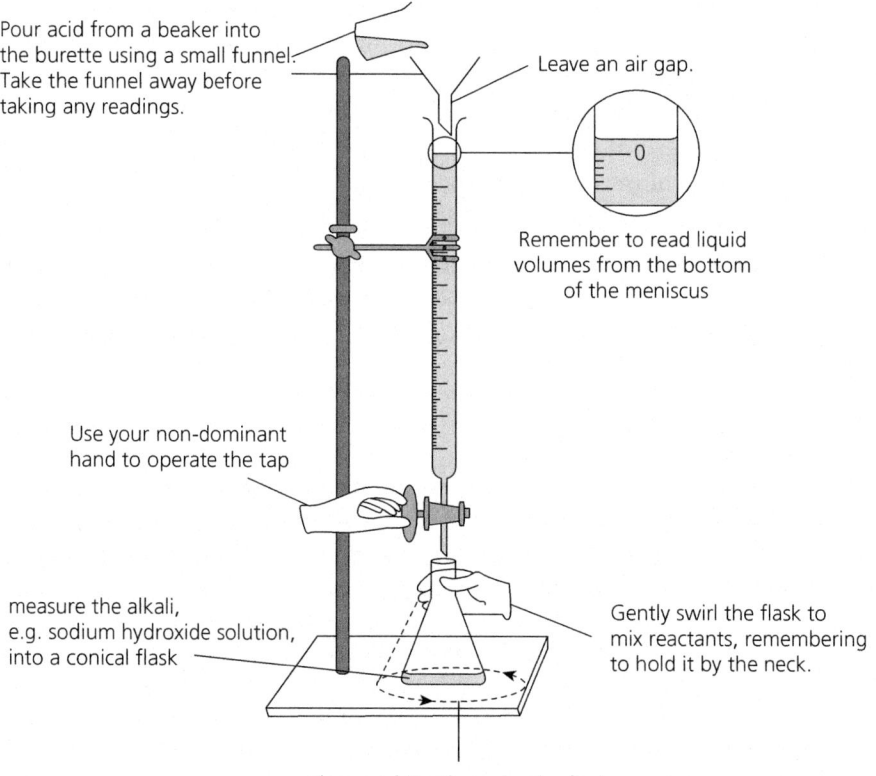

Figure 7.8 Titration apparatus and instructions

7.5 Neutralisation and titration

The steps in carrying out a titration are as follows:

1. Fill the burette, usually with acid, by pouring the solution through a small plastic funnel. The burette tap allows liquid to run through in a controlled way: release the tap so the liquid fills the glass below it to the tip. Remove the funnel. Run liquid through to a suitable start point on the burette scale. This does not need to be 0.00 cm³. Read the volume corresponding to the bottom of the meniscus.
2. Measure the alkali solution using a pipette, drawing the solution up with a pipette filler until the bottom of the meniscus is level with the marked line on the pipette. Run the alkali into the conical flask, touching the tip of the pipette to the inside of the flask to ensure the exact volume is dispensed. Add the indicator to the solution in the flask. Place the flask on a white tile under the tap on the burette.
3. Open the tap using the non-dominant hand. Use the dominant hand to swirl the flask gently as the solution runs in until the indicator changes colour, proceeding dropwise close to the end point.
4. Subtract the initial volume reading from the volume reading at the end point; the difference is the volume of acid required to neutralise the alkali in the conical flask.

During the first, rough titration, the acid is run in quickly. The volume used serves as a rough guide to the actual end point. In subsequent titrations, using clean conical flasks and a refilled burette, the acid is run through quickly until the volume is close to the rough reading, then dropwise to achieve an accurate endpoint. Once two consecutive volumes of acid are concordant, that is, they are within 0.2 cm³ of each other, the titration is complete. The volume of acid required to neutralise the alkali is taken as the mean of the two concordant values.

From these results the unknown concentration, whether of the acid or alkali, can be calculated. The volumes (V), concentrations (C) and the reacting ratio shown in the reaction equation (n) of the acid and alkali are related by the following equation:

$$n_1 \times C_1 \times V_1 = n_2 \times C_2 \times V_2$$

7 Acids and alkalis

 Maths

Titration: Worked example

A student carries out a titration, putting 25.0 cm³ of 0.200 mol dm⁻³ sodium hydroxide solution in a flask. They find it is neutralised by 12.50 cm³ of a dilute solution of hydrochloric acid. The equation for the reaction is:

$$HCl(aq) + NaOH(aq) \rightarrow NaCl(aq) + H_2O(l)$$

The reacting ratio (n) between the acid and alkali is 1:1 as the equation shows that 1 mole of acid reacts with 1 mole of alkali (see Chapter 2).

The units of concentration are mol dm⁻³. This means moles per decimetre cubed or moles per litre. To calculate concentration, volumes must be converted to dm³ by dividing values in cm³ by 1000.

So the data needed are:

For left-hand side, NaOH: $n_1 = 1$, $C_1 = 0.200$ mol dm⁻³, $V_1 = 25.0$ cm³ = 0.0250 dm³

For left-hand side, HCl: $n_2 = 1$, $C_2 =$ unknown mol dm⁻³, $V_2 = 12.50$ cm³ = 0.0125 dm³

Putting these in the equation:

$$1 \times 0.2 \times 0.025 = 1 \times C_2 \times 0.0125$$

Multiplying:

$$0.005 = 0.0125 \times C_2$$

Rearranging:

$$C_2 = \frac{0.005}{0.0125} = 0.4$$

So the concentration of the hydrochloric acid is 0.4 mol dm⁻³.

Cross-disciplinary

A burette gets its name from the French for 'small wine jug' and its precision makes it a key measuring instrument in chemistry. A pipette is named after the French for 'small pipe'.

Careers

Environmental scientists monitor and assess pollutants, including acids and alkalis, in water, on land and in the air. They collate data to provide information about environmental changes, chemical markers of these changes, and contribute to identifying potential causes and solutions. Individual environmental scientists gather water samples and undertake acidity and chemical composition tests. They gather and analyse data from sensors in a range of locations, including remote rural places and city centres. By sharing data, environmental scientists contribute to a deeper understanding of pollution and how this affects the environment.

An example of collaboration is the UK Uplands Waters Monitoring Network, founded by the UK Government in 1988, which monitors how acid rain affects the levels of sulfur compounds found in water and air and how this affects biodiversity. The group began by identifying potential sites where acid rain was the only significant source of pollution, monitoring changes to local ecology. By comparing sites over time, scientists have constructed a picture of the effect of acid pollution on local temperature, levels of acidic gases and incidence of types of plants and animals. Their data can be seen at https://uwmn.uk/. Their current work is focused on the impact of climate change on UK water acidification.

The UK Uplands Waters Monitoring Network has achieved policy change at the UK Government's Department of Farming, Environment and Rural Affairs (Defra). After demonstrating that nitrates and phosphates from fertilisers, washed off from soil in rainfall, add to water acidification, agricultural policies on fertiliser use changed.

Food scientists analyse food samples to make sure that they are safe for human consumption and conform to appropriate legal requirements. They assess the nutritional value of food to ensure that packaging reports food composition accurately. Food scientists inspect places where food is produced to ensure that the site of and finished products meet legal requirements. They propose new food sources, and improve food production methods, investigating new ways to process, preserve and package food.

Controlling acidity and alkalinity of food is integral to the food scientist's job. A change in pH may cause food to become unpalatable. Unexpected shifts in pH may indicate contamination by mould or bacteria. Some foods have acid added for taste, or to improve preservation; food scientists know what pH a food ought to be and measure this accurately. This is not always easy because food may consist of solid pieces suspended in a paste. The food scientist separates and tests each component precisely. They will assess results against expected standards and report their findings.

Food scientists solve problems with food and its manufacture. For example, carbon dioxide gas is used to extend shelf life of bread packaged in sealed wrapping. One manufacturer found that bread was going mouldy before the use-by date. A food scientist found that the packaging gas did not have a sufficiently high concentration of carbon dioxide.

In another example, sorbic acid is added to preserve soft drinks. A manufacturer received complaints that the drink tasted of petrol. A food scientist showed that the concentration of preservative was too low to kill mould. Small amounts of mould had used the sorbic acid for food, producing a chemical that tasted and smelled like petrol.

 Resources

References and further reading

Bennett, H. (2019) Magnificent molecules: Ethanoic acid. *Education in Chemistry*. Available at https://eic.rsc.org/magnificent-molecules/ethanoic-acid/3010392.article

Bennett, H. (2019) Magnificent molecules: Hydrochloric acid. *Education in Chemistry*. Available at https://eic.rsc.org/magnificent-molecules/hydrochloric-acid/3010539.article

Eason, M. (2016) Acids and bases: creating solutions. *Education in Chemistry*. Available at: https://eic.rsc.org/cpd/acids-and-bases/2000001.article

Hennah, N. (2018) How to teach acids, bases and salts. *Education in Chemistry*. Available at: https://eic.rsc.org/cpd/acids-bases-and-salts/3009612.article

Johnstone, A.H. (1991) Why is science difficult to learn? Things are seldom what they seem. *Journal of Computer Assisted Learning*, 7: 75–83

Kind, V. (2004) *Beyond appearances: Students' misconceptions about basic chemical ideas*. Available at https://edu.rsc.org/resources/beyond-appearances/2202.article

de Vos, W. and Pilot, A. (2001) Acids and bases in layers: The stratal structure of an ancient topic. *Journal of Chemical Education*, 78(4).

Websites

Making a Universal Indicator, microscale practical: http://science.cleapss.org.uk/Resource-Info/PP057-Making-a-Universal-Indicator-A-Microscale-Approach.aspx

Approaching acids, bases and indicators inclusively: https://edu.rsc.org/ideas/how-to-teach-acids-bases-and-indicators-for-all/3009995.article

Making red cabbage indicator: https://edu.rsc.org/resources/making-a-ph-indicator/422.article

Creating a rainbow in a burette: http://resources.schoolscience.co.uk/Salters/chemclub2_3.html

Creating a 'rainbow' in a glass tube with stoppers at each end: www.rsc.org/learn-chemistry/resource/res00000700/universal-indicator-rainbow?cmpid=CMP00005976

Burning magnesium in oxygen: https://edu.rsc.org/resources/the-change-in-mass-when-magnesium-burns/718.article

Microscale versions of burning magnesium in air using bottle caps instead of a crucible can be found on the CLEAPSS and SSERC websites at: http://science.cleapss.org.uk/Resource-Info/Finding-the-formula-of-magnesium-oxide.aspx

Reacting elements with oxygen: https://edu.rsc.org/resources/reacting-elements-with-oxygen/705.article

Demonstrating the displacement of hydrogen by magnesium: www.rsc.org/learn-chemistry/resource/res00001916/the-rate-of-reaction-of-magnesium-with-hydrochloric-acid?cmpid=CMP00006119

Practicals to prepare soluble salts using insoluble metal oxides and carbonates reacted with dilute acid. (The bases that are reacted here are copper(II) oxide and magnesium carbonate): https://edu.rsc.org/resources/preparing-salts-by-neutralisation-of-oxides-and-carbonates/1762.article

Microscale reaction of sulfuric acid and copper(II) oxide: http://science.cleapss.org.uk/Resource/Hydrated-copper-sulfate-preparation-Microscale-method.vid

Bath bomb instructions in the Salters' Chemistry Club Handbook, volume 2, experiment number 3: http://resources.schoolscience.co.uk/Salters/chemclub2_3.html

An interactive online activity to prepare students for carrying out a titration: www.rsc.org/learn-chemistry/resource/res00002077/titration-screen-experiment

Titrating sodium hydroxide with hydrochloric acid: https://edu.rsc.org/resources/titrating-sodium-hydroxide-with-hydrochloric-acid/697.article

UK Uplands Waters Monitoring Network: https://uwmn.uk/

Other resources

The STEM learning web site provides access to resources relating to acids, alkalis, bases and indicators: www.stem.org.uk

Instructions for making soaps and detergents: www.rsc.org/learn-chemistry/resource/res00001746/making-soaps-and-detergents?cmpid=CMP00005261 (Note: castor oil is only essential to make a detergent and a soap from the same oil, otherwise other oils can be substituted.)

8 Redox reactions and electrolysis

Nicklas Lindström and Katherine Aston

Introduction

Studying the reactions known as redox reactions and electrolysis enables learners to explore the relationship between chemistry and technology. Using these reactions, scientists have developed powerful techniques used globally to extract valuable elements from natural materials, especially mineral ores. Redox chemistry is applied in many other settings including protecting cars, ships, aeroplanes and many other metallic objects (such as lampposts, bolts, screws, nails and railings) from corrosion; and electrochemical cells, including hydrogen fuel cells and batteries, supply power to gadgets of all sizes. In organic chemistry, many biochemical processes in living organisms rely on redox reactions controlling metabolic pathways.

This chapter discusses redox reactions and electrolysis. Simply naming the topic introduces vocabulary which students have not met before, and language is an important theme in this chapter. Redox combines abbreviations of **red**uction and **ox**idation. The term emphasises that reduction and oxidation reactions always occur together. Redox reactions involve transfer of electrons between substances. Electrolysis (from *electro* generally meaning 'electric phenomena' and *lysis* meaning 'splitting up') is a type of redox reaction. In electrolysis, electricity brings about the redox reaction. Understanding redox reactions is a pre-requisite for understanding electrolysis.

The chapter begins with an introduction to reduction and oxidation, and to representing redox reactions using equations. It describes familiar reactions of metals with oxygen, water, acids and metal salts as redox reactions and describes how these reactions can be used to place metals in order of reactivity, in a reactivity series. Redox reactions for metal extraction require different techniques depending on the reactivity of the metal. Electrolysis is introduced as a redox reaction that requires an electrical current. Electrolysis is used to extract highly reactive metals such as aluminium. While electrolysis uses an electrical current to enable a redox reaction, the reciprocal process is also possible: redox reactions can be used to generate an electrical current – in fuel cells, for example.

8 Redox reactions and electrolysis

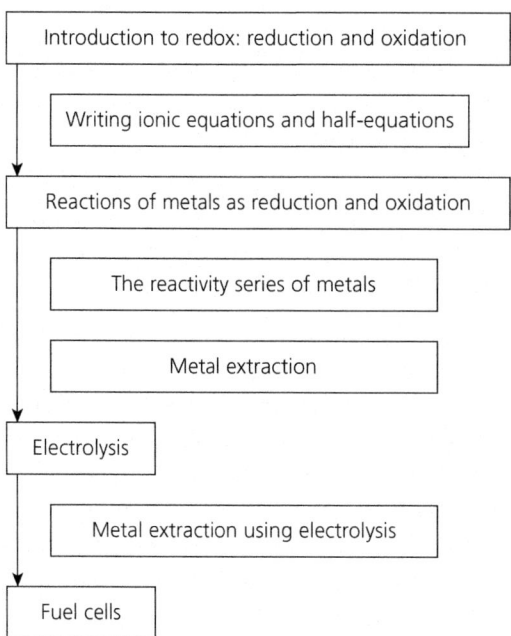

Figure 8.1 Redox reactions and electrolysis: teaching sequence overview

Redox reactions

Redox reactions involve transfer of electrons between substances. A substance *accepting electrons* has been 'reduced'. A substance *transferring electrons* is 'oxidised'. Therefore, oxidation and reduction always occur together, and neither can happen independently of the other. Chemists often represent oxidation and reduction in two half-equations. One half-equation shows the reduction process, while the other shows the oxidation process. Two half-equations add together to make an overall equation for a redox reaction.

Reduction and oxidation in reactions of metals

Reduction and oxidation occur in reactions of metals with oxygen, acids and water (among other reagents). In these reactions, metals are oxidised, meaning that each metal atom transfers one or more electrons to another atom or molecule, to create positive metal ions. A half-equation representing this, using M as the symbol for any metal transferring one electron only, is:

$$M(s) \rightarrow M^+(aq) + e^-$$

metal atom → metal ion + one electron

Extracting metallic elements from compounds of metals requires the opposite process: that is, reduction. Positively charged metal ions each accept one or more electrons from another atom or molecule to form metal atoms. A half-equation representing this, with M⁺ as an ion of any metal is:

$$M^+(s) + e^- \rightarrow M(s)$$

metal ion + one electron → metal atom

The choice of extraction technique is determined by the reactivity of a metal. We gain understanding of this by examining how readily a metal reacts with oxygen, water and acids.

When a highly reactive metallic element reacts with a compound containing ions of a less reactive metal, displacement may occur, due to energetics associated with the difference in reactivity. Ions of less reactive metals each accept one or more electrons from atoms of the more reactive metal, so are reduced. Simultaneously, atoms of the more reactive metal are oxidised, as they transfer one or more electrons (see half-equations above and below) to ions of the less reactive metal. Macroscopically, we observe the less reactive metal being displaced, which means it forms as an element. For example, when adding magnesium ribbon to copper(II) sulfate solution, magnesium dissolves (atoms oxidise to form ions), and copper metal is displaced (ions reduce to form atoms). Copper forms as an orange-red solid. Half-equations are:

$$Mg(s) \rightarrow Mg^{2+}(aq) + 2e^- \quad \text{OXIDATION}$$

magnesium atoms → magnesium ions + 2 electrons

$$Cu^{2+}(aq) + 2e^- \rightarrow Cu(s) \quad \text{REDUCTION}$$

copper ions + 2 electrons → copper atoms

Combining these into one equation gives:

$$Mg(s) + Cu^{2+}(aq) \rightarrow Mg^{2+}(aq) + Cu(s)$$

This equation is more detailed than is normally required for students aged 14–16 but illustrates the reaction that occurs. Sulfate ions are not included as they are not involved: they are spectator ions that remain unchanged. Electrons are not shown because charges balance between the positively charged ions and neutral atoms on either side of the arrow. A full equation for this type of reaction presented to 14–16-year-olds is:

copper sulfate + magnesium → magnesium sulfate + copper

$$CuSO_4(aq) + Mg(s) \rightarrow MgSO_4(aq) + Cu(s)$$

This equation requires care, as it gives the *completely incorrect* impression that copper and magnesium 'swap partners' during the displacement reaction, so molecules of copper sulfate change to molecules of magnesium sulfate. Sulfate ions remain separate from copper ions in the original solution and from magnesium ions after the reaction. No bond exists between copper ions and sulfate ions when they are in solution. No bond forms between magnesium ions and sulfate ions in solution. Copper sulfate, magnesium sulfate and all other salts *do not exist in solution as molecules* such as $CuSO_4$ or $MgSO_4$. Saying or implying that displacements involve metals swapping partners also implies that particles can make choices. Particles do not have brains, are not alive, cannot and do not 'hold hands' or make partnerships by choice (see Chapter 5). The reaction proceeds because energetic considerations make it favourable for magnesium atoms to oxidise, transferring electrons and to copper ions, which accept electrons, and are reduced. The reason why two electrons are involved here relies on the electron configurations of these elements (see Chapter 4).

> **Enrichment**
>
> Copper has been known since around 10 000 BCE: ancient societies shaped native copper into axes, arrowheads, shields, ornaments and jewellery. Extraction of metals from ores developed in ancient civilisations from around 5000 BCE. An archaeological site at Pločnik in Serbia revealed evidence of copper smelting by the Vinča culture from around 5400 BCE, the first society to produce copper from an ore. The method involved low-temperature furnaces with pipe-like vents and a chimney. The Vinča experimented with copper ores including green malachite, bright blue azurite (forms of copper carbonate) and red cuprite (copper(I) oxide). Extracting metals from ores involves reducing the metal by reacting the ore with a *reducing agent*, such as charcoal, to reduce metal ions in the ore to metal atoms. Copper is widely used today in, for example, wiring, piping, coinage, cooking equipment and for medical purposes.

Electrolysis

Electrolysis breaks down an ionic substance by passing an electrical current through it. Students are used to thinking of electrical current as a flow of electrons. However, in electrolysis, the current *through the ionic substance* is a flow of ions. Ions must be free to move to carry an electrical current, so ionic substances must be melted or dissolved for electrolysis to occur. Electrolysis involves redox reactions. The process is used to extract highly reactive metals from compounds.

Students' prior knowledge and experience

Concepts of redox reactions and electrolysis are introduced in chemistry curricula at 14–16. These topics build on students' knowledge of electricity from physics, and previous experiences of chemical changes (see Chapter 2 and Chapter 7).

As a pre-requisite, students need to understand that matter is made from particles that are invisible to the eye. That these particles can be atoms, ions or molecules and that atoms have sub-atomic structures comprising protons, neutrons and electrons is also necessary prior knowledge. Thus, the focus in this topic is on sub-microscopic understanding of matter (see Chapter 1). Particle theory is introduced to students aged 11–14 and developed in curricula for 14–16-year-olds. From early secondary physics, students know that objects with opposite charges attract each other, while objects with like charges tend to repel each other. They know that particles of liquids and aqueous solutions have greater freedom of movement than those in solids, which vibrate around a fixed position. These ideas support understanding of electrolysis, which occurs in liquid and aqueous states where ions are free to move. Students learn that electricity is a flow of sub-atomic particles known as electrons, and that electrical current is a flow of electrons through a complete circuit. In electrolysis, students encounter electrical currents in which ions carry electrical charge (see Section 8.6).

From early secondary chemistry, students learn to represent reactions using symbol equations (see Chapter 2). Students aged 11–14 describe reactions of metals with acids. Aged 14–16, they identify these as redox reactions and describe reduction and oxidation processes. Aged 11–14, students may use oxidation to describe substances bonding with oxygen when, for example, metals or fuels react with oxygen. Aged 14–16, the concept of oxidation is extended to mean loss of electrons.

Students will be aware of the periodic table and that chemical elements have different properties (see Chapter 4). Specifically, metallic elements, such as gold and silver, are very unreactive (that is, they do not react easily with oxygen or acids) while others, such as magnesium and zinc, react relatively easily. Students learn that metals can be arranged in a reactivity series based on their readiness to form compounds and predict how a specific metal may react with acid, water, oxygen and metal salts. They know carbon is a *reducing agent* used to extract some metals from metal ores. Aged 14–16, students describe these reactions in terms of reduction and oxidation.

Before teaching redox and electrolysis, students should be familiar with symbols and charges for common ions, and confident in writing balanced chemical equations (see Chapters 2 and 5). As redox and electrolysis build on these aspects of the 14–16 curriculum, a recommendation is to teach these topics towards the end of students' courses.

Progression

Beyond 16, chemistry students are introduced to a more sophisticated model for redox reactions, known as the oxidation state model. This allows for redox reactions involving covalent bonds, in which electrons partially transfer between atoms during a reaction. Note that Chapter 5 advises against using dot-and-cross diagrams to show ionic bonding, as in this case they tend to over-emphasise electron transfer, leading students to the impression that the transfer itself is the ionic bond. Here, the diagrams illustrate the principle of electron transfer occurring in redox reactions, which represent a broad range of contexts. It is not intended to represent ionic bonding between two individual ions. Students learn to combine half-equations to write an overall equation for a redox reaction.

Students build on their knowledge of electrolysis to construct, describe and draw electrochemical cells. In some cells, including those used for electrolysis, an electrical current causes a chemical reaction. Other cells, like those used in laptop computer batteries, use a chemical reaction to generate a current which can be used to power devices. The voltage a cell produces relates to differences in reactivity between the two metals used. Post-16 students learn that electrode potentials quantify the relative reactivity of metals. Students calculate expected voltages for a cell, or predict the direction of a redox reaction, by comparing the electrode potential for each metal.

In advanced organic chemistry, students learn standard chemical reactions for oxidation and reduction of carbon compounds, including hydrocarbons and carbonyl compounds. Redox reactions are central to many biochemical processes, including photosynthesis and respiration, which feature in advanced biology courses.

 ## 8.1 Introduction to redox reactions: reduction and oxidation

Magnesium reacts violently with atmospheric oxygen, producing a white smoke of magnesium oxide (see Chapter 2). The light is so bright that students should not look directly at burning magnesium. An equation representing the reaction is:

$$Mg(s) + O_2(g) \rightarrow 2MgO(s)$$

magnesium + oxygen → magnesium oxide

8.1 Introduction to redox reactions: reduction and oxidation

Chemists say this is an oxidation reaction in which magnesium is oxidised, meaning 'reacted with oxygen'. This is relatively easy for students to grasp as oxidation points to oxygen. However, chemists also describe magnesium as oxidised when it reacts with other elements, such as chlorine, which does not contain or involve oxygen. The terms oxidation and oxidised extend beyond reactions with oxygen. To understand this, consider what happens to two electrons in the outer shell of each magnesium atom. When magnesium oxide forms, in effect, two electrons transfer from a magnesium atom to an oxygen atom. In practice, the reaction does not proceed directly by electron transfer: there are complex steps involved. Nevertheless, the overall effect is that each magnesium atom becomes a magnesium ion, (formula: Mg^{2+}) by electron loss:

$$Mg(s) \rightarrow Mg^{2+}(s) + 2e^-$$

Oxidation describes *any reaction involving electron transfer or electron donation*. For ease of reference, this is often referred to as *electron loss*. In this example, magnesium atoms are oxidised, and act as a reducing agent, transferring electrons to oxygen atoms. Whenever a substance transfers or loses electrons to another particle, chemists say it is oxidised. Note that electrons are never completely lost but donated to another atom or molecule.

Oxidation **is L**oss of electrons

In the reaction, each oxygen atom accepts two electrons, becoming an oxide ion:

$$\tfrac{1}{2}O_2(g) + 2e^- \rightarrow O^{2-}(s)$$

Electron gain is reduction, so oxygen atoms *accepting electrons* are described as reduced. This language appears counterintuitive: atoms gain electrons while the number of protons and neutrons are unchanged, so nothing is reduced in number. Chemists say reduction because this term originally meant reducing the amount of oxygen in compounds. Reduction applied especially when metals were being extracted from ores, which are often compounds including oxygen. Reduction now means *electron gain or electron acceptance*. Whenever a substance gains or accepts electrons, chemists say it is reduced.

Reduction **is G**ain of electrons

Mnemonics may help students remember scientific concepts. In this instance:

OILRIG = **O**xidation **is L**oss, **R**eduction **is G**ain

8 Redox reactions and electrolysis

There are many ways of describing and representing reduction and oxidation. Table 8.1 shows these terms relating to oxygen and magnesium in the reaction above.

Table 8.1 Oxidation and reduction in the reaction of magnesium and oxygen

Oxidation	Reduction
Magnesium gains oxygen.	Oxygen gains magnesium
Magnesium is oxidised.	Oxygen is reduced
Magnesium is a reducing agent.	Oxygen is an oxidising agent
Magnesium atoms transfer electrons to oxygen atoms, becoming magnesium ions.	Oxygen atoms accept electrons to become oxide ions.
$Mg(s) \rightarrow Mg^{2+}(s) + 2e^-$	$½ O_2(g) + 2e^- \rightarrow O^{2-}(s)$

These descriptions emphasise the reaction in phrases that are broadly equivalent in meaning. Chemists move fluently between these descriptions (see Figure 1.1). To gain fluency, give students opportunities to practise describing reactions using these phrases, alternating between them. Redox reactions can be introduced by getting students to identify which chemical species (atom, ion or molecule) are reduced and oxidised, describing this as electron donation/acceptance, or gain/loss. These fundamental ideas help students to write half-equations for reduction and oxidation.

It is helpful to introduce redox reactions by using reactions familiar to students. This allows students to focus on developing descriptions of redox reactions, rather than simultaneously understanding new reactions. Examples are given in Chapter 2.

KEY ACTIVITY

Reaction between magnesium and oxygen

The reaction between magnesium and oxygen is an engaging demonstration that introduces redox reactions. Full details are available at https://chem.rutgers.edu/cldf-demos/1016-cldf-demo-burning-magnesium

The demonstration describes a familiar reaction using redox terminology. To check prior knowledge, demonstrate the reaction and ask students to write the equation.

$$Mg(s) + O_2(g) \rightarrow 2MgO(s)$$

magnesium + oxygen → magnesium oxide

Students can draw dot-and-cross diagrams representing the formation of magnesium and oxygen ions from their atoms. Note that in reality, all electrons are identical, and atoms do not have visible shells: these are representations only (see Chapter 5). These can be organised to show the half-equations in symbols underneath, as shown in Figures 8.1 and 8.2.

8.1 Introduction to redox reactions: reduction and oxidation

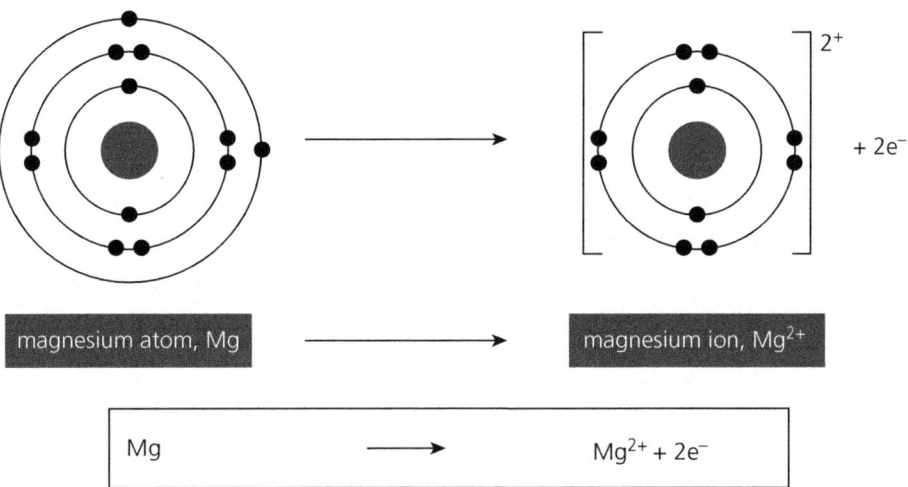

Figure 8.2 Formation of a magnesium ion and the half-equation for this process

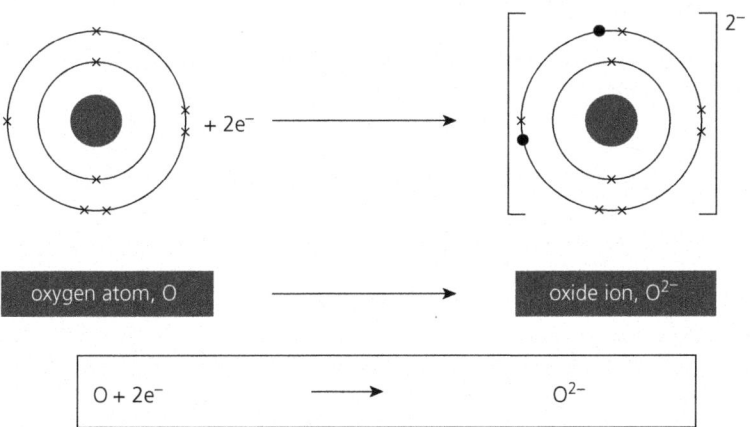

Figure 8.3 Formation of an oxide ion and the half-equation for this process

The reaction begins with oxygen molecules (O_2) rather than oxygen atoms. These split apart in an initial step:

$$O_2(g) \rightarrow 2O(g)$$

Each atom accepts two electrons from one magnesium atom, forming oxide ions, O^{2-}. The charge is −2 because the nuclei are unchanged, each with eight protons. The addition of two extra electrons means there are ten negative charges, an excess of −2. In equations, by convention, one oxygen atom may be shown as half of an oxygen molecule, $½O_2$.

8 Redox reactions and electrolysis

> Each magnesium atom donates two electrons forming an Mg^{2+} ion. Magnesium atoms each have twelve protons in their nuclei. Donating two electrons means that there are ten electrons surrounding the atomic nuclei, creating a +2 charge overall.
>
> ## Questions to ask
> - Which element is oxidised? How do you know?
> - Which element is reduced? How do you know?
> - Which element is the reducing agent?
> - Write equations showing what happens to magnesium atoms and oxygen atoms in the reaction.
> - What is the overall equation for the reaction? How does this show redox?
>
> Discussing the historical meaning of oxidation and reduction linked to oxygen (see Table 8.1) can help students see how these terms evolved. A more accurate representation is oxidation as electron loss (donation) and reduction as electron gain (acceptance). The OILRIG mnemonic can be applied to describe the reaction using oxidation and reduction.

As a second example, the reaction of magnesium with chlorine is also a redox reaction:

$$\text{magnesium} + \text{chlorine} \rightarrow \text{magnesium chloride}$$
$$Mg(s) + Cl_2(g) \rightarrow MgCl_2(s)$$

In this reaction, magnesium is oxidised, and chlorine is reduced.

Students must recognise which element loses (donates) electrons and which element gains (accepts) electrons, then describe a reaction stating which species are oxidised and reduced. Drawing the formation of ions from their atoms is a helpful step. For this reaction, diagrams showing the magnesium atom becoming an ion are above. Chlorine gas exists as Cl_2 molecules. These split apart in a first step:

$$Cl_2(g) \rightarrow 2Cl(g)$$

Chlorine atoms become chloride ions (symbol: Cl^-). Figure 8.3 shows diagrams representing this, along with the half-equation.

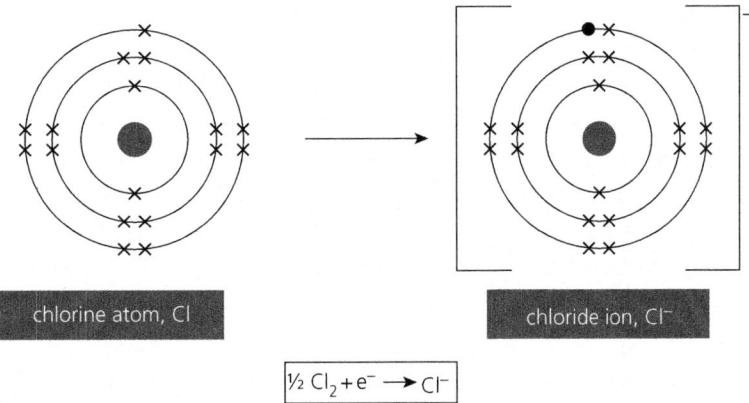

½ Cl$_2$ + e$^-$ ⟶ Cl$^-$

Figure 8.4 Formation of a chloride ion and the half-equation for this process

Each chlorine atom accepts only one electron, due to its electron configuration (see Chapter 4). Each magnesium atom transfers an electron to two chlorine atoms. Oxidation applies to electrons, so magnesium is oxidised, acting as a reducing agent. Reinforce these ideas using formation of other ionic compounds from their elements, for example, sodium chloride and aluminium oxide.

 Writing ionic equations and half-equations

Ionic equations show how ions form and react. In a redox reaction, chemists refer to ionic equations as half-equations. Half-equations represent a redox reaction as the sum of two 'halves', a reduction half and an oxidation half. For example, in the reaction between magnesium and oxygen, this ionic half-equation shows what happens to magnesium:

$$Mg(s) \rightarrow Mg^{2+}(s) + 2e^-$$

While this half-equation shows what is happening to oxygen:

$$\tfrac{1}{2}O_2(g) + 2e^- \rightarrow O^{2-}(s)$$

State symbols show precisely what happens. One 'half-reaction' cannot occur alone, so both half-equations are needed to express what is happening during a reaction. This emphasises that reduction and oxidation always occur together. There are four steps to writing a balanced ionic half-equation. This is exemplified for the reaction between zinc and silver nitrate in Table 8.2.

8 Redox reactions and electrolysis

Table 8.2 Writing balanced half-equations for a redox reaction

Step		Example	Prior knowledge and teaching notes
1	Identify ions present in the reagents and products. Write an ionic equation showing reactants and products that change.	$Zn(s) + 2Ag^+(aq) + 2NO_3^-(aq) \rightarrow Zn^{2+}(aq) + 2NO_3^-(aq) + 2Ag(s)$ $\cancel{NO_3^-(aq)}$ is unchanged, hence is ruled out. $Zn(s) + 2Ag^+(aq) \rightarrow Zn^{2+}(aq) + 2Ag(s)$	Nitrate ions appear on both sides. They are spectator ions that do not take part in the reaction. This is the overall ionic equation without spectator ions.
2	Identify ions that are oxidised and reduced in the reactants and products.	$Zn(s) \rightarrow Zn^{2+}(aq)$ $2Ag^+(aq) \rightarrow 2Ag(s)$	Consider each species in turn. Zinc atoms transfer electrons to silver ions, becoming positive zinc ions. This is oxidation. Silver ions accept electrons to become silver atoms. Gain of electrons is reduction.
3	Balance and simplify atoms/ions in the half-equations.	Reduction $2Ag^+(aq) \rightarrow 2Ag(s)$ $Ag^+(aq) \rightarrow Ag(s)$ Oxidation $Zn(s) \rightarrow Zn^{2+}(aq)$	The equation is balanced by coefficient 2 on both sides of the arrow. This simplifies to one ion becoming one atom. This equation is balanced showing one atom becoming one ion.
4	Balance charges in the reactions by adding in electrons.	Reduction $Ag^+(aq) \rightarrow Ag(s)$ Charge +1 0 Add one electron on the left to balance. $Ag^+(aq) + e^- \rightarrow Ag(s)$ Charge +1 −1 0 Oxidation $Zn(s) \rightarrow Zn^{2+}(aq)$ Charge 0 +2 Add two electrons on the right to balance. $Zn(s) \rightarrow Zn^{2+}(aq) + 2e^-$ Charge 0 +2 $2 \times (-1) = -2$	The number of particles is balanced, but the charges are unbalanced. Adding electrons to the positively charged side balances the charges. Electrons are negative, so adding electrons balances the positive charge.
The half-equations are		Oxidation $Zn(s) \rightarrow Zn^{2+}(aq) + 2e^-$	Reduction $Ag^+(aq) + e^- \rightarrow Ag(s)$

8.2 Writing ionic equations and half-equations

KEY ACTIVITY

Practising writing half-equations

This activity takes students through systematic practice of this task: https://edu.rsc.org/ideas/how-to-sequence-and-segment-your-teaching-of-ionic-equations/4011307.article

Step 1: Learn symbols and charges of ions

Students need to know symbols and charges of metal ions from the reactivity series (Section 8.4) and these non-metal ions: oxide, hydroxide, halides (chloride, bromide, iodide), sulfate, nitrate, carbonate and hydrogencarbonate.

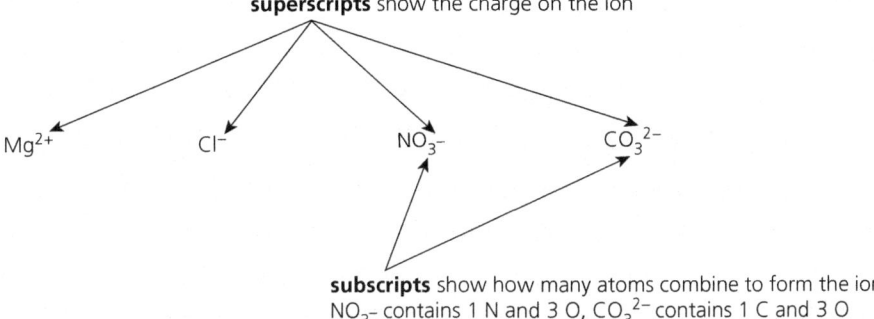

Figure 8.5 Understanding symbols for ions

Step 2: Learn to 'split' an ionic compound into its ions

Once students have secure knowledge of ion symbols and charges, the next step is learning to split the formula of an ionic compound into constituent ions. Examples (not exhaustive) are shown in Table 8.3.

Table 8.3 Ionic compounds and their constituent ions

Compound		Metal ion(s)	Anion(s)
Compounds with one metal ion and one anion			
Sodium chloride	NaCl	Na$^+$	Cl$^-$
Potassium bromide	KBr	K$^+$	Br$^-$
Lithium nitrate	LiNO$_3$	Li$^+$	NO$_3^-$
Silver nitrate	AgNO$_3$	Ag$^+$	NO$_3^-$
Calcium carbonate	CaCO$_3$	Ca^{2+}	CO$_3^{2-}$
Magnesium sulfate	MgSO$_4$	Mg^{2+}	SO$_4^{2-}$
Copper(II) oxide	CuO	Cu^{2+}	O^{2-}
Compounds with one metal ion and two anions			
Magnesium chloride	MgCl$_2$	Mg^{2+}	Cl$^-$ Cl$^-$
Zinc nitrate	Zn(NO$_3$)$_2$	Zn^{2+}	NO$_3^-$ NO$_3^-$

Calcium chloride	$CaCl_2$	Ca^{2+}	Cl^- Cl^-
Iron(II) nitrate	$Fe(NO_3)_2$	Fe^{2+}	NO_3^- NO_3^-
Compounds with two metal ions and one anion			
Sodium carbonate	Na_2CO_3	Na^+ Na^+	CO_3^{2-}
Lithium oxide	Li_2O	Li^+ Li^+	O^{2-}
Potassium sulfate	K_2SO_4	K^+ K^+	SO_4^{2-}
Compounds with one metal ion and three anions			
Aluminium nitrate	$Al(NO_3)_3$	Al^{3+}	NO_3^- NO_3^- NO_3^-
Iron(III) chloride	$FeCl_3$	Fe^{3+}	Cl^- Cl^- Cl^-
Aluminium hydroxide	$Al(OH)_3$	Al^{3+}	OH^- OH^- OH^-

Table 8.3 reminds students to look at subscripts when breaking down ions. For example, $AgNO_3$ becomes $Ag^+ + NO_3^-$, while $Zn(NO_3)_2$ becomes $Zn^{2+} + 2NO_3^-$. Ions such as OH^-, SO_4^{2-} and NO_3^- are polyatomic, meaning made of more than one atom. Students may incorrectly break down polyatomic ions. As a rule, a formula that is bracketed represents one ion, for example, nitrate (NO_3^-).

Students can practice this using ion cards such as those available here: www.rsc.org/Education/Teachers/Resources/Aflchem/resources/28/28%20resources/28-2%20Formula%20cards.pdf

Step 3: Apply the first two steps in writing half-equations

Once the process of writing half-equations (Table 8.2) has been demonstrated, students can be asked to write half-equations for familiar reactions (see Table 8.3). This process can be reinforced throughout the teaching of redox reactions and electrolysis by writing half-equations for each new reaction.

 The reactions of metals as reduction and oxidation

Reactions of metals with oxygen, acids and water

Aged 11–14, students describe characteristic reactions of metals with oxygen, acids and water (see Chapter 4 and Chapter 7). Aged 14–16, students learn these reactions are examples of reduction and oxidation. To reconsider these reactions in redox terms, students need to identify species that are reduced and oxidised and write

8.3 The reactions of metals as reduction and oxidation

half-equations representing these. Table 8.4 identifies reduction and oxidation processes in reactions of metals with oxygen, acid and water.

Table 8.4 Describing metal reactions as reduction and oxidation (redox) reactions

Reaction of metals	Overall reaction	Oxidation	Reduction
With oxygen	metal + oxygen → metal oxide		$O_2(g) \rightarrow O^{2-}(s)$
With acid	metal + acid → salt + hydrogen	$M(s) \rightarrow M^{n+}(aq) + ne^-$	$2H^+(aq) + 2e^- \rightarrow H_2(g)$
With water	metal + water → metal hydroxide + hydrogen		

Oxidation equations are given for a metal M which forms a positive ion of charge $+n$

Analysing these reactions as redox reactions reveals an underlying connection between them: they are all oxidations of metals, with the metals as reducing agents. In each reaction, metal atoms donate electrons to form positive metal ions. The oxidation half-equation is constant, whether the metal reacts with oxygen, an acid, or water. For example, magnesium is oxidised in reaction with oxygen, hydrochloric acid, sulfuric acid, or water:

$$Mg(s) \rightarrow Mg^{2+}(s) + 2e^-$$

Differences arise from species being reduced and compounds formed. When a metal reacts with oxygen, oxygen atoms are reduced. Oxygen molecules split apart, and each oxygen atom accepts two electrons to form negatively charged oxide ions (O^{2-}). The half-equation is:

$$O_2(g) + 4e^- \rightarrow 2O^{2-}(s)$$

Negatively charged oxide ions bond with positive metal ions, forming metal oxides.

When a metal reacts with acid, hydrogen ions (H^+) are reduced. Each hydrogen ion accepts one electron forming a hydrogen atom, and two hydrogen atoms bond together forming hydrogen gas (H_2) molecules. The half-equation for this reaction is:

$$2H^+(aq) + 2e^- \rightarrow H_2(g)$$

The positively charged metal ions exist in aqueous solution. This reaction is an example of a displacement reaction, as elemental hydrogen is displaced from the solution as a gas. When Group 1 metals react with water (see Chapter 4), water molecules act as a (very weak) acid, forming hydrogen ions (H^+) and hydroxide ions (OH^-). In the reaction, hydrogen ions (H^+) are reduced, producing hydrogen gas.

Questions to ask

→ What do you notice about what happens to metal atoms in these reactions?

8 Redox reactions and electrolysis

→ What happens to non-metal ions in these reactions?
→ Do all metals react in the same way?
  What do you notice about the pattern of reactions?

8.4 The reactivity series

Aged 11–14, students learn that magnesium and iron react easily with oxygen, water and acids, while other metals, like copper, do not. This leads to listing metals in reactivity order, in a reactivity series. The reactivity series of metals with oxygen, water, and acids are almost identical (Table 8.5). Carbon, a non-metal, is often included in reactivity series, because it is used in extracting metals that have low reactivity. Carbon does not react with dilute acids or water and will only react with oxygen under high temperatures and pressure, hence it is shown italicised and bracketed in Table 8.5.

Table 8.5 Reactivity series of metals and carbon with oxygen, water and dilute acids

Element	Reactivity	Reaction with oxygen	Reaction with water	Reaction with dilute acid
Potassium	most reactive	React quickly at room temperature	React with cold water	React violently with dilute acid
Sodium				
Lithium				
Calcium				
Magnesium		React readily when heated and slowly at room temperature	React readily with hot steam	React with dilute acid
Aluminium				
*(Carbon)**				
Zinc				
Iron			Less or no reaction with hot steam	
*(Hydrogen)**				
Copper				Do not react with dilute acid
Silver		Little or no reaction when heated.		
Gold	least reactive			

*Carbon and hydrogen are non-metals used in extracting metals that are low in reactivity. They are included in reactivity series by convention.

8.4 The reactivity series

The order of reactivity is consistent because all these reactions involve metal oxidations. One or more electrons are removed from metal atoms to form positively charged metal ions. For a metal M which forms a positive ion with charge $n+$ the half-equation for oxidation is:

$$M(s) \rightarrow M^{n+}(aq) + ne^-$$

Electrons are attracted by positively charged nuclei. Oxidising metal atoms by transfer of electrons requires energy. The amount of energy needed to transfer electrons differs for each metal. The energy required is one factor determining whether a reaction proceeds at a given temperature (Chapter 6). The reactivity series reflects the relative amounts of energy required to oxidise metal atoms. Metals higher in the reactivity series, like potassium and sodium, require relatively little energy to oxidise. They oxidise easily, reacting with oxygen, acids and water at room temperature. Metals low in the reactivity series, like silver and gold, require much more energy to oxidise. High amounts of energy are less readily available at room temperature and pressure, so these metals react with oxygen, acids and water slowly, at higher temperatures, or not at all.

The reverse is true for reduction of metal ions. If a metal atom easily donates electrons, becoming a metal ion, reversing this process by adding electrons to form metal atoms requires large amounts of energy. The reduction half-equation is:

$$M^{n+}(aq) + ne^- \rightarrow M(s)$$

The more reactive a metal is, the less easily its ions can be reduced.

The reactivity series can be used to predict how easily a metal will corrode, as corrosion always involves oxidation of a metal. Silver and gold are used for decorative objects like jewellery because they corrode very slowly or not at all, ensuring the jewellery stays shiny and retains its value. Alternatively, a thin layer of silver or gold may be used to cover a more reactive metal, forming a physical barrier to corrosion. This process is called plating.

The relative reactivity of a metal can be worked out by studying its reactions, with oxygen, water, dilute acids and the compounds of other metals. The feasibility of these reactions depends on the energy changes involved in reducing and oxidising both the metal reactant and the other reactant. For example, the same metal will react more readily with dilute acid than with water. This is because acids are more easily reduced than water.

Displacement reactions

Displacement reactions of metals involve reduction and oxidation. The literal meaning of 'displacement' is 'pushing out', as described above. Another example involves addition of iron filings to black copper(II) oxide powder. When heated, iron bonds with oxygen forming iron oxide, 'pushing out' copper which forms copper metal. Iron displaces copper.

$$Fe(s) + CuO(s) \rightarrow FeO(s) + Cu(s)$$

Displacement reactions involve two metals, one of which is reduced, and the other oxidised. Analysing this reaction shows that:

Iron is oxidised: $\quad Fe(s) \rightarrow Fe^{2+}(s) + 2e^-$

Copper is reduced: $\quad Cu^{2+}(s) + 2e^- \rightarrow Cu(s)$

From the reactivity series, we predict that iron requires less energy to oxidise than copper. In general, a metal is oxidised when reacting with a compound containing a metal below it in the reactivity series. The lower-placed metal will be reduced and displaced. For example, when iron filings are heated with magnesium oxide, no reaction occurs: magnesium is higher in the reactivity series than iron. However, if aluminium powder is heated with iron(III) oxide, a vigorous reaction occurs in which molten iron metal is released (see Enrichment).

Students need to use the reactivity series to predict how specific metals will react with oxygen, air and water, and other metal compounds, and to use observations from reactions to deduce a reactivity series.

Enrichment

The thermite (or thermit) reaction was discovered in 1893 by Hans Goldschmidt, a German chemist who studied with Robert Bunsen (inventor of the Bunsen burner). The name thermite means the dry, solid combination of two powders: a metal and a metal oxide. The metal (reducing agent) is normally aluminium, zinc or magnesium. In the oxide, a less reactive metal, such as iron or copper, is bonded with oxygen. Thermites are named after the metal oxide, hence iron thermite (iron(III) oxide), and copper thermite (copper(II) oxide). A promoter, often barium peroxide, is added to facilitate ignition via a magnesium-ribbon fuse. Iron thermite (iron(III) oxide with aluminium) is most frequently used. The reaction proceeds dramatically, producing flames, a huge amount of heat and molten iron. Instructions for carrying out this reaction are provided in Chapter 6.

The reaction has been used to weld railway lines using special apparatus that allowed molten iron to flow into the gap between two pieces of track. Thermite has military applications in grenades and artillery.

8.4 The reactivity series

KEY ACTIVITY

Metal reactions as redox reactions

Table 8.6 lists demonstrations and student experiments illustrating redox reactions involving metals. Students will have met many of these reactions in lower secondary chemistry, the challenge for students at 14–16 is to relate the patterns of reactions to the reactivity series and to analyse the processes of reduction and oxidation in each reaction. Students should experience a variety of these reactions, so they can practise making predictions from the reactivity series and conversely using observations to deduce a reactivity series. Students need to identify which reactants are reduced/oxidised, write half-equations, and examine patterns across reactions. Through discussion, students should recognise similarities in reactions, to guide them towards identifying that the metals are always oxidised. Half-equations are shown in the table. Chapters 2, 4 and 7 give further background information.

Some of these reactions are hazardous. Teachers intending to carry out these demonstrations must consult their employers' risk assessment (usually from CLEAPSS or SSERC), comply with all control measures and are strongly advised to practice before showing the reactions to a class.

8 Redox reactions and electrolysis

Table 8.6 Redox reactions of metals

Reactions of metals	Recommended activities	Specific oxidation/reduction	Full instructions
With oxygen	**Teacher demonstration** Reaction of lithium (or sodium) with air $2Li(s) + \frac{1}{2}O_2(g) \rightarrow Li_2O(s)$ $2Na(s) + \frac{1}{2}O_2(g) \rightarrow Na_2O(s)$ Reactions with chlorine emphasise that oxidation does not require oxygen.	Oxidised $Li(s) \rightarrow Li^+(aq) + e^-$ $Na(s) \rightarrow Na^+(aq) + e^-$ Reduced $\frac{1}{2}O_2(g) + 2e^- \rightarrow O^{2-}(s)$	https://edu.rsc.org/experiments/heating-group-1-metals-in-air-and-in-chlorine/732.article
With acid	**Teacher demonstration** Reaction of magnesium with hydrochloric acid $Mg(s) + 2HCl(aq) \rightarrow MgCl_2(aq) + H_2(g)$ **Class experiment** Reactions of magnesium, zinc, iron, tin and copper with acids (hydrochloric acid, nitric acid and sulfuric acid) $M(s) + 2H^+(aq) \rightarrow M^{2+}(aq) + H_2(g)$ Where M is the metal	Oxidised $Mg(s) \rightarrow Mg^{2+}(aq) + 2e^-$ Reduced $2H^+(aq) + 2e^- \rightarrow H_2(g)$ Oxidised $M(s) \rightarrow M^{2+}(aq) + 2e^-$ Reduced $2H^+ + 2e^- \rightarrow H_2$	https://edu.rsc.org/resources/the-reactivity-of-the-group-2-metals/409.article https://edu.rsc.org/resources/the-reaction-of-metals-with-acids/509.article
With water	**Teacher demonstration** Reactions of alkali metals with water $2Li(s) + 2H_2O(l) \rightarrow 2LiOH(aq) + H_2(g)$ $2Na(s) + 2H_2O(l) \rightarrow 2NaOH(aq) + H_2(g)$ $2K(s) + 2H_2O(l) \rightarrow 2KOH(aq) + H_2(g)$	Oxidised $Li(s) \rightarrow Li^+(aq) + e^-$ $Na(s) \rightarrow Na^+(aq) + e^-$ $K(s) \rightarrow K^+(aq) + e^-$ Reduced $2H^+(aq) + 2e^- \rightarrow H_2(g)$ Hydrogen ions come from water Hydroxide ions remain in solution	https://edu.rsc.org/resources/reactivity-trends-of-the-alkali-metals/731.article

8.4 The reactivity series

Reactions of metals	Recommended activities	Specific oxidation/reduction	Full instructions
With metal compounds (displacement)	**Teacher demonstration** Heating iron with copper(II) oxide and magnesium oxide Iron reacts with copper(II) oxide on heating $Fe(s) + CuO(s) \rightarrow FeO(s) + Cu(s)$ Iron does not react with magnesium oxide	Oxidised $Fe(s) \rightarrow Fe^{2+}(s) + 2e^-$ Reduced $Cu^{2+}(s) + 2e^- \rightarrow Cu(s)$	https://edu.rsc.org/resources/displacement-reactions-between-metals-and-their-salts/720.article
	Class experiment Reactions of metals and metal salts Copper, magnesium, zinc and lead are each reacted with copper(II) sulfate, lead nitrate, magnesium sulfate and zinc sulfate Example reaction $Mg(s) + CuSO_4(aq) \rightarrow MgSO_4(aq) + Cu(s)$	For example, oxidised $Mg(s) \rightarrow Mg^{2+}(aq) + 2e^-$ For example, reduced $Cu^{2+}(aq) + 2e^- \rightarrow Cu(s)$	https://edu.rsc.org/resources/the-position-of-iron-in-the-reactivity-series/1779.article
	Teacher demonstration Microscale thermite reaction $Fe_2O_3(s) + 2Al(s) \rightarrow 2Fe(l) + Al_2O_3(s)$ The reaction generates so much heat that the iron formed is molten	Oxidised $Al(s) \rightarrow Al^{3+}(s) + 3e^-$ Reduced $Fe^{2+}(s) + 2e^- \rightarrow Fe(l)$	http://science.cleapss.org.uk/Resource-Info/The-Microscale-Thermite-Reaction.aspx

8 Redox reactions and electrolysis

> **Questions to ask for each reaction**
> - Based on the reactivity series, what reaction do you predict will occur?
> - Which species are oxidised and reduced in this reaction?
> - What are the half-equations for oxidation and reduction?
> - What similarities are there to the other metal reactions you have learned about?
> - Of the metals you have seen reacting, which is the most reactive? Which metal is the least reactive? Why do you think metals react differently?

8.5 Metal extraction

Human societies have made and used metallic objects for thousands of years. Iron trinkets made from elemental iron found in iron meteorites have been found in Egyptian tombs dating back to 4000–3000 BCE, predating our ability to extract iron from its ores. In the future, rocks from space may be an important source. Metals like platinum and cobalt, which are relatively rare on Earth, are more abundant in asteroids. If asteroid mining becomes cost-effective, metals from asteroids could be used to replenish diminishing reserves on Earth. Meanwhile, learning terrestrial extraction methods is a valuable starting point. Most metal obtained globally is extracted from metal compounds by redox reactions. Metal extraction techniques have increased the quantity and variety of metals available.

Smelting was the first technique developed for extracting metals. Metal ores are heated with another substance, often carbon in the form of charcoal (from wood) or, more productively, coke, obtained by heating coal in the absence of air. Extraction of elemental metal from metal compounds is always a redox reaction. Metal ions are reduced to metal atoms, accepting electrons, while the other substance in the reaction is oxidised. For example, the equation for extracting iron from iron(III) oxide (haematite) is:

$$\text{iron(III) oxide} + \text{carbon} \rightarrow \text{iron} + \text{carbon dioxide}$$
$$2Fe_2O_3(s) + 3C(s) \rightarrow 4Fe(l) + 3CO_2(g)$$

In the reaction, as shown above, iron ions are reduced:

$$Fe^{3+}(s) + 3e^- \rightarrow Fe(l)$$

And carbon is oxidised:

$$C(s) + O_2(g) \rightarrow CO_2(g)$$

The element iron is molten when formed because the reaction reaches temperatures of nearly 2000 °C, which is above the melting point of iron (1538 °C). The ability to control this reaction to produce

8.5 Metal extraction

good quality iron provided the basis for the Industrial Revolution that transformed many societies from the eighteenth century onwards.

Electrolysis is also used to extract metals from ores. This requires passing an electrical current through the molten metal compound. The metal is reduced (see Section 8.6). Electrolysis is mainly used to extract metals above carbon in the reactivity series (see Section 8.3).

KEY ACTIVITY

Patterns in metal extraction

This activity helps students to relate the reactivity series to patterns in how metals are found in nature, and the different extraction techniques used to reduce metals. Table 8.7 shows metals, how they occur naturally, their earliest known extraction date and techniques used to obtain them. Metals higher in the reactivity series required development of new extraction techniques, so were first extracted from ores more recently than those lower in the reactivity series. These metals were first extracted in the early 1800s, following the start of the industrial revolution in Europe. Large-scale extraction of aluminium was possible from the late 1800s.

Table 8.7 Common metals, their ores and extraction techniques

Metal	Naturally found as	Date and location	Technique	Narrative
Gold (Au)	Elemental gold		Found as an element, so extraction is not needed	Flakes of gold have been found in caves inhabited by humans dating to 40 000 BCE.
Copper	Elemental copper (Cu) Malachite (copper(I) carbonate hydroxide, $Cu_2CO_3(OH)_2$)	Around 6000 BCE Middle East	Heat with carbon	Copper is found as elemental copper (Cu) and bound in copper compounds. While copper objects, such as jewellery, pre-date 6000 BCE, extracting copper increased the quantity of copper available (see Enrichment, Section 8.6).
Iron	Magnetite (iron(II, III) oxide, Fe_3O_4) Haematite (iron(III) oxide, Fe_2O_3)	Around 1500 BCE Middle East	Heat with carbon	Iron is extracted by heating with carbon. It was first extracted thousands of years after copper, partly because temperatures of above 1200 °C are required, whereas copper ores react with carbon at about 700 °C. Lining furnaces with clay created an environment that could reach higher temperatures. The clay lining reduced heat losses, allowing furnaces to maintain the high temperature needed to extract iron from its ores.

8 Redox reactions and electrolysis

Metal	Naturally found as	Date and location	Technique	Narrative
Potassium	Sylvite (potassium chloride, KCl)	1807 London, UK	Electrolysis	Potassium was first isolated as a metal by electrolysis from potassium hydroxide by Sir Humphrey Davy. Davy also isolated sodium by electrolysis in 1807.
Calcium	Limestone (calcium carbonate, $CaCO_3$) Gypsum (calcium sulfate, $CaSO_4$)	1808 London, UK		Sir Humphrey Davy first extracted calcium from lime (calcium oxide). By the end of 1808, he had also used electrolysis to extract magnesium, strontium and barium.
Aluminium	Bauxite (aluminium oxide, Al_2O_3)	1825, Denmark	Displacement reaction	Hans Christian Ørsted first isolated aluminium in small quantities using potassium to displace aluminium from aluminium chloride. potassium + aluminium chloride → aluminium + potassium chloride
		1886, USA and France	Electrolysis	The Hall-Héroult process provided a cost-effective method for large-scale aluminium production (see Section 8.6).

Give students the first four columns from Table 8.7. Ask students to look for patterns, and to compare this table with the reactivity series (Table 8.5).

Questions to ask

- What is the relationship between the reactivity series and how metals are found?
- What relationship occurs between the reactivity series and how metals are extracted?
- Is it possible to extract aluminium metal by heating an aluminium ore with carbon? Why/why not?
- What is the relationship between the reactivity series and the dates of extraction? Why is this?
- What technique would extract (a) zinc and (b) magnesium from their ores?

When teaching metal extraction, emphasise that most elements are bound in compounds. Generally, school chemistry focuses on elements as the building blocks of substances, and students see elements combine to form compounds. Elements are not present in

8.5 Metal extraction

compounds as elements (see Chapter 4). Students meet elemental metals, such as aluminium, tin, zinc, iron and copper in their everyday experiences. Students can form the impression that elements occur naturally, while compounds are synthetic. This is an opportunity for students to see that effort, perseverance and ingenuity were needed to extract elements in stable states from naturally occurring compounds by chemical reactions. This is illustrated by the stories of iron extraction and aluminium manufacture.

KEY ACTIVITY

Extracting iron from a match head

Practical work on metal extractions provides an engaging context for students to practise describing redox reactions. This class experiment requires about ten minutes. Students extract metal themselves by reducing iron(III) oxide with carbon present in a match head:

$$\text{iron oxide} + \text{carbon} \rightarrow \text{iron} + \text{carbon dioxide}$$

$$2Fe_2O_3(s) + 3C(s) \rightarrow 4Fe(s) + 3CO_2(g)$$

Reduction of iron: $\quad Fe^{3+} + 3e^- \rightarrow Fe(s)$

Oxidation of carbon: $\quad C + O_2 \rightarrow CO_2(g)$

The procedure is simple and requires little equipment. Students make their own sample of iron and show that it is magnetic. Full practical instructions are available at: https://edu.rsc.org/resources/extraction-of-iron-on-a-match-head/722.article

Emphasise that carbon is present in the match head. Sodium carbonate (Na_2CO_3) is used to bring iron(III) oxide into contact with the carbon, but does not participate in the reaction.

Students can describe the reaction in terms of reduction and oxidation, and write half-equations. They can compare the match-head extraction with large-scale iron extraction in a blast furnace.

Questions to ask
- What differences do you observe between the iron(III) oxide and the iron?
- What gas is produced in this reaction?
  How could we prove this?
- Which reagent is reduced and which is oxidised?
  How do you know?
- Write half-equations for (a) the reduction and (b) the oxidation.
- What is the overall equation for the reaction?

8.6 Electrolysis

In electrolysis, an electrical current passes through a substance, causing a chemical reaction. Electrolysis is the process of splitting, or breaking down, a substance by passing an electrical current through it. Conductors called electrodes connect the substance in an electrical circuit so that an electrical current can flow through it. The substance is called an electrolyte. Reduction and oxidation reactions happen at the electrodes. Chemists describe these reactions as electrochemistry because they are chemical reactions which require an electrical current. Electrolysis is an example of a redox reaction. Energy supplied by an electric current forces reduction and oxidation reactions to take place. Electrolysis cannot occur on its own: it is non-spontaneous.

> **Enrichment**
>
> Electrolysis has many practical applications including extracting reactive metals, such as aluminium, from their ores; extracting chlorine from sea water; and electroplating a metal object with a thin layer of a less reactive metal, such as copper, silver or gold. Gold- and silver-plating enhances the appearance and value of jewellery and other decorative objects. Electroplating is also used to protect vehicles and objects from corrosion.

Introducing electrolysis: zinc chloride

Electrolysis of a molten ionic compound is a good introduction to electrolysis as only one substance is present, making the reactions straightforward. To undergo electrolysis, a substance must conduct electricity. Ionic substances such as zinc chloride are usually solid at room temperature. They have high melting points due to ionic bonding, in which positive and negatively charged ions alternate in crystalline lattice structures, held in position by electrostatic attraction. Zinc chloride melts at 275 °C. Solid ionic compounds do not conduct electricity, as the ions are not free to leave the lattice structure. However, when the ionic lattice is destroyed, as it is in a molten liquid or when in solution, movement of positive and negative ions is unconstrained, so the substance conducts electricity. In electrolysis of zinc chloride, heat breaks down the ionic lattice, melting the compound and so freeing up zinc and chloride ions. Zinc and chloride ions can then carry electric current and move to electrodes. At the electrodes, zinc ions are reduced to zinc metal, and chloride ions are oxidised to chlorine gas.

8.6 Electrolysis

KEY ACTIVITY

Electrolysis of molten zinc chloride

Electrolysis introduces students to practical equipment they have not previously seen. A typical set of electrolysis apparatus is shown in Figure 8.6. A bulb can be included in the circuit to show that electricity is passing through the electrolyte.

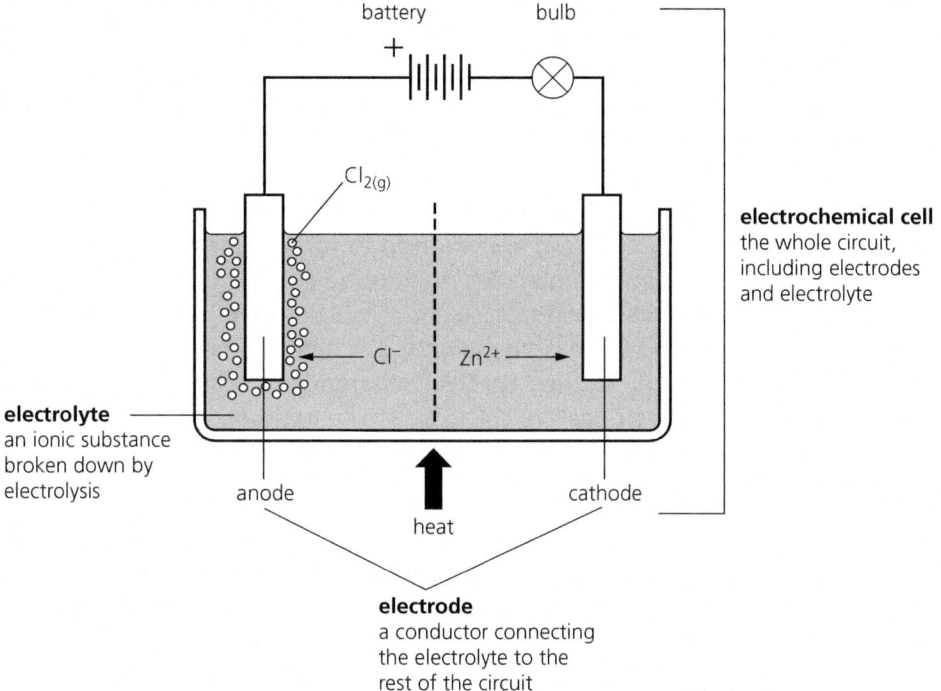

Figure 8.6 Electrolysis of molten zinc chloride

There are two electrodes in an electrochemical cell. The positively charged electrode is attached to the positive terminal of the battery. It is called the anode because the positive electrode attracts negatively charged anions. This can confuse students because anions (−) and the anode (+) are similar words but have opposite charges. Similarly, the negatively charged electrode is attached to the negative battery terminal. It is called the cathode because the negative cathode attracts positively charged cations.

This teacher demonstration must be carried out in a fume cupboard. Full practical instructions, including a video for teachers, are available at https://edu.rsc.org/resources/electrolysis-of-molten-zinc-chloride/826.article

A good starting point is to observe and describe the macroscopic features of electrolysis (Chapter 1). This includes helping students to label the unfamiliar equipment as the electrochemical cell (see diagram) is set up, and then to demonstrate the reaction so students can make observations.

8 Redox reactions and electrolysis

If a bulb is in the circuit, this will light when current is flowing through the molten electrolyte. Students will observe bubbles of chlorine gas forming at the anode (+). At the cathode (−), students may see metal crystals of zinc metal forming. Overall, the reaction is:

$$ZnCl_2(l) \rightarrow Zn(s) + Cl_2(g)$$

Once students can describe the practical equipment and observations, these can be explained by considering electrolysis on the sub-microscopic level. Zinc chloride ($ZnCl_2$) is an ionic lattice of positive zinc ions (Zn^{2+}) and negative chloride ions (Cl^-). When heated, the ionic lattice breaks apart, forming molten zinc chloride. This comprises positively charged zinc ions and negatively charged chloride ions that move freely. When an electrical current passes through the liquid, chloride ions move towards the positively charged electrode, the anode. At the anode, they transfer electrons into the electric circuit, and are oxidised to form chlorine atoms (Cl). The chlorine atoms combine to form molecules of chlorine gas (Cl_2). The gas is released as a yellow-green vapour that can be tested to show it is chlorine (see Chapter 9). The electrons are carried through the external circuit, through the battery (and light a bulb, if one is present in the circuit) to the negatively charged electrode, the cathode. Zinc ions (Zn^{2+}) move towards the negative cathode. At the cathode, each zinc ion accepts two electrons and is reduced, forming zinc atoms (Zn). Zinc metal is deposited on the cathode as a shiny grey coating.

Like all redox reactions, electrolysis involves oxidation and reduction: one substance loses or transfers electrons, which are gained or accepted by the other substance. In electrolysis, electrons transfer from one substance to another via the electrodes and electric circuit. Reduction occurs at the cathode, and oxidation at the anode. Half-equations for the electrolysis of zinc chloride are:

Oxidation at the anode: $2Cl^-(aq) \rightarrow Cl_2(g) + 2e^-$

Reduction at the cathode: $Zn^{2+}(aq) + 2e^- \rightarrow Zn(s)$

Questions to ask
- Why does the zinc chloride need to be heated?
- Why do ions move towards the electrodes?
- What happens to zinc ions? Write a half-equation to represent the change.
- What happens to chloride ions? Write a half-equation to represent the change.
- What is observed at each electrode, and why?
- Show that the electrolysis of zinc chloride is a redox reaction.
- Use the half-equations to write the overall equation for the reaction occurring in this experiment.

Learning the language of electrolysis

Electrolysis introduces new language to describe chemical reactions. Some electrolysis terminology will be familiar from learning about redox reactions: atoms, ions, cations, anions, reduction, oxidation and half-equations.

Students will need support with the specific terminology of electrolysis. For example, current includes the flow of ions through a solution. This is challenging, and students often assume that electrons, rather than ions, conduct electricity through the solution/molten substance. Specific vocabulary includes the following: electrolysis occurs in a *cell* comprising an *electrolyte* and two *electrodes*, the *cathode* and the *anode*. Electrodes are categorised as *inert* if they do not react, or non-inert if they change in the process (see below).

Electrolysis requires understanding of pairs of concepts with opposite meanings. Considering these words as pairs helps students to connect them.

- Ions can be positively charged (cations) or negatively charged (anions).
- In a cell, there is a positive electrode (anode) and a negative electrode (cathode).
- The electrodes are inert (not involved in the reaction) or non-inert (reactive).
- Substances can accept electrons (reduction) or transfer electrons (oxidation).

KEY ACTIVITY

Concept map for electrolysis

A concept map is a visual structure of vocabulary and ideas associated with a topic, showing how these interconnect. Drawing a concept map enables students to represent their ideas and develop links between different ideas. Concept-mapping is useful for topics with new vocabulary and challenging ideas.

Provide students with a large sheet of paper, and small pieces of paper about 5 cm square (sticky notes are ideal). They also need a pencil and an eraser.

Provide vocabulary for electrolysis: electrolysis, electrode, electrolyte, current, ion, cell, half-equation, atom, reduction, oxidation, anion, cation, anode, cathode, inert and non-inert electrodes.

8 Redox reactions and electrolysis

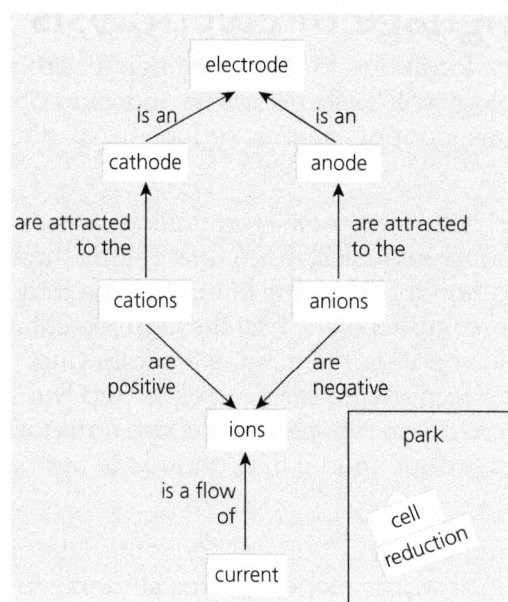

Figure 8.7: An introductory concept map for electrolysis

Students should rule a square of about 10 cm each side in one corner of the large sheet, and label this 'Park'. They then write the terms listed above on small pieces of paper, and arrange those they understand on the large sheet of paper, thinking of and labelling the connections between them with arrows (see Figure 8.7). Students may not get the connections correct: concept maps are a useful tool for revealing misconceptions. They can adjust the connections until they are satisfied they have included as many terms as possible. Terms they do not understand can be placed in Park.

Encourage students to share, develop and add to their concept map throughout the topic, taking terms from Park as their knowledge develops. Focusing on similarities and differences between students' concept maps over time is useful technique for promoting dialogic teaching and illuminates areas that students are finding challenging and require consolidating.

Questions to ask
- Which words seem to have the most connections to other words?
- Which words seem to have the fewest number of connections to other words?
- Find pairs of words which represent opposite meanings.
  Write down the pairs and the meanings of the words.

KEY ACTIVITY

The electrolysis of sodium chloride solution

Electrolysing sodium chloride solution (brine) introduces what happens when more than one ion is present at an electrode. This shows that water may participate in electrolysis. These are aspects which students find challenging. Students might predict that the electrolysis of sodium chloride solution produces sodium (at the cathode) and chlorine (at the anode). However, although chlorine forms as predicted, hydrogen forms at the cathode instead of sodium.

Sodium chloride solution contains Na^+ ions, Cl^- ions, and water molecules. Some water molecules dissociate to form hydrogen ions (H^+) and hydroxide (OH^-) ions (see Chapter 7). When a current flows through the solution, positively charged ions, namely hydrogen and sodium ions, move to the cathode. Negatively charged ions, namely chloride and hydroxide ions, (Cl^- and OH^-) migrate to the anode.

At each electrode, only one ion will react. Which ion will react depends on the energy changes of reduction or oxidation, where electrons are transferred to or from the ions to form atoms.

At the cathode (negative electrode), two reduction reactions are possible:

$$2H^+(aq) + 2e^- \rightarrow H_2(g)$$

$$Na^+(aq) + e^- \rightarrow Na(s)$$

As stated above, sodium ions require relatively more energy to accept electrons (Section 8.4) than hydrogen ions do. Ions of more reactive metals are more difficult to reduce by electrolysis than ions of less reactive metals. Sodium metal is more reactive than hydrogen gas. This means that hydrogen ions require less energy to accept electrons than sodium ions. Hydrogen ions each accept one electron, become hydrogen atoms and combine in pairs to form hydrogen gas. Sodium ions remain in solution.

At the anode (positive electrode), two oxidation reactions are possible:

$$2Cl^-(aq) \rightarrow Cl_2(g) + 2e^-$$

$$4OH^-(aq) \rightarrow 2H_2O + O_2 + 4e^-$$

In this electrolysis, we find that chloride ions are oxidised to chlorine atoms, which combine in pairs to form chlorine gas. Hydroxide ions remain in solution. This is unusual because in many electrolysis reactions hydroxide ions are more easily oxidised than chloride ions.

Electrolysis is a complex process. Substances produced depend on many factors including the ease of reduction/oxidation, relative concentrations of ions, rates of reaction and the type of material used as the electrode. However, the electrode reactions for any set of cations and anions can often be predicted by considering the relative reactivity of the ions. At the cathode, the cation which is most easily reduced is likely to react. Similarly at the anode, the anion which is most easily oxidised is likely to react.

8 Redox reactions and electrolysis

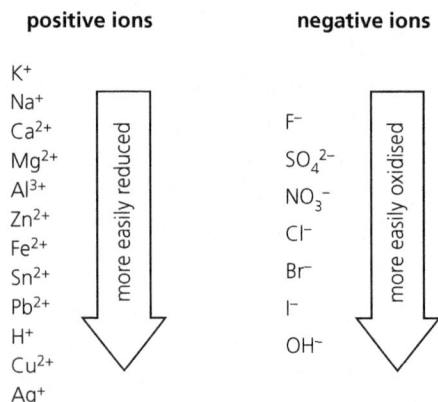

Figure 8.9 A reactivity series for ions

Sodium and hydroxide, which are unchanged, are called spectator ions, meaning that they are present but do not participate, like a spectator watching a sports game. Here the macroscopic observations are that two gases are produced: chlorine appears as a yellow-green gas at the anode, and hydrogen at the cathode. Both these can be identified via gas tests (see Chapter 9). Also, the pH of the solution rises due to increased concentrations of sodium and hydroxide ions (see Chapter 7).

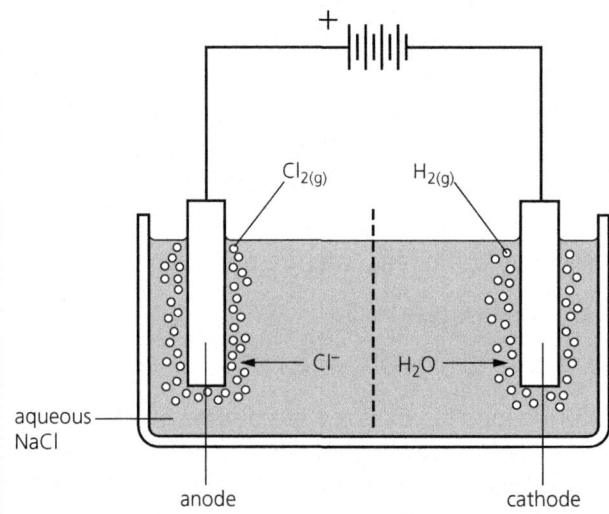

Figure 8.10 The electrolysis of aqueous NaCl

Electrolysis of sodium chloride solution makes a vital contribution to industry. The products all have important uses: hydrogen is used as a fuel and to manufacture ammonia, which is in turn used to make fertilisers; chlorine is used in water treatment and to make bleach; sodium hydroxide solution is used in paper- and soap-making.

Full instructions and teaching notes for demonstrating the electrolysis of sodium chloride can be found at https://edu.rsc.org/resources/colourful-electrolysis/735.article This demonstration uses Universal Indicator to track the reaction.

Questions to ask

- Which ions are present in the solution?
- Which ions will be attracted to (a) the cathode? (b) the anode?
- What do you observe at each electrode?
  What does that suggest about the products formed at each electrode?
- Write half-equations for the reactions occurring at each electrode.

KEY ACTIVITY

Animating electrolysis

Students need to draw and interpret visual representations, including movement of ions and reactions at electrodes. In this activity, students make animations of the electrolysis of sodium chloride solution. The activity uses dual coding, representing science ideas in pictures and words, helping students make connections and aiding memorisation. Full instructions and explanations can be found at https://edu.rsc.org/resources/colourful-electrolysis/735.article

Students need access to a stop-motion animation app. These are freely available for mobile devices including phones. To make a stop-motion animation, students illustrate the electrolysis process, photographing each illustration. The app animates the shots into a short film. Students provide a voiceover describing the animation. To illustrate the electrolysis process, students can use a range of media, such as drawings on paper or white boards, modelling material, plastic counters and coloured paper. In their choice of medium, students must represent the ions, atoms, electrodes, electrolyte particles and half-equations.

Questions to ask

Ask students to draw animations which answer these questions:

- Which ions are present at the start of electrolysis?
- What happens to the ions when the electrical current is switched on?
- What happens at the anode?
- What happens at the cathode?
- What are the half-equations for reduction and oxidation?

8 Redox reactions and electrolysis

KEY ACTIVITY

Electrolysis of copper(II) sulfate solution

Copper(II) sulfate solution comprises copper(II) ions (Cu^{2+}) and sulfate ions (SO_4^{2-}) in water (mainly molecules, formula H_2O). The ions and water molecules move freely. The solution conducts electricity: when copper(II) sulfate solution is placed in a beaker with electrodes connected to a 6 V power supply, electrons and ions move through the solution. The presence of an electrical current can be confirmed by including a bulb or an ammeter in the circuit. Redox reactions occur at electrodes. The reactions change depending on the material used for the electrodes.

Full instructions and teaching notes for this experiment are available at https://edu.rsc.org/experiment/electrolysis-of-copperii-sulfate-solution/476.article

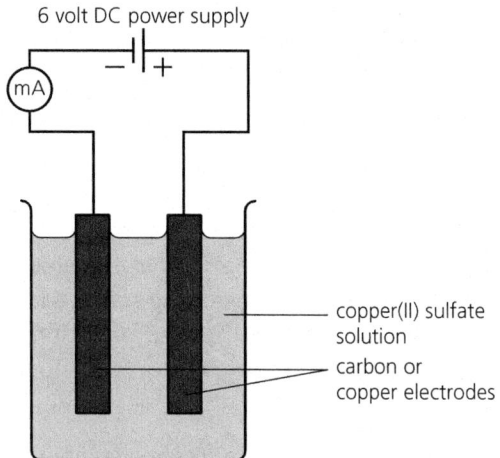

Figure 8.11 Apparatus for the electrolysis of copper sulfate solution

With inert electrodes, carbon rods

Carbon rods are inert electrodes, meaning they do not take part in either reaction at the anode or cathode. The carbon rods provide a surface on which reactions occur. For copper(II) sulfate solution, the reactions are:

At the anode, (+), carbon electrode: $\quad 2H_2O(l) \rightarrow O_2(g) + 4H^+(aq) + 4e^-$

This equation shows water molecules are oxidised, decomposing to produce oxygen, hydrogen ions and four electrons, which enter the external circuit. The hydrogen ions dissolve in solution and oxygen gas is liberated. The macroscopic observation is bubbles of clear colourless gas forming at the anode. This can be collected and tested to identify it as oxygen (see Chapter 9).

At the cathode, (−), carbon electrode: $\quad Cu^{2+}(aq) + 2e^- \rightarrow Cu(s)$

This equation shows copper ions are reduced, each gaining two electrons to become copper atoms. Over time, a layer of copper will build up on the cathode. The macroscopic observation is that a red coloured substance is deposited on the part of the rod submerged in the copper(II) sulfate solution, while the blue colour of the solution fades, ultimately to colourless.

Combining these two equations gives the overall equation for electrolysis of copper(II) sulfate with carbon electrodes:

$$2H_2O(l) + 2Cu^{2+}(aq) \rightarrow O_2(g) + 4H^+(aq) + 2Cu(s)$$

Notice that sulfate ions are spectator ions in this reaction.

With non-inert electrodes, copper rods

Copper rods are non-inert electrodes, meaning that these take part in the anode and cathode reactions. With copper electrodes, the half-equations for the electrolysis of copper(II) sulfate are as follows:

At the anode, (+), copper electrode: $\quad Cu(s) \rightarrow Cu^{2+}(aq) + 2e^-$

At the cathode, (−), copper electrode: $\quad Cu^{2+}(aq) + 2e^- \rightarrow Cu(s)$

These show that copper atoms in the anode are oxidised to form copper ions (Cu^{2+}). Copper ions move into solution and are attracted to the cathode. At the cathode, copper ions are reduced to form copper atoms. The macroscopic observations are that copper is deposited on the cathode, the anode gradually dissolves, and the solution remains a strong blue colour. Progress of the reaction can be determined by finding the masses of the electrodes: the anode loses mass and the cathode gains mass. However, combining the two half-equations shows there is no overall chemical reaction:

Overall equation with copper electrodes:

$$Cu(s) \rightarrow Cu(s)$$

Enrichment

Copper is valuable: it can be pulled into stable, non-corroding wires, so is used in electronics, buildings; it is easily shaped by hammering into many useful objects; and its warm colour is decorative and pleasing. The resale value of copper is high, so recycling the metal is economically viable.

Electrolysis is used to purify copper for re-use. Purifying copper by electrolysis is cheaper than finding deposits of copper ore, excavating them, and transporting them for extraction. The electrolyte used for purification is aqueous copper(II) sulfate solution. Impure copper is used as the anode (+), with pure copper as the cathode (−). Copper atoms deposited form layers of pure copper on the cathode. Over time, the anode loses mass and the cathode gains mass. Any impurities present fall to the base of the electrolyte tank.

Table 8.8 summarises electrolysis of zinc chloride, sodium chloride and copper(II) sulfate solutions.

8 Redox reactions and electrolysis

Table 8.8 Electrolysis reactions of aqueous solutions of common salts

Salt solution	Cathode (−) Reduction occurs	Anode (+) Oxidation occurs	Observations	Reference
Zinc chloride	$Zn^{2+}(aq) + 2e^- \rightarrow Zn(s)$	$2Cl^-(aq) \rightarrow Cl_2(g) + 2e^-$	Zinc metal is produced at the cathode. Chlorine gas is produced at the anode.	https://edu.rsc.org/resources/electrolysis-of-molten-zinc-chloride/826.article
Sodium chloride	$2H^+(aq) + 2e^- \rightarrow H_2(g)$	$2Cl^-(aq) \rightarrow Cl_2(g) + 2e^-$	Hydrogen gas is produced at the cathode. Chlorine gas is produced at the anode.	https://edu.rsc.org/resources/colourful-electrolysis/735.article
Copper(II) sulfate inert electrodes	$Cu^{2+}(aq) + 2e^- \rightarrow Cu(s)$	$2H_2O(l) \rightarrow O_2(g) + 4H^+(aq) + 4e^-$	Copper metal is produced at the cathode. Oxygen gas is produced at the anode.	https://edu.rsc.org/experiment/electrolysis-of-copperii-sulfate-solution/476.article
Copper(II) sulfate non-inert electrodes	$Cu^{2+}(aq) + 2e^- \rightarrow Cu(s)$	$Cu(s) \rightarrow Cu^{2+}(aq) + 2e^-$	The anode dissolves. A red–orange layer (pure copper) is deposited on the cathode. The anode loses mass and the cathode gains mass.	

The extraction of aluminium by electrolysis

The extraction of aluminium relates electrolysis back to the extraction of metals and the reactivity series. Metals which are more reactive than carbon cannot be extracted by heating with carbon. Electrolysis, developed in the 1800s, was a powerful chemical tool which enabled reactive metals, including aluminium, to be extracted from their compounds (see Section 8.5).

Aluminium is the most abundant metal on Earth. However, most aluminium is found bound in aluminium oxide, in the ore bauxite. Electrolysis is used to extract aluminium (see Table 8.7). A challenge is that bauxite only melts at temperatures over about 2000 °C. This very high melting point meant that several decades were required to find a cost-effective process to electrolyse aluminium. In 1886, two 22-year-old scientists, Charles Hall and Paul Héroult, separately realised almost simultaneously that bauxite dissolved in molten cryolite, sodium aluminium fluoride (Na_3AlF_6). This forms an electrolyte that is molten at about 1000 °C, much lower than the melting point of pure bauxite, thus requiring much less energy to electrolyse the ore. The electrodes are inert coke (carbon). The overall reaction is:

$$2Al_2O_3(l) \rightarrow 4Al(l) + 3O_2(g)$$

At the cathode, aluminium ions are reduced: $Al^{3+} + 3e^- \rightarrow Al(l)$

Molten aluminium collects at the bottom of the electrolysis cell and is tapped off.

At the anode, oxide ions are oxidised: $2O^{2-} \rightarrow O_2(g) + 4e^-$

Oxygen formed reacts with the carbon anode forming carbon dioxide. This degrades the anodes, which have to be replaced regularly.

Video clips can be used to illustrate the electrolysis of aluminium oxide (for example, https://edu.rsc.org/resources/aluminium-extraction/16.article). Learning about the extraction of aluminium also allows students to draw together their understanding of redox reactions, electrolysis, and metal extraction.

Questions to ask
→ Why is the bauxite dissolved in cryolite?
→ Why do the anodes need replacing?

8　Redox reactions and electrolysis

- → Write half-equations for the reactions at the cathode and anode.
- → Compare the electrolysis of aluminium oxide with the electrolysis of zinc chloride.
- → Compare the reactions for extracting aluminium and iron.
  Why is carbon not used to extract aluminium?
  Why is electrolysis not used for extracting iron?

 Fuelling the future: fuel cells

In electrolysis, an electrical current initiates redox reactions. A fuel cell is an electrochemical cell that uses *spontaneous* redox reactions to generate an electrical current. The fuel is commonly hydrogen. In the fuel cell, hydrogen combines with oxygen from the air. Hydrogen is oxidised to form hydrogen ions, and oxygen is reduced to form oxide ions. Hydrogen and oxide ions combine to form water. During reduction and oxidation an electrical current forms by the flow of electrons through an external circuit. This current can power electrical devices.

German physicist Christian Friedrich Schönbein and Welsh physicist William Grove first developed hydrogen fuel cells in the 1830s. Schönbein and Grove worked separately but corresponded about their ideas. Engineer Francis Thomas Bacon developed the first fuel cell capable of powering a vehicle in the 1930s. The US National Aeronautics and Space Administration (NASA) used these in the 1960's Apollo missions. Diminishing fossil fuel reserves and the impact of carbon dioxide emissions on global climate change means fuel cells can contribute significantly to addressing our energy needs.

KEY ACTIVITY

Electrolysis of acidified water – the Hoffmann voltameter

The electrolysis of water illustrates the principle of producing hydrogen for fuel cells. The Hoffmann voltameter comprises a set of fine glass tubing, connected to a low voltage supply (Figure 8.12). The arrangement is used to electrolyse water. This activity shows how water can be separated into hydrogen and oxygen gases, and then recombined.

8.7 Fuelling the future: fuel cells

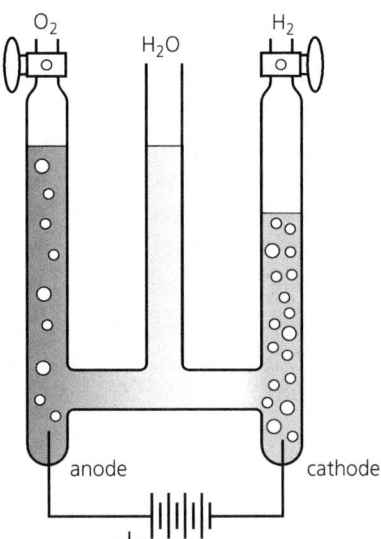

Figure 8.12 A Hoffmann voltameter used to demonstrate the electrolysis of water

Before demonstrating the electrolysis of water, review prior understanding of electrolysis. For example, discuss with students whether pure water can be electrolysed. Electrolysis requires the presence of ions, to conduct electricity. Water molecules are covalently bonded. This means that pure (see Chapter 9) water cannot be electrolysed. Invite students to identify ions present in a bottle of mineral water, using the label. A high concentration of ions makes electrolysis possible.

Prior to the lesson, wearing eye protection, fill the voltameter by opening the taps and adding dilute sulfuric acid through a funnel placed in the central tube, taking care not to overfill. (If the solution overflows, rinse the equipment with distilled water and leave it to dry.) Close the taps: this is vital, otherwise gases formed will escape into the atmosphere! Leave some empty volume in the central tube, as liquid displaces when the current is flowing. The voltameter is now ready.

In the lesson, turn on the power supply. If time is short, switch on the current beforehand. Macroscopic observations are of bubbles of clear, colourless gas forming in both outer tubes, but this takes time to develop. The gases collect beneath each tap (see Figure 8.12), displacing liquid up the central tube. The volume of gas formed at the cathode is twice that formed at the anode. The overall reaction is that water molecules decompose to produce oxygen and hydrogen gases:

$$2H_2O \rightarrow 2H_2 + O_2$$

Reduction at cathode: $2H^+ + 2e^- \rightarrow H_2$

Oxidation at anode: $2O^{2-} + 4e^- \rightarrow O_2$

8   Redox reactions and electrolysis

> The ratio of hydrogen: oxygen is 2:1; two hydrogen molecules form for every oxygen molecule. Measure the volumes of hydrogen and oxygen using gradations on the tubes or a ruler.
>
> When reasonable volumes of the gases have formed (say about 15–30 cm^3), wearing eye protection, attach rubber tubing to each tap to collect each gas separately into an upturned test tube, set under water. Gas tests are described in Chapter 9. Alternatively, collect both gases together, by running the rubber tubing into soap solution in a trough. This forms bubbles of hydrogen and oxygen on the surface. Ignite a bubble with a lighted splint attached to a metre rule, keeping students at a safe distance behind a safety screen. A pleasing and surprisingly loud pop is produced, as hydrogen and oxygen recombine to form water. It is useful to practise this technique before trying it in front of students.
>
> ### Questions to ask
> - What would happen if the voltameter were filled with pure water? Why?
> - Which gas forms at each tap? How do you know?
> - Write half-equations for these reactions, and an equation for the overall reaction.
> - Why are the volumes of gas different? Explain the ratio of gas volumes.
> - What reaction occurs in the explosion? Write an equation representing the reaction.

Fuel cells

A number of car manufacturers including Hyundai, Toyota and Honda (market leaders at the time of writing) have developed cars powered by hydrogen fuel cells. A major challenge for fuel cell cars is achieving refuelling. Like cars run by diesel or petrol, hydrogen cars need a constant supply of fuel, so to be practically useful, a huge network of hydrogen filling stations is needed, requiring extensive national investment. Hydrogen is explosive so is difficult to store and transport in large quantities, leading to technical challenges with filling stations. However, some countries and regions are developing infrastructure to support hydrogen-fuelled cars. By 2020, the state of California had 40 hydrogen stations. An overview of hydrogen fuel cell technology suitable for teachers can be found at https://pubs.rsc.org/en/content/articlelanding/2019/ee/c8ee01157e

Miniature cars powered by hydrogen fuel cells are available from school scientific suppliers. These are equipped with a solar panel to electrolyse water. The hydrogen gas produced is used in its hydrogen fuel cell. The reaction is the reverse reaction of that in the Hoffmann voltammeter (above).

8.7 Fuelling the future: fuel cells

Figure 8.13 A toy car powered by a hydrogen fuel cell. The cell is highlighted in the black box.

Figure 8.14 A toy car powered by a hydrogen fuel cell. The cell is producing hydrogen gas. This particular version can use a battery or solar panel to electrolyse water.

KEY ACTIVITY

How can we power our vehicles?

Many countries are committed to reducing or eliminating vehicles powered by diesel or petrol, as part of global efforts to reduce carbon emissions. Electric batteries and hydrogen fuel cells are two developing technologies which can be used to power vehicles including lorries, cars, buses and ships.

Students can research how electric batteries and hydrogen fuel cells work in vehicles, and the advantages and disadvantages of each technology. This research can inform a class debate and discussion on which technology is more promising for powering vehicles. As the technology is developing quickly, look for up-to-date articles and references. Some possible sources of information are listed below:

- Alternative fuels: https://driveclean.ca.gov/
- Electric batteries: https://energysavingtrust.org.uk/advice/electric-vehicles/
- Hydrogen fuel cells: www.bmw.com/en/innovation/how-hydrogen-fuel-cell-cars-work.html

Questions to investigate

- Why are there concerns about petrol and diesel engines?
- Why are electric batteries and hydrogen fuel cells preferred to petrol/diesel engines?
- What vehicles are currently available which use these new technologies?
- What are the advantages of powering vehicles using (a) electric batteries (b) hydrogen fuel cells?
- What are the biggest challenges for each technology?

Careers

Electrochemical processes underpin many technologies and their associated careers. These technologies include development of batteries and fuel cells, and applications of electrolysis in electroplating, food preservation and metal extraction.

Scientists and engineers are constantly developing battery technology, creating batteries which last longer and have lower environmental impact. Many scientists and engineers develop batteries for particular uses. The automotive industry is developing electric cars which use either hydrogen fuel cells or conventional batteries. Manufacturers of laptops, tablets and mobile phones compete for batteries with a longer life and shorter recharge time. Advances in renewable energy depend on batteries to store the power generated by solar panels and wind turbines.

Electroplating is an application of electrolysis. Electroplating deposits a thin layer of metal on a surface, changing the properties of the surface.

Engineers use electroplating to manufacture surfaces with specific properties. For example, the gold-plated telescope mirror on the Mars Orbiter Laser Altimeter was able to map the mountains of Mars (see https://spinoff.nasa.gov/spinoff1997/hm2.html). The plating made the mirror exceptionally reflective, corrosion-resistant and cleanable. The same patented electroplating process is used to gold-plate the Oscars awarded annually by the US Academy of Motion Picture Arts and Sciences. While large-scale electroplating processes are fully automated, manual electroplating is common for creating decorative artwork. Restoration specialists use electroplating to restore treasures like silver-plated cutlery and classic cars to their original splendour. Jewellery makers electroplate natural objects such as leaves and acorns, preserving their beauty for many years.

Food scientists develop and test preservatives, which keep food fresh for longer. Preservatives called antioxidants inhibit oxidation reactions in food, which would otherwise spoil the food. In many countries, preservatives must be approved by the government for use in food. This involves scientists testing the preservatives and specifying the maximum quantities which can be used in food. Antioxidants are also used to extend the shelf life of other products, such as cosmetics.

8.8 Resources

References and further reading

De Jong O and Treagust D. (2002) The teaching and learning of electrochemistry in Gilbert J.K., De Jong O., Justi R., Treagust D.F., Van Driel J.H. (editors) *Chemical education: Towards research-based practice*, Springer. Available at https://doi.org/10.1007/0-306-47977-X_14

Kind, V. (2002) *Contemporary Chemistry for Schools and Colleges*, Royal Society of Chemistry.

Schmidt, H., Marohn, A., Harrison A. (2006) Factors that prevent learning in electrochemistry. *Journal of Research in Science Teaching*, 44(2): 258–283. Available at https://doi.org/10.1002/tea.20118

Staffell, I., Scamman, D., Velazquez Abad, A., Balcombe, P., Dodds, P. E., Ekins, P., Shah, N. and Ward, K. R. (2019) The role of hydrogen and fuel cells in the global energy system. *Energy and Environmental Science*, 12(2): 463–491. Available at https://pubs.rsc.org/en/content/articlelanding/2019/ee/c8ee01157e

Websites

Teaching redox chemistry: https://edu.rsc.org/cpd/redox-chemistry/2000011.article

Teaching displacement reactions: https://edu.rsc.org/cpd/displacement-reactions/4011106.article

Teaching the extraction of metals: https://edu.rsc.org/cpd/the-extraction-of-metals/4010857.article

The history of metallurgy: https://www.3mvet.eu/en/news/history-metallurgy

Five ways to explain electrolysis: https://edu.rsc.org/ideas/5-ways-to-explain-electrolysis/4012108.article

Burning magnesium in air: https://chem.rutgers.edu/cldf-demos/1016-cldf-demo-burning-magnesium

Extraction and mining of copper: www.ase.org.uk/resources/copper-mining-and-extraction

Sequencing and segmenting the teaching of ionic equations: https://edu.rsc.org/ideas/how-to-sequence-and-segment-your-teaching-of-ionic-equations/4011307.article

The extraction of iron on a match head: https://edu.rsc.org/resources/extraction-of-iron-on-a-match-head/722.article

Electrolysis of molten zinc chloride: https://edu.rsc.org/resources/electrolysis-of-molten-zinc-chloride/826.article

Electrolysis of sodium chloride: https://edu.rsc.org/resources/colourful-electrolysis/735.article.

Electrolysis of copper (II) sulfate: https://edu.rsc.org/experiment/electrolysis-of-copperii-sulfate-solution/476.article

Electrolysis of aluminium: https://edu.rsc.org/resources/aluminium-extraction/16.article

Alternative fuels for vehicles: https://driveclean.ca.gov/

Electric vehicles: https://energysavingtrust.org.uk/advice/electric-vehicles/

Hydrogen fuel cell cars: https://www.bmw.com/en/innovation/how-hydrogen-fuel-cell-cars-work.html

Electroplating the Mars Orbiter laser altimeter: https://spinoff.nasa.gov/spinoff1997/hm2.html

Other resources

Ion cards for practising ionic formulae: www.rsc.org/Education/Teachers/Resources/Aflchem/resources/28/28%20resources/28-2%20Formula%20cards.pdf

Hydrogen fuel cell toy cars are available from school science suppliers.

9 Chemical analysis
Sheila Curtis

Introduction

Chemical analysis is, along with synthesis of substances, one of the main purposes for doing chemistry. Analysis is summarised by questions such as, 'What is this material?', 'How pure is this sample?' and 'How much of this substance is there?' Scientists and non-scientists use analysis techniques to make sense of specific chemical reactions occurring in everyday settings, including health, forensic science, food, cosmetics, materials science and many industries. Chemical analysis combines technological advances with understanding of chemical reactions, molecular structures and chemical properties of specific substances. Analysts apply techniques to test materials, investigate unknown materials and develop new materials. Automated instrumentation has mainly replaced 'wet' chemistry tests relying on hand-held experiments using test tubes and pipettes that were previously commonplace.

Students' familiarity with television dramas in which scientists apply chemical analysis techniques to solve problems in fictional or genuine crime scenarios may lead to a view that these always generate an unequivocal answer leading to a conviction or proof of innocence. In reality, chemical analysis applies in many situations, involves high-level thinking, application of prior ideas and often a combination of techniques to solve problems. Results may be inconclusive, generating uncertainty and a need for further tests, or application of contextual information.

In school science, this area of study offers a great opportunity for students to be involved in problem-solving which models real-world science. Students apply knowledge of chemical reactions and become familiar with wet (in solution) tests they can carry out in the laboratory. Adopting an inquiry-based approach to efficient chemical analysis is useful. However, take care to ensure that the degree of openness of inquiry tasks matches the complexity of concepts students must recall to solve a problem. Teachers' expectations for problem-solving should be made clear to students embarking on inquiry through clear outcomes and scaffolding for the less experienced students.

Students' prior knowledge and experience

Chemistry in primary education features physical properties of materials, changes of state, dissolving and that new materials can form from other materials. Children's understanding of chemical reactions develops through experience of everyday substances, such as reacting vinegar (aqueous ethanoic acid) with baking powder (sodium hydrogen carbonate) and baking cakes or bread. They learn about electrical conductivity, heat insulation and floating or sinking in water. In early secondary school, students develop understanding about chemical elements and compounds including:

→ Chemical reactions: precipitation, crystallisation, acid/base and decomposition.
→ Patterns of reactions from the periodic table: reactivity and trends in groups.
→ Acid–alkali reactions leading to neutralisation and formation of salts.

In England, the science curriculum introduces 'working scientifically' (Department for Education, 2013) from ages 7–11, which comprises skills and knowledge relevant to analysis. These include planning scientific enquiries to answer questions; using scientific equipment; recording data and knowing when to take repeat readings; using classification keys, tables and graphs; setting up comparative and fair tests; reporting and presenting findings; explanations of and degree of trust in results; identifying scientific evidence used to support or refute ideas or arguments. Students aged 11–14 apply sampling techniques; use mathematical concepts to calculate results; present results appropriately; interpret observations and data; offer explanations for patterns and in relation to predictions and hypotheses; and show awareness of error. 14–16-year-olds learn chemical analysis explicitly.

Research (see Section 9.9) suggests secondary students find inorganic chemical analysis challenging. Reasons include poor understanding of the purpose of procedures and lack of mastery of practical skills required to carry out tests accurately. Working memories may become overloaded by carrying out multi-step chemical analysis tests; reading and executing instructions; preparing additional tests, such as for gases; making observations; recording and interpreting results; and taking note of time left to tidy up and complete their reports. Teachers expect students to 'get the right answer', so students may expect this to be obvious and definitive, rather than inconclusive. Appreciating that tests may yield inconclusive answers is valuable. Teachers should clarify lesson outcomes, for example, applying a chemical concept to interpret an observation, or learning a practical skill. Achieving both in a short lesson is unlikely to produce meaningful learning.

Teaching pre-16 students to combine analytical practices with knowledge of chemical reactions and properties of substances at the sub-microscopic level is vital. This avoids reversions to macroscopic observations that constrain reaching suitable conclusions. Complex analytical situations require recognising relevant cues, and systematic application to problem-solving (Warren, 2019; Ngai et al, 2014). Investigative processes are fundamental to analytical chemistry: deciding what to test, which observations are relevant and what conclusions to draw. The topic provides opportunities for students to consolidate ideas and find significance in chemical ideas that appear disparate. Thus, practising analytical chemistry problems supports revision for summative examinations.

Progression

From their primary education (ages 7–11) children may have macroscopic level understanding of changes of state, labelling them as boiling, melting, evaporation, condensation and solidification, associating these with heating and cooling. 11-year-olds may know a simple particle model and describe state changes using particle terminology. Purity may be understood non-scientifically as 'unadulterated' or 'natural' rather than meaning one single substance. Students' understanding of this concept develops through learning that melting and boiling points are substance-specific and rely on purity. Filtration, distillation and chromatography illustrate that mixtures can be separated into 'pure' substances. Aged 14–16, students learn techniques including those discussed in this chapter. The concept of purity is a basis for undertaking calculations of percentage yield for a product of a chemical reaction.

Post-16 students use physical and chemical properties and spectroscopy to identify materials, including organic compounds and transition metal ions. At university, students studying archaeology, biology, pharmacology, biomedical, Earth and environmental sciences apply chemical analysis to study materials of interest in laboratories equipped with analytical instruments, such as spectrometers and chromatographs. Some rely on simple portable tests, such as for carbonates in rocks, that can be conducted during fieldwork.

Teaching sequence overview

This chapter focuses on qualitative analysis. To be successful chemical analysts, students need to recall properties of materials, have knowledge of chemical reactions and apply these to analyse results of tests. This chapter considers these three aspects of analysis.

Physical properties	Chemical reactions	Spectroscopic techniques
density, melting point/boiling point, chromatography.	'Wet' chemistry or test-tube tests to identify gases from reactions. To identify ions (cations and anions) by precipitation of metal hydroxides, silver halides and reaction with acids.	Interaction of materials with light and other parts of the electromagnetic spectrum, using flame tests in the laboratory and instrumental techniques such as flame emission spectroscopy.

Figure 9.1 Chemical analysis: teaching sequence overview

To introduce the topic, students need to understand the chemical definition of the term pure and its importance for chemical analysis. An approach for developing this understanding is discussed next.

Stuff, pure substances and mixtures

Students should know the difference between pure substances and mixtures, together with techniques for refining purity, including filtration, evaporation, distillation and chromatography. In analysis, purity is important as contamination of chemicals used in tests can cause false or inconclusive results. Impure samples demonstrate lower melting points, higher boiling points and alteration of density. Knowing if a sample is pure or impure is useful information for analytical chemists seeking to identify a material.

In chemistry, pure means a single substance free from contamination from any other substance. Students associate pure with 'natural', 'untreated', 'uncontaminated' or 'unadulterated' substances. For example, juice squeezed directly from an orange is pure, as this is only juice, without additional water, preservatives or heat treatment. However, orange juice is a mixture comprising mostly water, but also containing vitamin C, citric acid, amino acids, carbohydrates, fibre and flavonoids. We often talk about 'pure air', measure air purity, and may purchase an air purifier that removes 'micro-particles' including bacteria and fungal spores. Yet chemically, air is a mixture comprising about 78% nitrogen, 20% oxygen, 1% argon and 0.04% carbon dioxide with other gases such as water vapour making up the rest.

Chemical purity is more complex than it first appears. A chemical element is often defined as a single substance made of atoms that are all of the same type, so it fits the description of pure. In practice, isotopes exist. An isotope has one or more additional neutrons in the nucleus. Although these atoms behave chemically as atoms of the element, they have different atomic mass values, meaning that atoms in a sample of an element are not all identical. Isotopes are accounted for in calculating atomic mass values (A_r values, Chapter 4). Some elements with very common isotopes have fractional A_r values: for example, chlorine, $A_r = 35.45$, and copper, $A_r = 63.55$. Chemicals are prepared to grades of purity, as removing every single particle of all other substances is difficult and expensive. For example, sodium chloride (formula: NaCl) is available as rock salt, a naturally occurring, impure mineral (formal name: halite); table salt, which is about 97% sodium chloride; and in chemical grades, which are 99.0–99.5% sodium chloride. Degree of purity or relative purity are more accurate terms than absolute purity. Nonetheless, emphasising that a pure substance is one type of material *as far as we are able to test it* is important in starting to understand chemical analysis.

KEY ACTIVITY

What is stuff like?

To introduce chemical analysis and purity, invite students to consider 'stuff' that chemists investigate. Knowing the type of stuff helps chemists decide analytical techniques for an investigation, and how to improve chemical purity. Figure 9.2 describes key terms used to describe stuff.

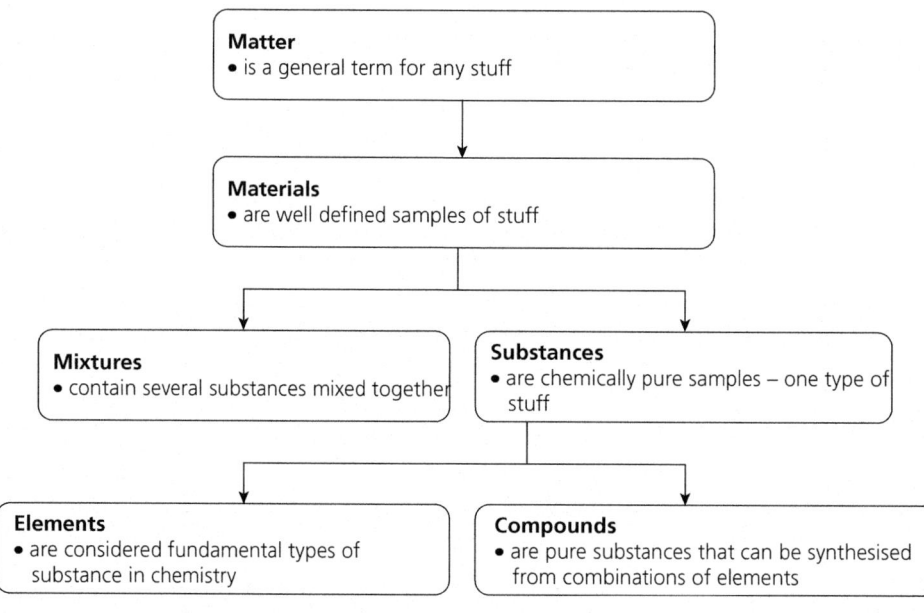

Figure 9.2 How some key terms are understood in chemistry

9.1 Stuff, pure substances and mixtures

Ask students to name examples of stuff: exhibit a wide range including some that students may not recognise. Work through Figure 9.2 to decide where examples fit in the flowchart.

- Everything can be regarded as matter, so all stuff passes the first box.
- Materials are stuff we work with, such as wood, glass, diamonds, sea water, fibreglass, paint, china, washing soda (sodium carbonate), flour, mayonnaise and poly(ethene). It is unlikely examples would not pass this box.
- Next, the flowchart divides into mixtures and substances. A mixture contains several substances mixed together; a substance is chemically pure, that is, one type of stuff. Ask students to consider how to recognise if a material is 'chemically pure' or a mixture. A general (but not absolute) rule can be that a chemical-sounding name indicates a single substance, while a general name implies a mixture. From the list earlier in the paragraph, sodium carbonate and poly(ethene) are substances, while the others are mixtures. Placing diamond may be difficult, as this is a general name, but comprises a single element, carbon, so is a substance.
- Substances sub-divide into elements (diamonds, once agreed as a substance) and compounds (sodium carbonate and poly(ethene)). This is discussed further in Chapter 4.
- Mixtures (not shown on Figure 9.2) sub-divide into composites (wood, fibreglass, china), emulsions and suspensions (paint, mayonnaise), and solutions (sea water), among other groups.

If uncertainty arises about a specific material, ask students to suggest confirmatory tests. These tests can include melting point, boiling point, density and separation by filtration, chromatography, evaporation or distillation. At this stage, there is no need to be too precise about correct and incorrect responses: what matters is that students start to think as chemical analysts, considering an identification process and making an assessment of 'purity'. Some materials may have more than one possibility and/lead to an inconclusive outcome. This is fine, as analysis often produces inconclusive results. Encourage students to accept that multiple/unclear results are part of the process. Information about handling inconclusive results is provided later in the chapter.

The activity leads to the significant point that many common materials are not 'substances' in the chemical sense. They are mixtures of substances that can, at least in theory, be separated into their component substances. Chapter 3 may help with this.

Questions to ask

- Name examples of natural materials that are 'pure'.
- What does pure mean to a chemist?
- What does pure mean in everyday life?
- Why is it useful to know if a material is a pure substance or a mixture?
- What tests would you carry out to find out where to place an unknown sample of material on the flowchart? How would the results help you decide?
- What particles are present in air?
  What particles are present in a sample of pure oxygen?
- Investigate how an analyst could test a sample of a gas to find out if it is air or pure oxygen.

9 Chemical analysis

KEY ACTIVITY

More about mixtures

The previous activity offers an opportunity to revisit particle ideas, leading to developing students' understanding of mixtures. Figure 9.3 shows images of particles of two pure substances and a mixture. These reinforce the point that a pure substance comprises one type of particle, while a mixture has more than one.

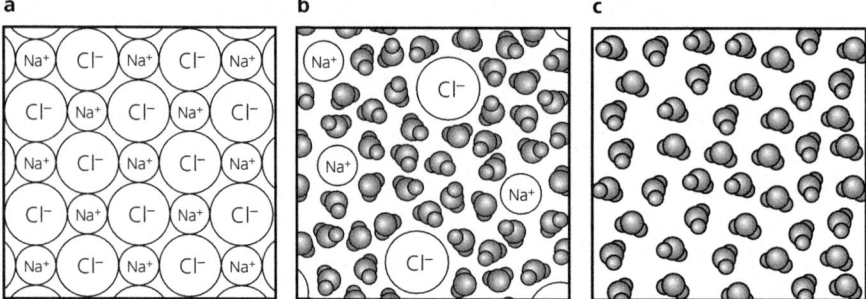

Figure 9.3 Two pure substances and a mixture

Figure 9.3a shows sodium chloride. This is a pure substance, as the sodium and chloride ions are bonded in a lattice structure (see Chapter 5) and the composition is constant. Figure 9.3b shows a solution of sodium chloride (brine, or 'sea water'), a mixture of water molecules, sodium and chloride ions. Figure 9.3c is a pure substance. Based on Figure 9.3b, this can be considered as pure water. Note that these are only representations and are not actual images of these particles (see Chapter 1).

Apply these particle principles in considering materials that could be pure substances or mixtures. First, choose materials in everyday use that are granular, with grain sizes big enough to handle. Examples are:

- Steel ball bearings that are the same size and colour (P for pure)
- Marbles of the same size, inserts and glass colour (P)
- Marbles of different sizes and colours (M for mixture)
- A mixture of steel ball bearings and marbles of the same size, inserts and glass colour (M)
- A fruit and nut mixture, such as raisins, currants, peanuts, brazil nuts and other dried fruit (M)
- Mixed nuts (M)
- One variety of unsalted, un-spiced nuts, for example, peanuts or cashews (P)
- One variety of salted nuts, for example, peanuts or cashews (M)
- Honey-coated banana chips (P)

Label jars with numbers or letters, not names. The letters in brackets are suggested correct responses (P for pure and M for mixture) and should not appear on labels.

Working in groups, students consider the contents of each jar. In analysis, making observations about a material is important, so ask students to observe what they think is in the jars. Then, for each jar:

9.1 Stuff, pure substances and mixtures

- How many components are present in each jar? What are they?
- Decide if the contents of each jar are a mixture or a pure substance.
- Give reasons for these answers.
- How could students separate the mixtures into pure substances (other than by tasting)?

In a plenary, discuss students' answers. They may not notice salt on salted nuts and miss the difference between raisins and currants. Students may disagree about some of the contents. For example, are honey-coated banana chips a mixture, because they contain honey and banana chips, or a pure substance because they are the only item in the jar? We cannot easily separate the honey from the banana chips, so these are a pure substance, roughly equivalent to a chemical compound. Salted and unsalted nuts may also raise discussion. Salt can be removed by placing nuts in water, dissolving the salt, then drying the nuts, so salted nuts are a mixture, while unsalted nuts are a pure substance. Other separation techniques that could be used are tweezers, fingers, sieves and crude filters. Reinforce which are 'pure substances' and the difference between these and mixtures.

Next consider a group of materials that are 'chemical', for example:

- Sand mixed with salt
- Sugar mixed with salt
- Salt solution
- Black ink
- Air
- Iron filings mixed with sulfur powder (flowers of sulfur)
- Iron filings mixed with copper turnings
- Water

Label the jars with letters or numbers. Students may benefit from using magnifiers or hand lenses to establish if more than one type of grain is present. Students should discuss in groups:

- If they think the contents are a mixture or a pure substance (all these are mixtures).
- Reasons for their answers.
- What tests could they carry out to investigate if a sample is a mixture or a pure substance.

In discussion, students may miss that sand/salt and sugar/sand are mixtures; and fail to observe differences between salt solution and water. Consider how to test a sample to find out if it is a mixture or a pure substance. For solid mixtures, sieve through fine sieves and/or use a magnet. Test to see if one or both components dissolve by making a solution. If some solid is left, filter to remove this, then evaporate to dryness to leave the dissolved solid. If both dissolve, then another test is needed to separate the solids before dissolving: a fine sieve or a magnet may help. For liquids, evaporation to dryness will reveal any solids, so will identify if a sample is a mixture; simple distillation will separate a pure sample of solvent from a solution.

Ask students to think *as analysts* to assess if a sample comprises one or more substances. Eventually, they may arrive at the following:

- Sand/salt – add water, dissolve the salt, filter, evaporate the solution to dryness, dry the sand in the filter paper
- Sugar/salt – a fine sieve could be used to separate the grains of sugar and salt
- Salt solution – evaporate to dryness
- Black ink – use chromatography (see below) or simple distillation to separate the water and pigment
- Air – use fractional distillation (see Section 9.9 Websites)
- Iron filings/sulfur – use a magnet
- Iron filings/copper – use a sieve and a magnet
- Water – use simple distillation to separate pure water and any dissolved solids

Regard each sample as an unknown to be tested to determine if it is pure or a mixture. Students are not expected to learn one 'correct' answer for each mixture. Separation tests do not identify substances precisely, so an analyst would carry out further tests (see below) to find out exactly what is present. The categorisation of some samples may be hard to resolve, leading to discussion about inconclusive tests and the need for further information. For example, how do we know air is a mixture? Is our water pure? Quite a few experiments over many years established that air is a mixture of gases (see Section 9.9 Websites). The closest we come to pure water is distilled, deionised water prepared for experiments. Most drinking water contains dissolved salts, such as magnesium and calcium salts and fluorides. Non-potable water may contain bacteria, sewage and algae among other substances.

Questions to ask

- Should all samples be analysed in the same way? Why/why not?
- What else might a chemist want to know about these samples?

An extension to this activity is separating these mixtures. Full instructions for methods, including sieving, filtration, evaporation, simple and fractional distillation, crystallisation and chromatography are available via this resource https://edu.rsc.org/cpd/separation-techniques/3009787.article

Investigating formulations

Formulation chemists work in many industries making mixtures for specific purposes. In contrast to chemists who combine substances in the expectation of a chemical reaction, formulation chemists make mixtures from substances that are not expected to react. Mixtures are used in many applications, including pharmaceuticals, cosmetics, perfumes, household products, foods, paints, fuels, fertilisers, pesticides, insecticides and animal foods.

9.1 Stuff, pure substances and mixtures

To make a formulation for a specific application, a chemist needs good understanding of the chemical behaviours of substances they intend to use. They test formulations for effectiveness, stability over a reasonable time period, discoloration, leaks from packaging, and check that products function in various environmental conditions.

In this activity, students examine formulations of everyday household products: shampoos and shower gels. They compare chemicals used and develop understanding of why each substance is needed in the mixture. Students analyse formulations, assessing if claims made about a product (and its cost) are justified based on the ingredients. Provide samples of shampoos and shower gels, with price per item and price per 100 ml. If possible and appropriate, students could bring samples from home, or undertake the task at home, then bring results to compare. Ask students to draw up a table with columns as follows: product name, cost, marketing claims, cost per 100 ml, ingredients, role of ingredient.

Full instructions are available at https://edu.rsc.org/resources/afl-what-are-shampoos-and-how-do-they-work/99.article The activity includes a cosmetics ingredient database. Prompt students to identify a surfactant, preservative, emulsifier, salt, thickener and perfume.

Legally, components must be listed in order of mass from highest to lowest. Manufacturers must list all ingredients but are not obliged to state quantities. An expensive product may have very small amounts of very expensive ingredients listed at the end. Marketing may state 'includes herbal extracts' which at very low concentrations make no difference to product efficacy.

Many products are complex mixtures, so investigating all ingredients is time-consuming. In a lesson, students can take one product per group, investigating ingredients between them. Some ingredients are listed as abbreviations and many have complicated names. Water is almost always the first listed ingredient, often as aqua.

Further activities linked to this task are provided in the list of resources in Section 9.9.

KEY ACTIVITY

Separating mixtures using chromatography

Chromatography involves passing a mixture of substances in solution across a *medium* so that individual substances travel separately. The medium is referred to as the *stationary phase* while the solution is the *mobile phase*. As a separation technique, chromatography relies on the affinity each substance has for the chosen medium. The name chromatography is derived from Greek *chroma* meaning 'colour' and *graphein* meaning 'to write'. The product of a chromatography experiment is a *chromatogram*.

Chromatography is the basis for highly sophisticated and sensitive analytical techniques. These include thin layer chromatography (TLC), in which the stationary phase is usually silica gel or aluminium oxide fixed to a metal or plastic plate; gas-liquid chromatography

(GLC) in which an inert gas is used to force a solution of substances through a narrow tube packed with an inert material; and high-performance liquid chromatography (HPLC) in which liquid solvent containing the sample mixture is pressurised and pumped through a column of solid adsorbent material. Further information about chromatography techniques is available in this resource: https://edu.rsc.org/download?ac=13852 Applications of chromatography are discussed in the Careers box.

In school, the stationary phase is usually paper. Post-16 students may use TLC. A variety of investigations are possible using paper chromatography. Younger students enjoy separating mixtures of dyes present in coloured inks and the food colours used in the hard coloured casings of some brands of sweets. For older students, separating pigments in leaves is a good experiment, requiring an extraction step prior to setting up the chromatography stage. This resource provides full instructions: https://edu.rsc.org/cpd/practical-chromatography/2500327.article

Students should learn to calculate R_f values of separated substances on a chromatogram. R_f is an abbreviation of retardation factor, the ratio of distance moved by a substance and distance moved by the solvent from the starting position. Figure 9.4 shows a chromatogram of three substances.

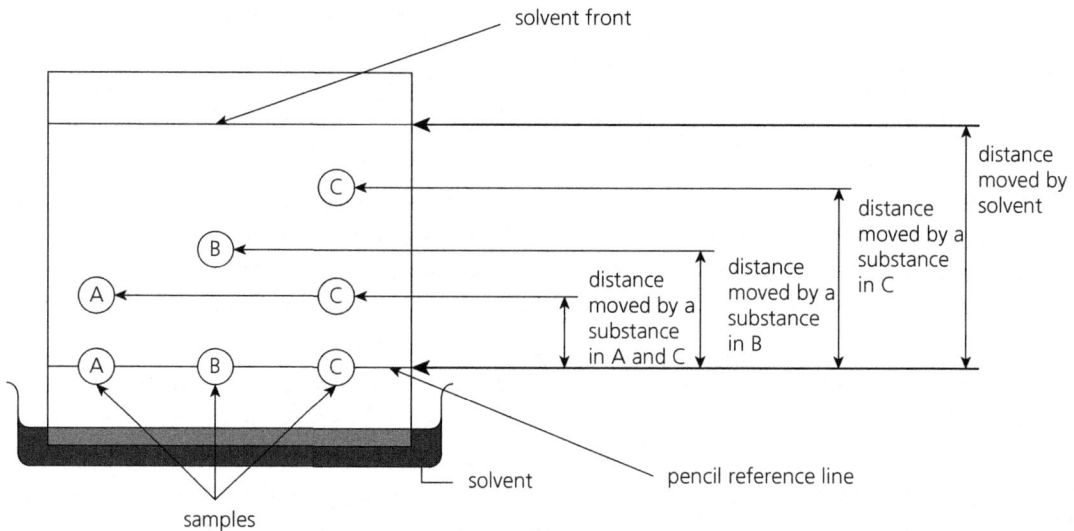

Figure 9.4 Chromatogram of three substances

The solvent front in this chromatogram has moved 4.2 cm from the reference line. R_f values for four substances are shown in Table 9.1, using the centre of each spot to measure distances.

R_f = distance travelled by spot/distance travelled by solvent

9.2 Identifying substances using physical properties

Table 9.1 R_f values for substances in the chromatogram in Figure 9.4

Substance	Distance moved from reference line/cm	R_f value based on solvent distance = 4.2 cm
A	1.3	0.31
B	2.0	0.48
C lower spot	1.3	0.31
C higher spot	3.2	0.76

Based on these data, A and B are either pure substances, or mixtures that cannot be separated using this solvent. Sample C is a mixture of a substance that could be A and another substance not present in A or B.

These results were obtained using a specific solvent. Separation patterns may change using a different solvent. Each spot could be a mixture that may separate into one or more substances if a different solvent is used. Note that R_f values vary from solvent to solvent.

Questions to ask
- What can we conclude about samples A, B and C?
- How do we know if the samples and spots on the chromatogram are pure substances?
- What other tests may be needed to confirm what is present in the samples and spots?
- How could the chromatography experiment be changed to get a different result using the same samples?

Resources that provide instructions for investigations using chromatography include:

- Why do leaves look red in autumn (what are the pigments in leaves)? https://edu.rsc.org/resources/chromatography-of-leaves/389.article
- Who wrote the love letter? https://edu.rsc.org/resources/chromatography/11333.article This practical begins with a simple explanation and video of paper chromatography but goes on to look in more detail at a more technical approach using column chromatography suitable for older students.
- What colourings are used in Smarties? https://edu.rsc.org/resources/chromatography-of-sweets/455.article

Identifying substances using physical properties

Substances can be identified or distinguished using differences in physical properties such as melting point, boiling point and density. Lessons should clarify if students are learning about physical properties or principles of analysis, noting these are not the same. Help students develop confidence by teaching properties separately before applying the technique to identify unknowns.

How can we identify substances by changes of state?

Students benefit from revisiting conceptual ideas behind each technique. For example, revisit particle theory (Chapter 3), to consolidate understanding of energy changes and bonds between particles. Relate these to melting and boiling a substance and the temperatures at which these changes occur.

An impure sample of a substance (a mixture) melts/freezes at a temperature below that of a pure sample. Particles of impurities disrupt intermolecular bonds between particles of the substance. This leads to less energy being needed to increase particle movement to the level needed for the substance to melt. Higher levels of impurities disrupt the structure further, causing greater variation in the types of bond present between particles. A melting point test helps an analytical chemist to determine if a sample is pure or impure. Further tests will determine any impurities present, and the proportion of impurities in a sample.

Conversely, boiling points increase when impurities are present. Impurities form intermolecular bonds with particles that require more energy to disrupt than those between particles of the substance alone. This means more energy is required to enable particles to break free of the liquid and transfer to the gas state. This increases the boiling point.

To develop students' understanding, carry out melting point determinations of first a pure substance then an impure sample of the same substance, plotting both on the same graph. The results should resemble Figure 9.5. Next, give students unlabelled samples of a different substance, one pure and the other impure. Ask them to test melting points to determine which is which. Dodecanoic acid (or lauric acid), used in soap, is suitable for this. Comparing solid soap (impure) with a pure sample of lauric acid allows students to observe smell and colour, using melting point as a confirmatory test. The technique for determining melting point is described next.

9.2 Identifying substances using physical properties

KEY ACTIVITY

Melting point determination for stearic acid

Students heat stearic acid to determine its melting point. Full instructions are available at https://edu.rsc.org/resources/melting-and-freezing-stearic-acid/1747.article

Using a data-logger to record the temperature over time allows a heating/cooling curve to be plotted automatically, so that students can focus on interpreting the graph. If a data-logger is not available, record the temperature at regular time intervals and plot these on a graph to show changes over time. For a pure substance, the heating curve features a horizontal line at the melting point temperature (69 °C for stearic acid, see Figure 9.5). A cooling curve shows a line in the same position as the freezing point (see Chapter 3). The melting/freezing point depends on purity. Impure stearic acid produces a sloping line under the melting/freezing point, as the sample melts/freezes over a range of temperatures just below 69 °C.

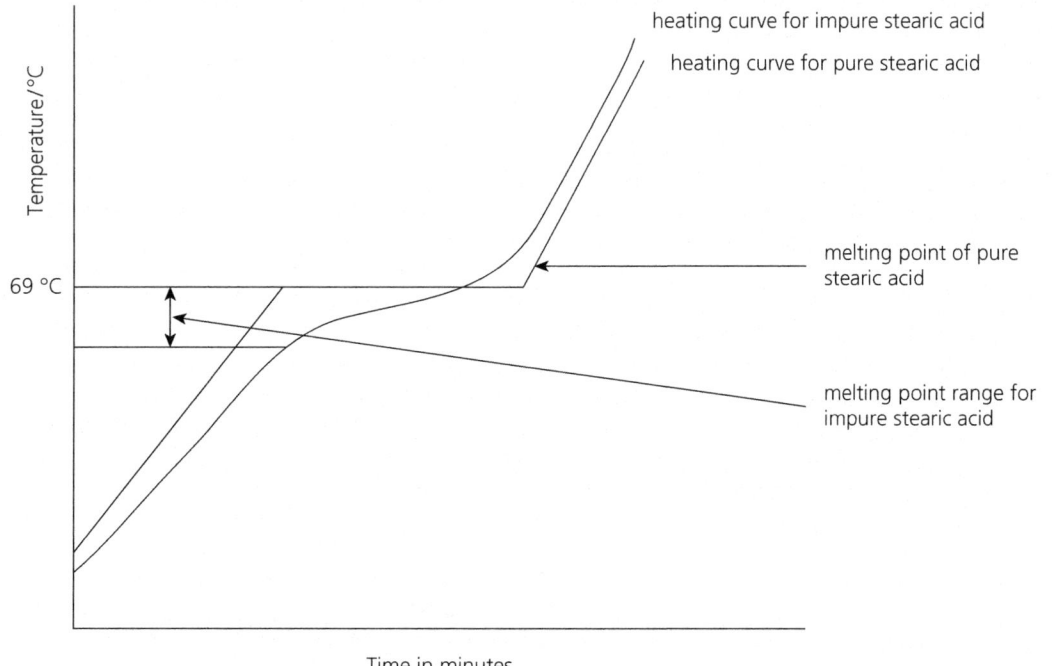

Figure 9.5 Heating curves for pure and impure stearic acid samples

Source: V. Kind, permission for reuse granted

Questions to ask
- At what temperature does stearic acid solidify/freeze?
- How do we know if the sample of stearic acid is pure?
- What might the curve look like if the stearic acid is impure?

9 Chemical analysis

- Why do impure and pure substances have different heating/cooling curves?
- Explain how a precise melting/freezing point helps to identify a substance.
- What will a melting/freezing point not tell us?

A heating curve for ice gives a similar-shaped curve, with the horizontal line at the melting point of ice (0 °C). However, a heating curve for wax (generally a mixture of polymers) does not have a clear horizontal line as different polymers in the mixture melt at different temperatures.

KEY ACTIVITY

Identifying a substance by boiling point determination

Clear colourless liquids, such as ethanol, propanone and water, can be identified from their boiling points. Full instructions for carrying out this demonstration safely are at https://edu.rsc.org/cpd/separation-techniques/3009787.article Boiling tubes containing propanone and water in a hot water bath at about 60 °C may stimulate cognitive conflict, as the colourless liquids appear identical but only the propanone boils. Revise why this is the case: Figure 3.9 shows differences in melting and boiling points for several substances. Solvents in any solution can be identified by boiling point, for example water from ink during distillation.

Water samples can be investigated using the context that access to potable water is essential for health. Test boiling points of water samples from various sources. Investigate whether all impurities have the same effect on boiling point: boil water samples dry and test any resulting solids (see below) to determine impurities present. Invite students to bring water samples for testing, including tap water, bottled water and water from streams, rivers and ponds. This may prompt discussion about how water is cleaned for drinking, and differences between pure and drinkable (potable) water.

How can we identify a substance using density?

Density measures closeness of packing of particles, as well as particle mass. A substance with closely packed, heavy particles is denser than a substance with particles that are the same mass but less closely packed. The unit of density is mass per unit volume: in chemistry, usually $g\,dm^{-3}$, grams per decimetre cubed. One cubic decimetre is one litre. Differences in density can establish the material from which an object is made. Measuring density is important for determining purity, especially of precious metals such as gold, silver and platinum (see Science in

9.2 Identifying substances using physical properties

context). Density identifies counterfeit coinage, both historically and today. Measuring density requires making measurements of an object's mass and volume. If the object is irregular, volume can be measured using a Eureka can. The can is filled with water and the object, attached to a thread, is lowered in. Water displaced by the object leaves the can via a spout arranged above a measuring cylinder. The volume of the object is read from the measuring cylinder scale. The following animation allows students to explore known and unknown blocks in a flotation tank https://phet.colorado.edu/sims/html/density/latest/density_en.html (Flash player required).

> ### Enrichment
> Gold, symbol Au, has been valued as a precious metal symbolising immortality, power and wealth for thousands of years. The purest samples of mined elemental gold contain up to 5% silver. Knowing the purity of this metal is vital to its value. The Egyptians devised a fire assay in around 1500BCE. This involved heating gold in a bone-ash crucible with lead metal. Impurities were absorbed by the crucible while lead and silver were removed by reaction with nitric acid. The pure gold remaining was weighed and its mass compared with the starting mass. Archimedes (287–212BCE) knew that the specific gravity (density) of gold was altered with base metals, such as lead. This principle was used to detect counterfeit coins made from a mixture of gold and cheaper metals such as silver.

KEY ACTIVITY

Investigating plastics for recycling using sink/float density separation

Students are likely to be familiar with setting aside plastics for recycling. This activity illustrates one step in the recycling process, in which a float–sink test is used to assess plastic density: https://edu.rsc.org/resources/identifying-polymers/385.article

The activity involves testing whether samples of plastics sink or float in a range of liquids. Each liquid has a different density. The plastics also differ in density, so a distinctive sink/float pattern emerges for each one. Using one or two liquids means some plastics cannot be identified, as more than one plastic behaves the same way. The aim is that students identify plastics from results of the tests. A table of expected outcomes for seven types of plastic (listed by plastic recycling number 1–6 and EPS for expanded polystyrene), a set of density data and suggested sources of plastic samples are provided. The activity could be extended to investigate the ease with which specific plastics are recycled and recycling costs. EPS is usually not collected for recycling but goes to landfill: students can find out

> why. Organisations such as the British Plastics Federation have additional information to support discussion (see Section 9.9 Other resources).
>
> **Questions to ask**
> - Why are the plastics tested in more than one liquid?
> - Why is the sink–float test useful for recycling plastics?
> - Investigate the practicality and cost of recycling plastics.
>   Are some easier/more expensive to recycle than others?
>   Are any impossible/very difficult to recycle?

 Identifying substances using chemical reactions

Qualitative inorganic analysis provides information about what is present in a substance sample, not how much of any substance is present. Qualitative tests rely on macroscopic observations such as colour changes, formation of precipitates and evolution of gases. The tests often involve substances in aqueous (water-based) solution, known colloquially as wet tests. Standard health and safety procedures need to be followed when undertaking tests. In general, this includes wearing eye protection; knowing how to deal with spillages; smelling gases by wafting a small sample gently towards the nose and inhaling cautiously; heating gently; and ensuring stock bottles of reagents and pipettes are not contaminated. Many reactions can be done with low risk and good results using the microscale approach (https://microchemuk.weebly.com/).

Developing students' understanding and expertise

When testing unknowns, students need to work accurately while simultaneously understanding the results. This can be challenging, so a stepwise approach to developing their skills and knowledge is recommended. Support students' learning by making explicit connections to analysis from other chemistry topics. For example, flame tests can be considered with the chemistry of Groups 1 and 2 (see Chapter 4) and tests for halide (chloride, bromide and iodide) ions with Group 7. This enables students to amass information over time, revisiting knowledge as needed when undertaking analysis.

9.3 Identifying substances using chemical reactions

Give students opportunities to practice tests before applying these to unknown substances. This develops mastery of techniques and understanding of results. Practice manipulative skills, so students become used to using small amounts of chemicals and samples; avoiding contamination by using clean spatulas and pipettes; keeping their working area tidy; being systematic. Provide examples that give inconclusive results (see Table 9.6), so students learn that not all tests give definitive answers. For example, several anions may give white precipitates in the same test, so another test is needed for precise identification. Tests are described below together with flowcharts to help students work stepwise. Encourage students to make accurate observations and describe these precisely. For example, 'a white precipitate forms' is preferred to 'the solution goes cloudy'; and 'no visible change' rather than 'nothing happens'.

Once students have developed confidence, training on unknowns can be undertaken. The simple flowchart in Figure 9.6 is helpful in considering the steps.

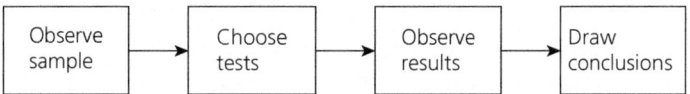

Figure 9.6 Steps in analysis

Observations of samples give important information about the type of substance. Is the sample a powder, crystalline or a solution? What colour is it? Based on these observations, students choose tests to identify, if the sample is assumed to be a salt, the cation (positive ion) and anion (negative ion) present. These tests are described below. If a gas test is required, the gas should be identified. A flame test colour may indicate a possible cation.

Drawing conclusions is challenging, as students may be reluctant to commit to a specific substance. Encourage them to apply skills and knowledge developed in previous stages (see above) and not to be afraid of being wrong! Focus on the process of analysis rather than emphasising obtaining one single correct answer. The flowchart below (Figure 9.7) summarises the tests for ions described. To build students' knowledge, present this flowchart when tests for cations (positively charged ions) and anions (negatively charged ions) have been learned. Separate flowcharts are provided for each of these ion groups.

9 Chemical analysis

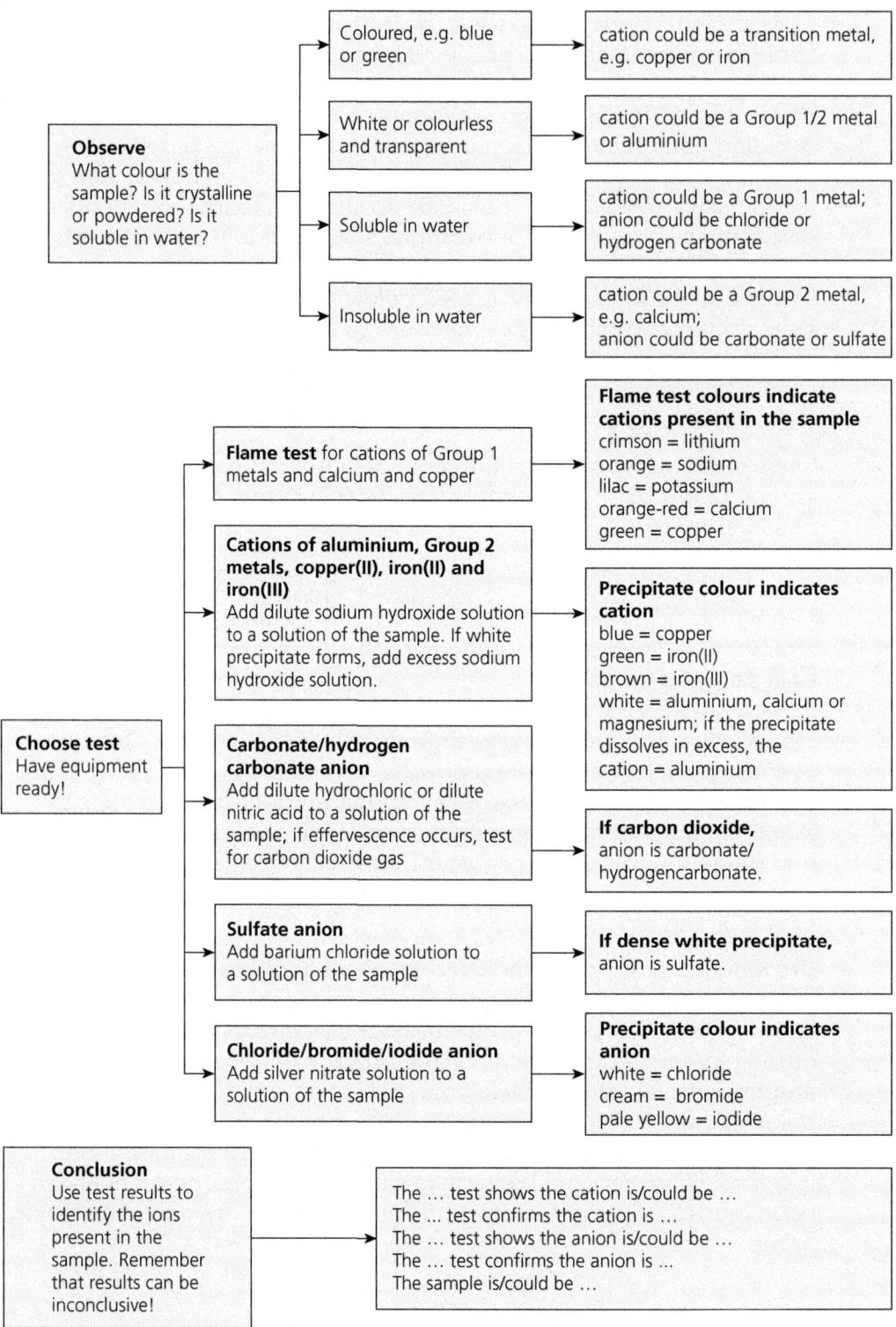

Figure 9.7 Qualitative analysis flowchart

 How to identify gases produced in reactions

Pre-requisite knowledge includes characteristics of gases, including the difference between bubbles formed in a reaction and those in a boiling liquid; and that gases have mass (Chapter 3). Familiarity with chemical reactions and associated practical work increases students' confidence in predicting which gas may be produced and testing techniques. Table 9.2 summarises reactions that produce gases.

In addition, awareness of the characteristics of sulfur dioxide and nitrogen dioxide are helpful, to connect to industrial equilibrium reactions (sulfur dioxide, Contact process) and burning fossil fuels (sulfur dioxide, nitrogen dioxide). The chemistry of ammonia can be explored through a microscale practical in a Petri dish: http://science.cleapss.org.uk/Resource-Info/PP064-Ammonia-chemistry-in-a-petri-dish.aspx

Table 9.3 shows tests for hydrogen, oxygen, carbon dioxide, chlorine and ammonia. Explanations and advice for carrying out each test are included. When these gases are mentioned in topics, such as electrolysis, acids and alkalis, and the Haber process, consolidate knowledge about how to test for them. Develop students' manipulative skills with repeated practice.

In general, to test for gases, each student, pair or group requires: a test-tube holder; test tubes and a test-tube rack; red and blue litmus paper; de-ionised water in a wash bottle; limewater (dilute calcium hydroxide solution); $1\,\text{mol}\,\text{dm}^{-3}$ hydrochloric acid; wooden splints; Pasteur pipettes; Bunsen burner.

The two most commonly used tests are for hydrogen gas and carbon dioxide gas. Either may be produced when a dilute acid is added to an unknown solid. Figure 9.8 shows how to test for hydrogen gas. In Step 1, observations suggest a grey powder is likely to be a metal. This should prompt the hydrogen gas test. In Step 5, remember hydrogen is lighter than air, so will escape if Step 3 is not done quickly.

9 Chemical analysis

Table 9.2 Examples of reactions producing gases

Gas	Reaction(s)	Example
Hydrogen H_2	acid + metal → salt + hydrogen	magnesium + hydrochloric acid → magnesium chloride + hydrogen $Mg(s)$ + $2HCl(aq)$ → $MgCl_2(aq)$ + $H_2(g)$
Oxygen O_2	Thermal decomposition of nitrates Decomposition of hydrogen peroxide	magnesium nitrate → magnesium oxide + nitrogen dioxide + oxygen $2Mg(NO_3)_2(s)$ → $2MgO(s)$ + $4NO_2(g)$ + $O_2(g)$ hydrogen peroxide → water + oxygen $2H_2O_2(l)$ → $H_2O(l)$ + $O_2(g)$
Carbon dioxide CO_2	acid + metal carbonate → salt + water + carbon dioxide acid + metal hydrogen carbonate → salt + water + carbon dioxide	hydrochloric acid + sodium carbonate → sodium chloride + water + carbon dioxide $2HCl(aq)$ + $Na_2CO_3(s)$ → $2NaCl(aq)$ + $H_2O(l)$ + $CO_2(g)$ hydrochloric acid + sodium hydrogencarbonate → sodium chloride + water + carbon dioxide $HCl(aq)$ + $NaHCO_3(s)$ → $NaCl(aq)$ + $H_2O(l)$ + $CO_2(g)$
Chlorine Cl_2	Electrolysis of sodium chloride solution	Anode (+) reaction: chloride ions → chlorine gas + electrons $2Cl^-(aq)$ → $Cl_2(g)$ + $2e^-$
Ammonia NH_3	ammonium salt + sodium hydroxide solution → ammonia + sodium salt + water	ammonium nitrate + sodium hydroxide → sodium nitrate + water + ammonia $NH_4NO_3(s)$ + $NaOH(aq)$ → $NaNO_3(aq)$ + $H_2O(l)$ + $NH_3(g)$

9.4 How to identify gases produced in reactions

Table 9.3 Tests identifying gases

Gas	Test	Explanation	Advice/further information
Hydrogen H_2	**Squeaky pop test** Lighted splint in the mouth of a test tube of hydrogen ignites the gas producing a pop sound.	Hydrogen gas reacts with oxygen gas in the atmosphere to produce water vapour $$2H_2(g) + O_2(g) \rightarrow 2H_2O(g)$$	Hold the test tube and splint at arm length. The squeaky pop is a very distinctive sound. A test for hydrogen is used when an unknown substance is reacted with dilute acid. See sections 8.3 and 8.4. A positive test confirms that the unknown substance is a metal (see Table 9.2).
Oxygen O_2	**Glowing splint test** Place a glowing splint into a test tube of oxygen. A bright yellow flame is produced.	The heat of the splint and high concentration of oxygen is sufficient to ignite the wood. This is a complex reaction summarised as: Wood + oxygen → carbon dioxide + water vapour + ash	Collect the gas by placing a thumb or finger over the mouth of the tube. Push the splint well into the test tube.
Carbon dioxide CO_2	**Limewater test** Bubble gas into limewater, which is calcium hydroxide solution. If carbon dioxide is present, a white precipitate appears as fine wisps of solid. Limewater does not contain lime juice! The name originates from using calcium hydroxide as lime or mortar for constructing buildings.	Carbon dioxide reacts with calcium hydroxide to produce calcium carbonate which is the precipitate: $$Ca(OH)_2(aq) + CO_2(g) \rightarrow CaCO_3(s) + H_2O(l)$$	Collect the gas by squeezing a plastic pipette then holding this above the reaction mixture, gradually releasing the squeezed bulb (see Figure 9.9). Students should pipette the gas repeatedly, and look closely for the precipitate. **The test identifies an anion as a carbonate or hydrogencarbonate.** To tell the difference, add barium nitrate solution to a solution of the solid: a white precipitate confirms the carbonate anion; no precipitate confirms hydrogencarbonate ions.
Chlorine Cl_2	**Blue litmus test** Dampen a piece of blue litmus paper. Chlorine gas turns the litmus paper red then white.	Chlorine is soluble in water forming an acidic solution. The low pH causes litmus to turn red. Chlorine solution has a bleaching effect so the paper becomes white.	Chlorine gas denser than air and green-yellow in colour. It bleaches litmus paper white. Smell with caution: chlorine is highly toxic; the smell is associated with swimming pools.
Ammonia NH_3	**Red litmus test** Dampen a piece of red litmus paper. Ammonia gas turns this blue.	Ammonia is highly soluble in water forming an alkaline solution. The high pH causes red litmus to turn blue.	Ammonia gas is highly toxic and less dense than air. Smell with caution: ammonia has a strong pungent smell.

9 Chemical analysis

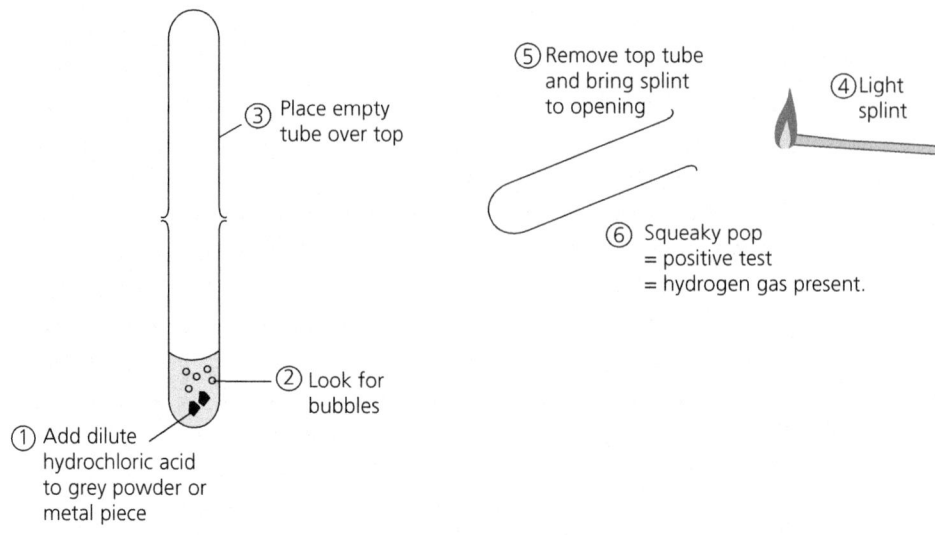

Figure 9.8 How to test for hydrogen gas

Figure 9.9 shows how to test for carbon dioxide. A test tube containing limewater placed beside the test tube used to make gas must be available. A difficulty is Step 5, as students find pipetting something they cannot see perplexing. Carbon dioxide is denser than air so remains in the tube. In Step 7, wisps of white precipitate show the presence of carbon dioxide. This is a positive test for carbonate (CO_3^{2-}) and hydrogen carbonate (HCO_3^-) ions in the solid used in Step 3. This test will not distinguish between these two anions, so the result should be described as showing the solid is a carbonate or hydrogencarbonate. To distinguish between them, make a solution

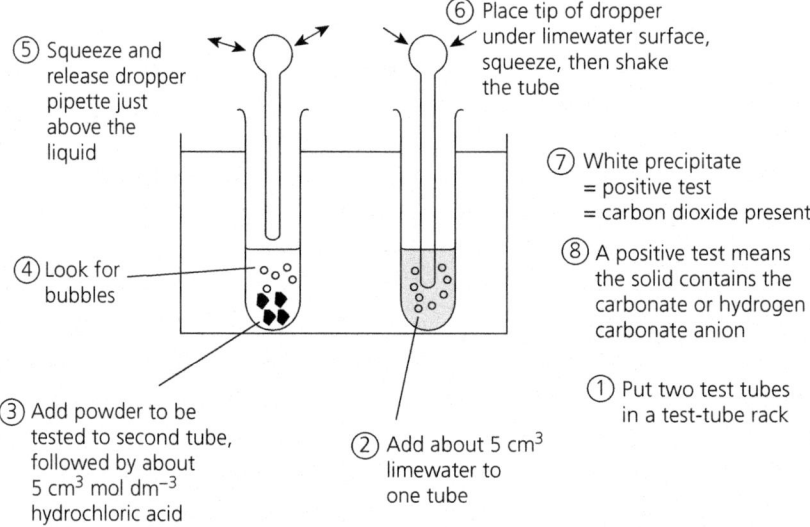

Figure 9.9 How to test for carbon dioxide

9.4 How to identify gases produced in reactions

of the salt that gives a positive carbon dioxide test. Add a few drops of barium nitrate solution: a white precipitate indicates that the salt contains the carbonate anion. If there is no precipitate, the salt contains the hydrogencarbonate anion.

Using the correct acid is crucial. Sulfuric(VI) acid reacts with carbonates and hydrogencarbonates. Carbon dioxide will be produced, but another product is a sulfate. Many sulfates are insoluble so reactions stop after a few minutes. For example, calcium carbonate (marble chips) reacts with sulfuric(VI) acid to form insoluble calcium sulfate which prevents further reaction between calcium carbonate and the acid:

$$CaCO_3(s) + H_2SO_4(aq) \rightarrow CaSO_4(s) + H_2O(l) + CO_2(g)$$

Figure 9.10 provides a flow chart to help students decide which gas test to choose. This starts with observations about the sample,

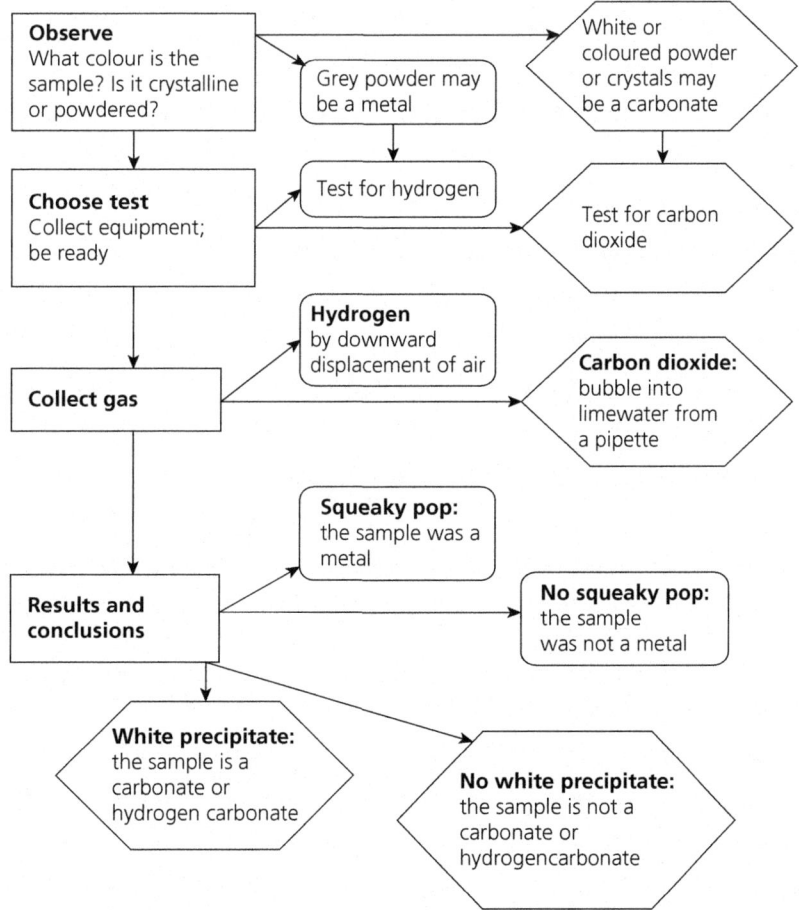

Figure 9.10 Testing for hydrogen or carbon dioxide

proceeds to selecting a test, considering the equipment, carrying out the test and drawing conclusions.

 How to identify ions in salts

Salts contain oppositely charged ions, cations (+) and anions (−). These ions form crystalline lattices. Ionic bonds form between oppositely charged particles (see Chapter 5). The properties of ionic compounds vary: they may be coloured, soluble/partially soluble/insoluble, and form coloured complex ions with water or other small molecules (see Chapter 4 and Chapter 5).

When analysing a salt, cation(s) and anion(s) are considered separately, as each ion has specific characteristics. For example, sodium chloride comprises the sodium ion (Na^+), which gives a characteristic bright orange flame test colour. The chloride ion (Cl^-) is identified by adding silver nitrate solution to a solution of the salt. A white precipitate of silver chloride forms. To identify a salt, students need to carry out at least two tests, one each for the cation and anion present.

The flowchart in Figure 9.7 (above) shows the steps. Focus on attention to detail, using small amounts of reagents and proceeding carefully, making observations at each stage. Microscale activities allow students to see results using minimum apparatus and chemicals, reducing risk and expense. Detailed practical instructions for carrying out cation and anion tests are widely available from websites such as https://www.bbc.co.uk/bitesize/guides/zxtvw6f/video, https://www.youtube.com/watch?v=fCZztwJmAl0 and https://edu.rsc.org/resources/qualitative-analysis-quizzes/2201.article

Identifying cations

Two tests are commonly used to identify cations: the flame test (described below) and the sodium hydroxide test. Students need to consider how to use these tests and what conclusions to draw from results. Figure 9.11 shows both tests in a flowchart. Start by making observations about the sample and using these to consider possible outcomes of tests.

9.5 How to identify ions in salts

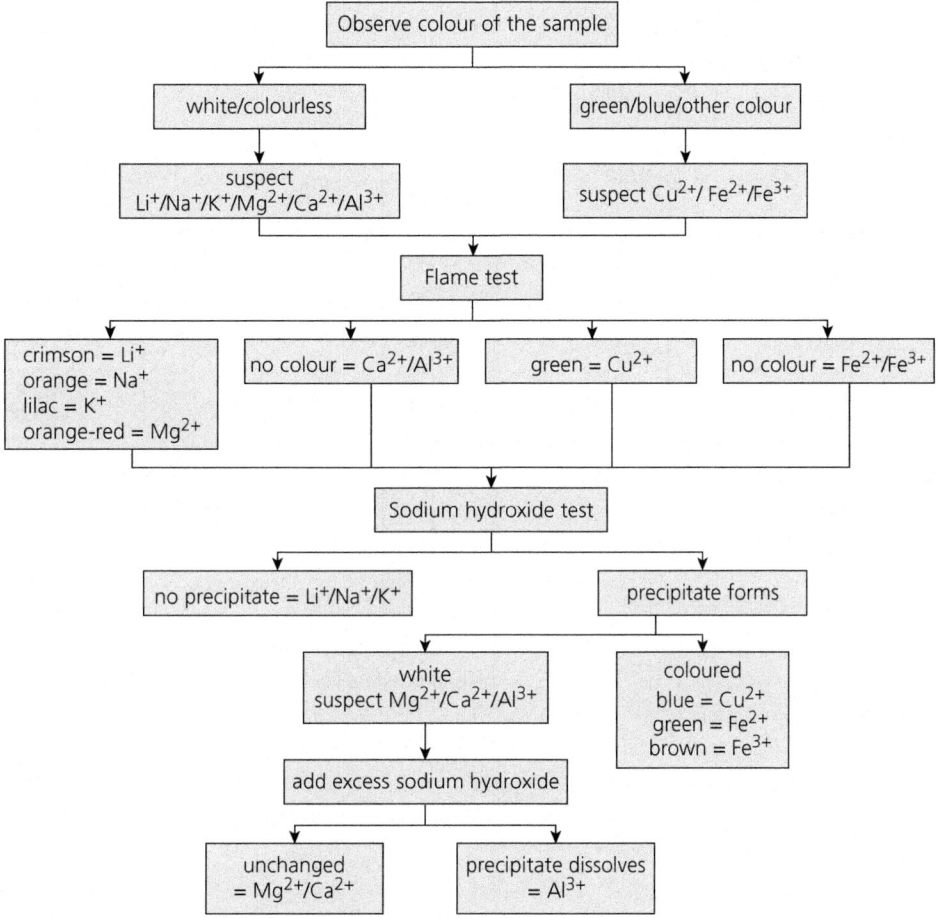

Figure 9.11 Flow chart for identifying cations

The flame test identifies Group 1 metal cations and copper(II) ions. The sodium hydroxide test identifies copper(II), iron(II), iron(III) and aluminium ions, and indicates if a Group 2 metal cation may be present.

The sodium hydroxide test

The steps in the sodium hydroxide test are shown in Figure 9.12 which also summarises the most likely outcomes of the test.

The test involves making a solution of the unknown sample, adding dilute sodium hydroxide solution, observing the result, then drawing conclusions about the cation present in the sample. Table 9.4 describes

9 Chemical analysis

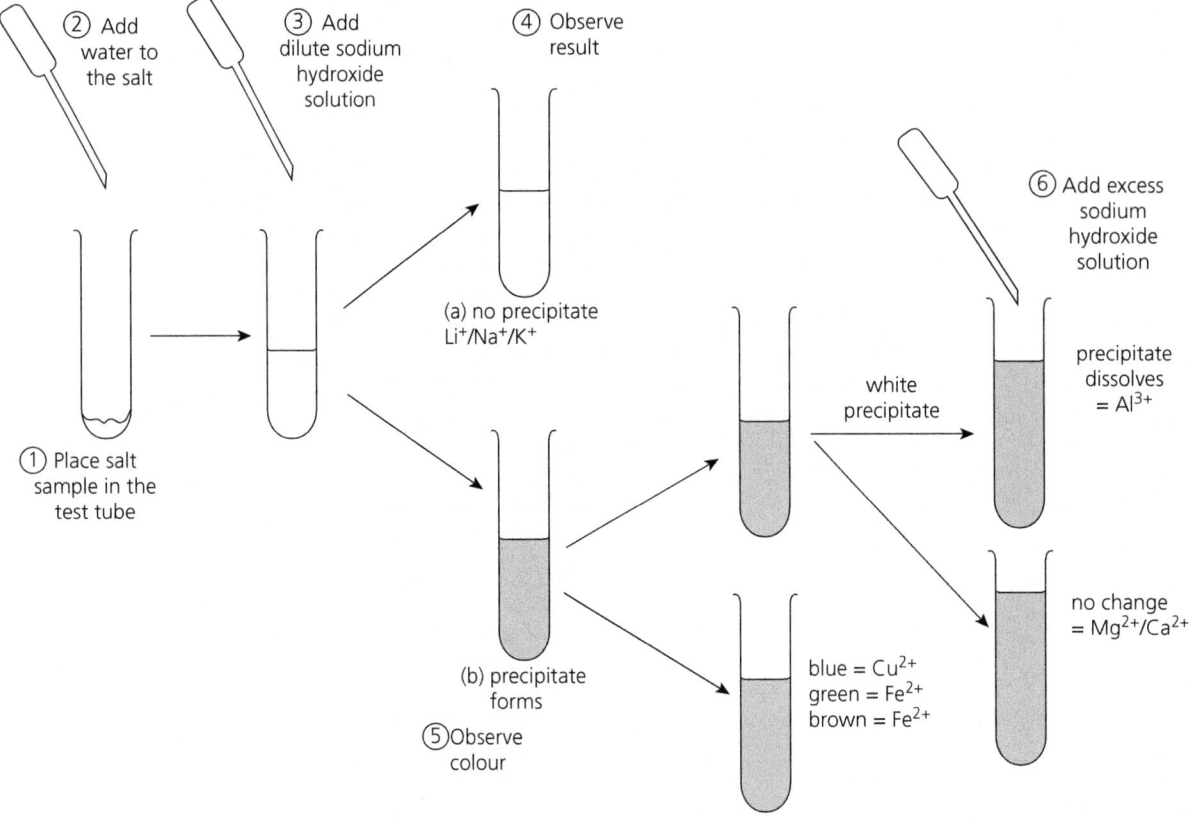

Figure 9.12 The sodium hydroxide test

the most common results that students aged up to 16 are expected to know.

Identifying anions

Anions are negatively charged, non-metallic ions. Students need to know tests for carbonate, sulfate and halide ions (chloride, bromide and iodide). These tests are summarised in Table 9.5 and Figure 9.13.

9.5 How to identify ions in salts

Table 9.4 Tests for cations

Cation test	Results		Chemical explanation
Add 1 mol dm^{-3} sodium hydroxide solution dropwise to an aqueous solution of a metal ion	Many cations in aqueous solution react with hydroxide ions to produce the metal hydroxide as a precipitate. metal cation + hydroxide ions → metal hydroxide The metal hydroxide precipitate colour is characteristic of the metal cation:		Sodium hydroxide is a strong alkali with a large number of hydroxide ions available in a small volume of solution. No metals of Group 1 give a positive result (a precipitate) with this test. All compounds of Group 1 metals, including hydroxides, are soluble in water.
	Fe^{2+} green	iron(II) hydroxide $Fe(OH)_2$	Iron(II) hydroxide turns brown on standing, as iron(II) ions are oxidised.
	Fe^{3+} brown/rust	iron(III) hydroxide $Fe(OH)_3$	Hydroxides of transition elements may dissolve in excess hydroxide molecules, forming complex ions. Complex ions are soluble and may be colourful.
	Cu^{2+} blue	copper(II) hydroxide $Cu(OH)_2$	
	Zn^{2+} white	zinc hydroxide $Zn(OH)_2$	Zinc hydroxide precipitate dissolves in excess sodium hydroxide solution to produce a clear, colourless solution.
	Ca^{2+} white	calcium hydroxide $Ca(OH)_2$	Compounds of Group 2 metals, including hydroxides, tend to be partially soluble. This means precipitates appear as fine wisps of solid rather than opaque clouds.
	Mg^{2+} thin, white	magnesium hydroxide $Mg(OH)_2$	
	Al^{3+} white gelatinous	aluminium hydroxide $Al(OH)_3$	Aluminium hydroxide dissolves in excess hydroxide ions to form a clear, colourless solution.

9 Chemical analysis

Table 9.4 Tests for cations (continued)

Cation test	Results	Chemical explanation
	Inconclusive results: four cations produce a white precipitate, so further tests are required.	
	To tell the difference between zinc and aluminium, examine the precipitate: a jelly-like consistency means aluminium ions are present. A powder-type precipitate means zinc ions are present.	
	Add excess hydroxide ions: if the precipitate dissolves, the cation in the solid is aluminium or zinc. No change indicates the cation is magnesium.	
	Calcium ions give an orange-red colour in a flame test on the solid. The other cations do not give a distinctive flame test colour.	
Add ammonium hydroxide solution to a test tube containing a metal hydroxide precipitate	Fe green — iron(II) hydroxide $Fe(OH)_2$	Ammonia reacts with water to produce ammonium ions and hydroxide ions. These react with iron(II) and iron(III) ions. Iron(II) ions form a green precipitate which does not react with excess sodium hydroxide solution or ammonia solution.
	Fe brown — iron(III) hydroxide $Fe(OH)_3$	A brown precipitate indicates iron(III) ions.
Add the solid to a test tube containing sodium hydroxide solution and warm gently	Warming releases ammonia gas (NH_3). The gas has a pungent smell and turns damp red litmus paper blue.	**This test is useful when all other tests are negative.** Ammonium compounds contain the NH_4^+ (ammonium) ion, which dissolves in sodium hydroxide solution. Warming produces ammonia gas and water. Ammonia gas is toxic so should be smelt with care by wafting gently.

9.5 How to identify ions in salts

Table 9.5 Tests for anions

Anion test	Results	Chemical explanation
For sulfates and carbonates Add aqueous barium nitrate or barium chloride solution dropwise to an aqueous solution of the unknown compound. Then add a few drops of dilute nitric acid.	Solutions of sulfate (SO_4^{2-}) and carbonate (CO_3^{2-}) ions produce white precipitates on addition of barium ions: barium ion + sulfate ion $\rightarrow$ barium sulfate $Ba^{2+}(aq)$ + $SO_4^{2-}(aq)$ $\rightarrow$ $BaSO_4(s)$ barium ion + carbonate ion $\rightarrow$ barium carbonate $Ba^{2+}(aq)$ + $CO_3^{2-}(aq)$ $\rightarrow$ $BaCO_3(s)$ Barium carbonate reacts with nitric acid: $BaCO_3(s) + 2HNO_3(aq) \rightarrow Ba(NO_3)_2(aq) + CO_2(g) + H_2O(l)$	All nitrates are soluble so barium nitrate is a source of aqueous barium ions. Barium sulfate is a dense white precipitate. The test confirms the presence of sulfate ions. Barium carbonate is insoluble and appears as a white precipitate. Carbonates react with dilute nitric acid to produce carbon dioxide. A fizz (effervescence) occurs; carbon dioxide can be identified by the limewater (calcium hydroxide solution) test described in Table 9.3.
For chlorides, bromides and iodides (halides) Test the aqueous solution of a compound with dilute silver nitrate solution.	Silver nitrate solution produces precipitates with chloride, bromide and iodide (halide) ions: silver ion + halide ion $\rightarrow$ silver halide The silver halide colour is characteristic of the halide ion present. AgCl silver chloride White AgBr silver bromide Cream AgI silver iodide Yellow Silver nitrate produces a precipitate with carbonate ions: silver ion + carbonate ion $\rightarrow$ silver carbonate $2Ag^+(aq)$ + $CO_3^{2-}(aq)$ $\rightarrow$ $Ag_2CO_3(s)$	Anions formed by halogens are known as halide ions. All have a 1− charge. All fluorides are soluble in water, so this test does not apply to them. Each halogen produces a coloured silver halide that is slightly darker than the preceding halide in Group 7. $Ag^+(aq)$ + $Cl^-(aq)$ $\rightarrow$ $AgCl(s)$ $Ag^+(aq)$ + $Br^-(aq)$ $\rightarrow$ $AgBr(s)$ $Ag^+(aq)$ + $I^-(aq)$ $\rightarrow$ $AgI(s)$ Silver halides darken in sunlight. Carbonates form a white precipitate of silver carbonate. This reacts with dilute hydrochloric acid to produce carbon dioxide (see above). Silver carbonate does not darken in sunlight.

9 Chemical analysis

Carbonates and sulfates form white precipitates with barium ions in solution. Addition of dilute nitric acid distinguishes between them. If the precipitate is barium carbonate, carbon dioxide is produced, identifiable using the test described above. Barium sulfate does not react with dilute nitric acid. Nitric acid is chosen because all nitrates are soluble, so no additional precipitates mask the result. Silver ions form precipitates with chloride, bromide, iodide and carbonate ions; silver sulfate(VI) is sparingly soluble and a precipitate may not be formed if concentrations of the reagents used are low.

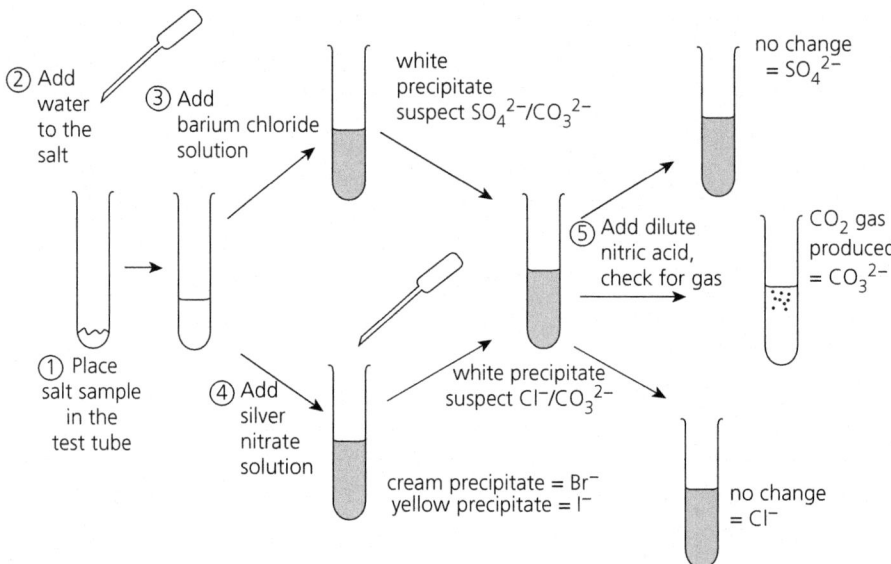

Figure 9.13 Testing for anions

Questions to ask

→ Why is nitric(V) acid the best acid to use?
→ Could the nitric(V) acid be added to the unknown solution before barium nitrate(V) solution?

9.6 Handling inconclusive results

Usually, for students aged up to 16, an unknown solid gives a distinctive result for the flame test (described below) and/or the sodium hydroxide test (see Figure 9.12 and Table 9.4) and an anion test (Figure 9.13 and Table 9.5). However, students should not expect to get positive outcomes for every test but *consider what inconclusive results mean and how to handle them*. Train students to think of identification as a process of steps, as the flowcharts (Figures 9.6 and 9.7) suggest. They should not panic if a test does not seem to give a positive result but think systematically about what to do next.

9.6 Handling inconclusive results

Table 9.6 suggests salts that give inconclusive results for one or more tests described above, with reasoning that helps to identify the ions present. When students are confident in their skills in performing tests, include one of these in a pair of unknowns (see case studies, below) to develop their ability in handling inconclusive outcomes.

Table 9.6 Salts that give inconclusive results in cation or anion tests

Salt		Cation identification		Anion identification	
Cation	Anion	Flame test	Sodium hydroxide test	Barium chloride solution	Silver nitrate solution
magnesium	hydrogen carbonate	No distinctive colour	Fine white precipitate that does not dissolve in excess hydroxide ions confirms the cation is magnesium.	No precipitate. The anion is not sulfate.	No precipitate. The anion is not carbonate, chloride, bromide or iodide.
					Add dilute nitric acid to a solution of the solid. Gas produced. Test for carbon dioxide gas is positive. This confirms that the anion is hydrogencarbonate.
zinc	chloride	No distinctive colour	Fine white precipitate that dissolves in excess hydroxide ion confirms the cation is zinc.	No precipitate.	White precipitate that darkens on standing confirms the anion is chloride.
aluminium	carbonate	No distinctive colour	Gelatinous white precipitate that dissolves in excess hydroxide ions confirms the cation is aluminium.	White precipitate forms.	White precipitate forms that does not darken on standing.
				Add dilute nitric acid to a solution of the solid. Gas produced. Test for carbon dioxide gas is positive. This confirms that the anion is carbonate.	
calcium	iodide	Orange-red colour indicates calcium ions	Fine white precipitate that does not dissolve in excess hydroxide ions confirms calcium ions.	No precipitate. The anion is not sulfate.	Pale-yellow precipitate that darkens on standing confirms the anion is iodide.

> **Enrichment**
>
> The first breathalysers used wet tests to identify if vehicle drivers had been drinking alcohol. The idea of a test to measure the amount of alcohol a person had consumed began in 1920s America, during the time of Prohibition, when selling, importing and transporting alcoholic beverages was illegal. In 1938 Rolla N. Harger, a professor of biochemistry and toxicology at the University of Indiana medical school devised a machine that could be operated by the police. Harger's machine relied on the colour

> change occurring when a sample of breath containing ethanol vapour passes over acidified potassium permanganate: it decolourises, going from purple-pink to colourless. When asked what the machine was called, Harger jokingly replied, 'The Drunkometer' which somehow stuck. Today, hand-held breathalysers have sensors that detect changes using semi-conductors or fuel cells, while table-top infra-red spectrometers are used in police stations.

 Spectroscopy

Analysis of light and other types of radiation absorbed or emitted by materials is known as spectroscopy. The word comes from Latin *specere* meaning 'to look at' and Greek *skopia*, which means 'to see'. Spectroscopy has provided scientists with extensive information about atomic structure. Chemical elements emit light of specific wavelengths when heated in a Bunsen flame or when an electrical current is passed through the vaporised substance. The emitted light is a mixture of light wavelengths. These are separated in a spectrometer, which passes light through a slit to form a fine beam, then through a prism or diffraction grating to split the beam into separate wavelengths of light, forming a spectrum. The pattern of lines, known as spectral lines, corresponds to light of specific wavelengths. Spectra for each chemical element are unique, so act as 'fingerprints' that identify elements. Examples of spectra are shown in Figure 9.15, below.

Modern spectroscopy extends to many types of radiation, including X-rays, nuclear magnetic resonance (NMR), ultra-violet (UV) and infra-red (IR), leading to advanced techniques including flame emission spectroscopy (FES) discussed below.

Flame tests

Flame tests are a form of spectroscopy. Colours emitted in flame tests identify metal cations. In a flame test, a sample of a compound is placed in a hot Bunsen flame and any colour observed. While this sounds straightforward, encouraging students to develop good technique is important. There are excellent videos and instructions available online showing details of how to carry out flame tests using nichrome wire or wooden splints (for example, https://edu.rsc.org/resources/flame-tests-using-metal-salts/1875.article). Alternatively, flame tests make a good teacher demonstration. In this format, solutions of compounds containing metal ions can be prepared in spray bottles and sprayed into a blue Bunsen flame. Alternatively, ethanol-based solutions of solids can be burned on watch glasses placed on a heatproof mat. Appropriate safety precautions should be taken (see https://edu.rsc.org/exhibition-chemistry/the-rainbow-flame-demonstration/3009399.article).

Specific chemical elements generate flames with characteristic colours as shown in Table 9.7.

Table 9.7 Flame test colours given by specific cations

Cation	Colour of flame test
sodium	bright orange/yellow
potassium	lilac
lithium	red
calcium	brick red/dark orange
strontium	red/crimson
barium	pale green/apple green
copper	green
magnesium	no colour

The colours are visible in fireworks, which are made to create specific effects on burning. Careful choice of chemicals creates specific colours, for example, purple fireworks use strontium and copper.

Flame emission spectroscopy

Modern analytical processes use instrumental methods to identify substances. These include mass spectrometers, flame emission spectrometers and NMR spectrometers. These produce fast, accurate and often quantitative results, identifying how much of a substance is present as well as what the substance is.

Old and new breathalyser tests illustrate differences between wet chemistry tests and instant electronic measurement. Old breathalysers detect the presence of alcohol in exhaled breath by reaction with potassium dichromate (potassium chromate(VI)) crystals, which change colour from orange to green. Modern digital devices incorporate a mini spectrometer (see https://www.youtube.com/watch?v=gMNCoJbq1cI) which give an accurate, sensitive and reliable quantitative readout.

Students can experience flame emission spectrometry using a hand-held spectrometer. These produce characteristic fingerprint pattern of lines for given elements. An example of such a spectrum is shown in Figure 9.15.

9 Chemical analysis

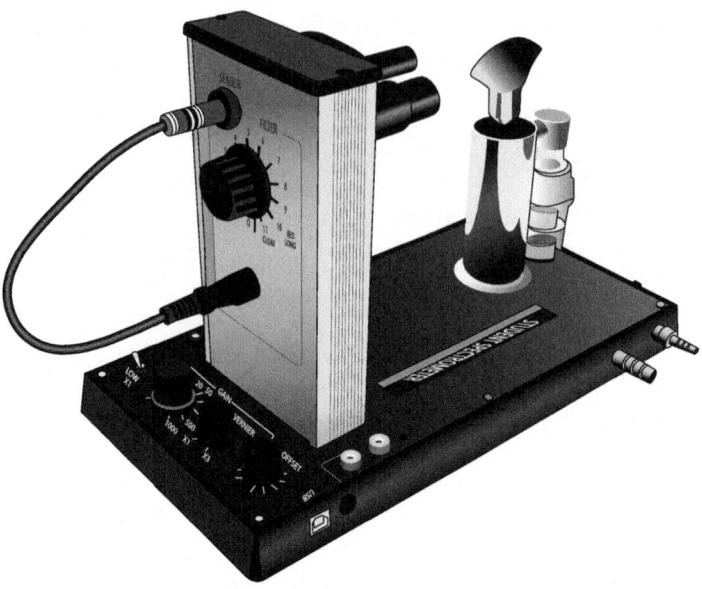

Figure 9.14 Flame emission spectrometer

Calcium line spectrum

Figure 9.15 Flame emission spectra for calcium

Flame emission spectroscopy (or photometry) is explained in the video https://www.youtube.com/watch?v=Iq-wmm_9S9E The underlying principles are also shown in Figure 9.16.

Flame emission spectroscopy and atomic emission spectroscopy are essentially the same. This video explains how an atomic emission spectroscopy instrument works https://www.youtube.com/watch?v=9DYV25Ki-kc

Analysis of the Sun's composition is illustrated in this video https://www.youtube.com/watch?v=0GUIKD0QneU

9.7 Spectroscopy

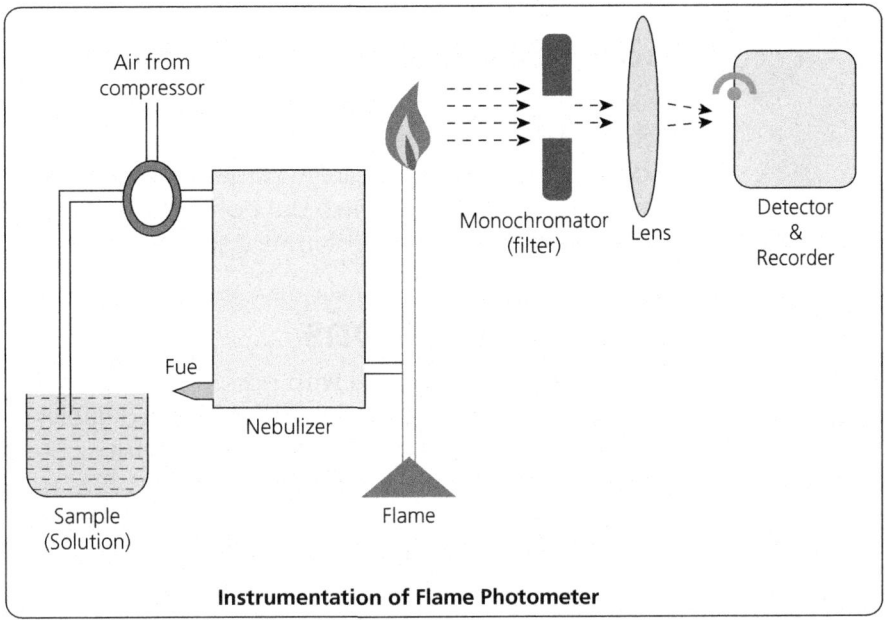

Figure 9.16 Principles of flame emission spectrometry

Enrichment

The seventeenth-century English scientist Isaac Newton is credited as the first person to pass sunlight through a prism to show multi-coloured light of different wavelengths, the most basic spectroscopy. Advances in optics by the mid-nineteenth century led to German chemists Gustav Kirchhoff (1824–1887) and Robert Bunsen (1811–1899) designing the first spectrometer to investigate spectra of chemical elements. They were the first to show that chemical elements produce characteristic spectral patterns. They discovered two new elements, caesium (symbol: Cs) and rubidium (Rb). The Bunsen burner developed by Robert Bunsen achieves a hot, clean gas flame. Their spectrometry experiments were amongst the first to use the Bunsen burner.

9.8 Case studies

This section suggests case studies that apply analytical techniques discussed above. The examples move from consolidating expertise in carrying out tests to applying these in contexts. Care is required as these may sound contrived, but context can be an engaging way of developing students' problem-solving skills.

Staged questions

1. A student finds a bottle with half the label missing: only 'sulfate' is visible. The green powder in the bottle dissolves in water forms a green precipitate with dilute sodium hydroxide. On standing, this precipitate turns brown at the surface. What is it?
2. A chemist shop is selling a white crystalline substance as a water softener and cleaning agent. You think it might be sodium carbonate.
   - **a** What result would you expect to see if you added dilute hydrochloric acid?
   - **b** What would happen when barium nitrate is added?
   - **c** What would the results from a and b mean?
   - **d** How would you prove this is not sodium sulfate?
   - **e** Why would you not be able to prove this is a sodium salt using sodium hydroxide?
   - **f** How could you identify a sodium ion in a sample?
3. You have been asked to identify a white crystalline solid. Use the information in Table 9.8 to identify the solid.

Table 9.8 Testing an unknown white solid

Test	Observations
Dilute sodium hydroxide solution is added to a solution of the solid	A white precipitate forms which then dissolves on addition of excess sodium hydroxide
Dilute hydrochloric acid is added to the solid	No reaction is seen
Silver nitrate solution is added to a solution of the solid	A white precipitate is seen

Identifying stuff

Examples of unusual materials can be assessed using the flowchart (Figure 9.7) and tests proposed/carried out to identify substances present. These could include seashells, eggshells, leaves, samples of natural fibres (for example, cotton, linen), rocks or animal bone.

Students should note the environment in which an object was found, which may give clues about the nature of the material.

With experience, students can undertake context-based investigations. For example, for 11–14s, investigations may use chromatography to determine colours in ink to establish the author of a mystery letter (for a suggested context, see https://serc.carleton.edu/sp/mnstep/activities/26438.html); use melting point tests to solve a mix up in a cake factory about butter and margarine needed to make a batch of cakes; test water samples to find out why people became ill after drinking river water on a trip to local beauty spot.

Identifying salts

To develop students' expertise, start with two substances that look similar and ask students to identify anions or cations only. Pairs of salts with different cations but the same anions are suitable, for example: potassium chloride and sodium chloride; magnesium carbonate and calcium carbonate; lithium nitrate and strontium nitrate; copper(II) sulfate and sodium sulfate.

For anions, choose pairs with the same cation and different anions, for example: sodium chloride and carbonate; magnesium nitrate and sulfate; potassium bromide and iodide.

When students can identify one ion in a pair of salts, introduce two unknowns in which all four ions (two cations and two anions) need identification. Suggested pairs are: sodium chloride and potassium carbonate; lithium bromide and magnesium sulfate; copper(II) sulfate and potassium iodide; lithium carbonate and iron(II) chloride.

Next, introduce contexts in which students identify one or more mystery substances. For example, a baker looking for bicarbonate of soda (sodium hydrogencarbonate) amongst flour, sugar and salt (sodium chloride); a geologist analysing samples of rocks, for example, halite (rock salt, sodium chloride), gypsum (calcium sulfate) and potash (potassium carbonate); a water-board analyst determining the minerals present in water samples from various sources.

Object from the beach

Table 9.9 shows two objects from a beach with tests to identify them. This approach can be used to consider other objects from different environments. The question to ask is 'What inferences can be drawn about the composition of the object?'

9 Chemical analysis

Table 9.9 Objects from the beach

Observation	Choose test	Results
Object found Hard shiny crystalline object with holes throughout. Look closely at the top edge of the second picture. This is an oyster shell. Limestone is made from shells fallen to the bottom of the sea and made into sedimentary rock	**Test for cation** Calcium via flame test – grind a sample of the object to test. **Test for anion** Add hydrochloric acid to the sample. Gas collected in a pipette and bubbled through limewater. Solution filtered off and sodium hydroxide added dropwise.	Flame test on sample gave brick red flame. Surface of the sample produced bubbles. Limewater turned milky. Initially no reaction; slight white precipitate formed.

What's that rock?

Students can test rocks using the information provided in the chapter, with a view to identifying the ions present. Images in the table, together with observations and inferences illustrate this activity.

9.8 Case studies

Table 9.10 What's that rock?

Image	Observations	Inferences
	Red colour in rock. No fizzing with acid but reacts to form a red-brown solution which gives a brown precipitate with sodium hydroxide. No precipitate forms with barium chloride or silver nitrate. Rock reacts with carbon at very high temperatures to produce carbon dioxide and a dark silver metal.	Iron (III) hydroxide formed with sodium hydroxide. Anion cannot be detected by any of the given tests but rock reacts with carbon to make carbon dioxide so it must be an oxide. Rock is iron (III) oxide; haematite rock.
	Green coloured rock. Fizzes with hydrochloric acid, producing a gas that turns lime water cloudy. Remaining solution produces a light blue precipitate with sodium hydroxide. Flame test gives bright green flame.	Colour suggests transition element. Flame test and sodium hydroxide suggest copper is cation. Acid test is positive test for carbonate. Rock is copper carbonate; malachite.
	Light grey/whitish rock. Dissolves in water when ground to produce a colourless solution and some sandy grains. Solution gives bright orange/yellow flame test Solution produces a white precipitate with silver nitrate.	White colour suggests Group 1 or Group 2. Solubility suggests Group 1. Flame test confirms sodium. White precipitate with silver nitrate suggests anion is chloride. Rock is sodium chloride; desert rose.
	Pink rock. Insoluble and resistant to acid. Characteristic pink colour.	Pink colour could be manganese. Not possible to identify rock as unreactive. Possibly quartz rock made from sand-type chemical silicon dioxide in giant lattice.

More complex contexts and activities are listed in the Resources.

9 Chemical analysis

> **Careers**
>
> Chemical analysis offers a wide range of career options. Analysts work in a wide variety of occupations. Specialist medical laboratory scientists run hospital and private laboratories carrying out many different types of tests on body tissues to establish if patients have specific diseases, illnesses or medical conditions. Forensic scientists prepare evidence for criminal court cases (*forensic* means 'for the court') that enables lawyers and juries to determine cause and effect. An anti-doping analyst tests athletes' urine and blood samples for banned substances. A water quality engineer tests if a water sample is safe to drink. Trading standards officers can commission tests to establish if items on sale are authentic and of consistent quality. Astro-chemists study the abundance and reactions of chemicals present in space. Chemical analysis techniques enable investigation of these and many other questions.
>
> Formulation chemists work in many industries, making and testing household products including washing powder, fabric conditioner, detergents, soaps, shampoos, hair conditioners, toothpastes, sunscreens and cleaning fluids; paints, varnishes and coatings; foods such as mayonnaise, sauces, packet soups and instant/ready meals; garden products such as fertilisers, pesticides, insecticides, and animal and plant foods. Cosmetic chemists make formulations for facial and body make-up, while perfumers combine their sense of smell and knowledge of oils to make scents and perfumes.
>
> Chromatographic techniques such as GLC and HPLC are common in food, medical and pharmaceutical industries as these permit finely details analysis of products. Jobs often combine these techniques with spectroscopy so technicians are skilled in both. Using two techniques enhances the ability to identify molecules precisely.

9.9 Resources

References and further reading

Bauer, M., Lehner, M., Schwabl, D., Flachberger, H., Kranzinger, L., Pomberger, R. and Hofer, W. (2018) Sink/float density separation of post-consumer plastics for feedstock recycling. *Journal of Material Cycles and Waste Management*, 20: 1781–1791. Available at https://link.springer.com/article/10.1007/s10163-018-0748-z

Ngai C., Sevian, H. and Talanquer, V. (2014) What is this substance? What makes it different? Mapping progression in students' assumptions about chemical identity. *International Journal of Science Education*, 36(14): 2438–2461. DOI: 10.1080/09500693.2014.927082

Department for Education (2013) National Curriculum for England: Science. https://www.gov.uk/government/publications/national-curriculum-in-england-science-programmes-of-study

Green, J. (2019) How to teach separation techniques. *Education in Chemistry*, November 2018. Available at https://edu.rsc.org/cpd/separation-techniques/3009787.article

Harger, R.N. (1950) Debuking the drunkometer. *Journal of Criminal Law and Criminology*, 40(4): 497–506.

Tan, D.K.C. Goh, N.K., Chia, L.S., and Treagust, D.F. (2004) Qualitative analysis practical work: An instructional package. *School Science Review*, 85(313): 97–102.

Warren, D. (2019) Applying practical knowledge to unfamiliar contexts. *Education in Chemistry*, September 2019. Available at https://edu.rsc.org/cpd/applying-practical-knowledge-to-unfamiliar-contexts/3010780.article

Websites

RSC learn chemistry videos on qualitative analysis: https://edu.rsc.org/resources/qualitative-analysis-quizzes/2201.article

Fractional distillation of air: https://www.bbc.co.uk/bitesize/guides/zt6g87h/revision/2

Use of float/sink density test to assess coffee bean quality: https://mill47.coffee/blogs/story-mill/green-bean-density

What shampoos do people use and why? Formulation activities about shampoo: https://www.rsc.org/education/teachers/resources/aflchem/resources/21/index.htm

Making soaps and detergents: https://edu.rsc.org/resources/making-soaps-and-detergents/1746.article

Know your poison. RSC article with activities to support understanding of the need for quick reliable instrumental analysis of low concentrations of material based on analysis of illegal drugs at music festivals: https://edu.rsc.org/feature/know-your-poison-the-festival-chemical-safety-net/3007847.article

Testing for the sulfate ion – a challenge: https://thescienceteacher.co.uk/tests-for-ions/

Use of instrumental non-destructive instrumental analysis to analyse ancient artefacts an article: https://www.chemistryworld.com/news/chemical-analysis-reveals-origin-of-pompeian-mosaic-tiles/3010127.article#/

Experiments that established air is a mixture of gases: https://www.chemicool.com/elements/composition-of-air.html

Other resources

Formulation: infographic on cosmetic chemistry, red lipstick: www.compoundchem.com/wp-content/uploads/2014/08/Cosmetic-Chemistry-of-Red-Lipstick.png

British Plastics Federation: https://www.bpf.co.uk/Sustainability/Plastics_Recycling.aspx

School Zone: https://www.bpf.co.uk/polymer-zone/schoolszone.aspx

The Royal Society of Chemistry has produced case studies of chemical analysts:

- an analytical chemist https://edu.rsc.org/job-profiles/analytical-chemist/4010854.article
- an analytical technician in plastics https://edu.rsc.org/job-profiles/analytical-technician-plastics/4010921.article
- and a laboratory analyst involved with water quality https://edu.rsc.org/job-profiles/laboratory-analyst-and-higher-degree-apprenticewater/4010930.article

10 Organic chemistry

David Paterson

Introduction

This chapter provides an introductory summary that should help teachers teaching organic chemistry for the first time. It presents essential knowledge, ideas and concepts in a logical sequence with suggestions for practical work. Organic chemistry refers to compounds containing the chemical element carbon and, usually, a combination of hydrogen, oxygen and other elements such as nitrogen and sulfur. The term organic implies compounds that are naturally occurring in plants and animals. A major source of organic compounds is crude oil, which formed over time from the remains of animals and plants. So organic also means compounds made from crude oil. There is a vast range of naturally occurring and manufactured organic compounds, so potentially this topic is very broad. The references and websites listed in Section 10.4 will support extension of the information provided in the text.

Introducing organic chemistry

The eighteenth-century Swedish chemist Carl Wilhelm Scheele (pronounced 'Shaylur') was probably the first person in Europe to isolate organic compounds from plant and animal sources. He showed that milk turns sour because of the compound lactic acid (2-hydroxypropanoic acid, formula $CH_3CHOHCOOH$). In the eighteenth century, chemists believed organic compounds could only be made in reactions in cells, by means of a life force. This belief changed after 1828, when the German chemist Friedrich Wöhler showed that organic compounds could be synthesised in laboratories. He made urea, the compound excreted in animal urine, by heating a solution of the inorganic compound ammonium cyanate:

$$\text{ammonium cyanate} \rightarrow \text{urea}$$

$$NH_4CNO \rightarrow CO(NH_2)_2$$

Organic chemistry relies on each carbon atom forming up to four bonds. Although other chemical elements also form four bonds per atom, bonds involving carbon atoms have relatively high bond enthalpy values (see Chapter 6), so require input of relatively large amounts of energy to break them. Carbon atoms bond readily with each other, a property called *catenation*, so carbon–carbon bonds are commonplace. These factors mean that many organic compounds can be formed based on chains and rings of carbon atoms. Carbon atom chains can be thousands or millions of atoms long.

Many organic compounds are grouped together in families called *homologous series*. Members of each series all have a small, common set of atoms called a *functional group* that provides the general chemical characteristics of the family. In a homologous series, each member differs from the previous member by addition of a $-CH_2-$ unit. Properties such as melting and boiling point gradually change across the series as the relative molecular mass increases.

Students' prior knowledge and experience

Organic chemistry is explicitly introduced at age 14–16. However, the products of organic chemistry are ubiquitous, so students are likely to have encountered examples of these substances in everyday contexts. Students may know names and practical uses of organic products. They are less likely to know systematic names, characteristic composition, structure and reactions.

Fuels, particularly petrol, diesel and natural gas, will be familiar to students, and they are likely to know these are obtained from crude oil. They are unlikely to refer to these substances as hydrocarbons. From their 11–14 science education, they will have met combustion reactions and know that burning fuels is exothermic. They should know that carbon dioxide gas produced in combustion is a greenhouse gas which contributes to global warming and climate change. They may know that fossil fuels, including crude oil, are finite resources, and that alternative energy sources are becoming commonplace. This prior knowledge is relevant for learning about alkane- and alcohol-based fuels.

Students will have used many natural products in daily life. They should know that food comprises carbohydrates, fats, proteins, vitamins, minerals and fibre. They will have encountered alcohols, vinegars and sugars and sweet-smelling substances in fruits, sweets, flowers and perfumes. Students learn that yeast is used to ferment sugars to make bread and alcohol. They may know that lemon juice and vinegar are acidic, from their taste and reaction with sodium hydrogencarbonate (commonly known as bicarbonate of soda). From biology, students will know that plants produce their own food (sugars) using sunlight.

Many everyday objects are plastic-based. Students will know that plastic recycling helps conserve finite resources. Secondary school technology lessons may have given them opportunities to design and make plastic items leading to knowledge of physical properties and uses for plastics. They are unlikely to know the molecular composition and structure of plastics and other polymers.

Ideas and techniques which students meet in their chemistry education are crucial for understanding organic chemistry. These include understanding particles and structure, chemical bonding, properties of substances, properties of acids, combustion and catalysis. Work on the periodic table and in biology means students should be familiar with classification, grouping substances (or organisms) with similar properties and features. Classification underpins secondary organic chemistry, which is taught by groups of substances with characteristic structures, properties and reactions. This organising principle, once grasped, allows students to make predictions about unfamiliar molecules.

Some practical chemistry techniques taught in lower secondary school apply to organic chemistry. For example, distillation is utilised in separating components of an ink and water mixture. They may also be aware that distillation is used to produce alcoholic drinks. Students may have experience of simple identification tests such as distinguishing acidic and alkaline substances by chemical indicators, and starch by addition of iodine solution.

Progression

Post-16 chemistry curricula feature a wider range of organic chemistry compounds, synthesis and analysis techniques. For example, the carbonyl group, C=O, is found in aldehydes and ketones, taught in Year 12, and in esters and amides, taught in Year 13. Many of these substances are strong-smelling, with a wide variety of applications as solvents and scents, and are used to manufacture pharmaceuticals, food supplements, materials and for environmental uses. Stepwise synthesis techniques are introduced in post-16 courses, together with advanced calculations of yield and further considerations of atom economy.

A major area of study involving organic chemicals is spectroscopy, in which substances interact with electromagnetic radiation to yield information about structure and bonding. For 14–16-year-olds, visible light spectroscopy is introduced via flame tests of metal ions, which show characteristic colours. Post-16 organic spectroscopy is introduced through IR spectroscopy, NMR spectrometry (which uses radio waves), and mass spectrometry, in which substances

10 Organic chemistry

are broken down into smaller parts for analysis. At university, these techniques are discussed in much greater depth, and extended to spectroscopy using microwaves and X-rays.

Teaching sequence overview

Start the teaching sequence with hydrocarbons, as this introduces families of organic substances, naming conventions and similarities of chemical properties. After this, either natural products or polymers can be studied. The natural products section introduces substances that may be familiar to students from their previous knowledge. Polymers follows from discussion of cracking of hydrocarbons and chemistry of alkenes. In the Figure 10.1, dotted and block lines indicate these alternative routes.

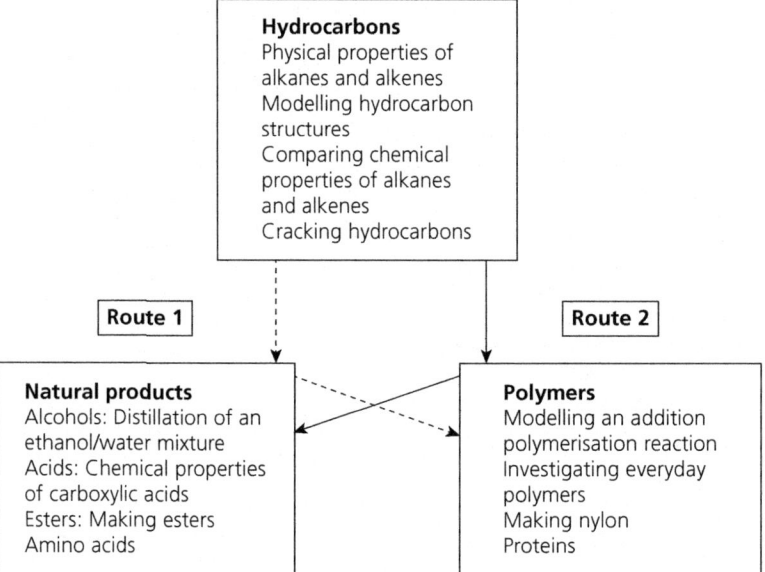

Figure 10.1 Organic chemistry: teaching sequence overview

10.1 Hydrocarbons

Hydrocarbons are the simplest organic compounds, comprising hydrogen and carbon only. There are three main homologous series of linear hydrocarbons, namely *alkanes*, *alkenes* and *alkynes*. In general, hydrocarbons are known formally as alkyl compounds. For students aged 14–16, only knowledge of the first four alkanes and alkenes is required. Tables 10.1 and 10.2 provide details of the structure and properties of first few members of the alkane and alkene homologous series.

Prefixes and suffixes (see above) are commonly used in chemistry to create systematic names. In a hydrocarbon series, the prefixes *meth-*, *eth-*, *prop-*, *but-* and *pent-* indicate the number of carbon atoms in the main chain, corresponding to 1, 2, 3, 4 and 5 carbon atoms respectively. The suffix indicates the type of bonding present: *-ane* shows that the carbon–carbon bonds are all single bonds; *-ene* means that one (or more) carbon–carbon double bond is present. Students should be able to predict formulae of additional members of a homologous series from basic information about the first five members, prefixes and suffixes.

All bonds in alkanes are single covalent bonds: each bond comprises one electron pair shared between two atoms. These are referred to as *saturated* structures. In alkenes, there is one double covalent bond between two carbon atoms in which two electron pairs are shared between two atoms. Note that carbon–carbon triple bonds exist: members of the alkyne series each have a triple covalent bond in their molecules, in which three electron pairs are shared between two atoms. No knowledge of alkynes is expected at age 14–16. All molecules with double or triple bonds are referred to as *unsaturated* structures.

Data in Tables 10.1 and 10.2 introduce structures and physical properties of alkanes and alkenes. Identifying patterns in boiling points may be challenging, as numbers range from negative to positive values in degrees Celsius. Students need to understand that substances such as short-chain alkanes and alkenes with boiling points below room temperature will be gases at room temperature (see Chapter 3). The low boiling points arise because there are weak intermolecular bonds between the molecules (see Chapter 5). These bonds require relatively little amounts of energy to break. As the number of carbon atoms increases, the boiling point increases. For larger molecules, greater surface area contact between adjacent molecules occurs, so more intermolecular bonds can form, increasing the energy required to separate them. Alkanes and alkenes with 10 carbon atoms are liquids at room temperature. Those with 20 and 30 carbon atoms are solids.

10 Organic chemistry

Table 10.1 Names, molecular formulae, boiling temperatures and structural formulae of some alkanes

Number of carbon atoms	Prefix	Alkanes			
		Suffix	Molecular formula	Structural formula(e)	Boiling temperature/ °C
1	Meth-	-ane	CH_4	H–C(H)(H)–H	−162
2	Eth-		C_2H_6	H–C(H)(H)–C(H)(H)–H	−89
3	Prop-		C_3H_8	H–C(H)(H)–C(H)(H)–C(H)(H)–H	−42
4	But-		C_4H_{10}	H–C(H)(H)–C(H)(H)–C(H)(H)–C(H)(H)–H	−1
5	Pent-		C_5H_{12}	H–C(H)(H)–C(H)(H)–C(H)(H)–C(H)(H)–C(H)(H)–H	36

Table 10.2 Names, molecular formulae, boiling temperatures and structural formulae of some alkenes

Number of carbon atoms	Prefix	Alkenes			
		Suffix	Molecular formula	Structural formula(e)	Boiling temperature/ °C
1	Meth-	-ene	—	—	—
2	Eth-		C_2H_4	$H_2C=CH_2$	−104
3	Prop-		C_3H_6	$H_2C=CH-CH_3$	−47
4	But-		C_4H_8	$H_2C=CH-CH_2-CH_3$	4
5	Pent-		C_5H_{10}	$H_2C=CH-CH_2-CH_2-CH_3$	30

KEY ACTIVITY

Physical properties of alkanes and alkenes

Provide a data table of boiling points of methane, ethane and propane and the equivalent alkenes and ask students to plot an appropriate graph of the number of carbon atoms (x-axis) against boiling point in °C (y-axis). Note that the y values run from negative to positive. Students should discuss any trends they see in their graph and then predict boiling points of butane and butene from their graphs. The addition of a line showing room temperature will help to provide context for these figures.

Questions to ask
- Describe and explain the general patterns for boiling point as the number of carbon atoms increases.
- Predict the physical states at room temperature of alkanes and alkenes with 10, 20 and 30 carbon atoms.
- What is the physical state at room temperature of an alkane or alkene that has a boiling point with a negative value? Explain this in terms of intermolecular bonds.
- Compare the patterns for alkanes and alkenes, noting similarities and differences.
- Predict the boiling points of butane and butene from the graph, and compare these.

KEY ACTIVITY

Making models of hydrocarbons

Creating models of organic compounds gives students an impression of molecular shape. Most schools are likely to have Molymod® sets, or similar kits which use colour-coded balls for atoms and have some means of connecting these to form organic compounds. Building models with up to five carbon atoms in the chain is useful. This task can be completed alongside the activity on physical properties described above. Students can compare models of alkanes and alkenes, examining their structures. Questions to prompt students thinking about the models and their structure are helpful. All models have limitations: some of these are discussed in Chapter 1.

Carbon atoms make a maximum of four single covalent bonds. Any double bonds count for two of these bonds, while triple bonds use three. A carbon atom can make a combination of single and double or/and triple bonds. For example, carbon in methane forms four single bonds, whereas two carbon atoms in ethene each form one double bond and two single bonds. Hydrogen atoms always form only one bond. In alkanes, bonds are arranged in a tetrahedral shape. In alkenes, carbons that are double bonded form a flat triangle shape. Emphasise that alkenes can have more than one double bond. Some students will identify that in butene the double bond can be between the first and second, or the second and third carbon atoms. These two structures are *isomers* of butene, named but-1-ene and but-2-ene respectively.

10 Organic chemistry

Each member of a homologous series differs by one –CH$_2$– group, known as a methylene group. The general formula of a homologous series shows the mathematical relationship between the numbers of carbon and hydrogen atoms. For alkanes, the general formula is C$_n$H$_{2n+2}$. This means the number of hydrogen atoms is twice the number of carbon atoms, plus two. The first member of the alkane series is methane, formula CH$_4$. For alkenes, the general formula is C$_n$H$_{2n}$, meaning the number of hydrogen atoms is twice the number of carbon atoms. The first alkene is ethene, formula C$_2$H$_4$. Students will recognise the name from polythene or poly(ethene) which is a polymer of ethene molecules. Methene cannot exist, as a carbon–carbon double bond requires two carbon atoms.

Questions to ask
- What is the maximum number of bonds that one carbon atom makes?
- What shape is formed by bonds around a carbon atom?
- How many bonds does one hydrogen atom make?
- Alkanes and alkenes use the same prefixes and suffixes.
  - What is a prefix? What is a suffix?
  - What do these terms mean in relation to alkanes?
  - What do these terms mean in relation to alkenes?
- How many carbon and hydrogen atoms add to an alkane or alkene to form the next series member?
- What is the pattern of the number of carbon atoms and number of hydrogen atoms in:
  - an alkane molecule?
  - an alkene molecule?
  - Write down a formula that applies to all alkanes in the form C$_n$H____.
  - Write a formula that applies to all alkenes in the form C$_n$H____.
- What is the first member of
  - the alkane homologous series
  - the alkene homologous series?

  What do you notice? Explain the differences in terms of bonding present in alkanes and alkenes.

KEY ACTIVITY

Chemical properties of alkanes and alkenes

A class practical will illustrate differences in properties of alkenes and alkanes. Bromine water is separately added to each hydrocarbon and any colour changes are observed. Each hydrocarbon is burned and observations of flame colours and smoke are made. Models of cyclohexane (Figure 10.2) and cyclohexene (Figure 10.3) are useful here.

Before the practical, check that students understand how many carbon atoms are in each molecule and that both molecules are ring-shaped, hence the *cyclo-* prefix. Ask them to

10.1 Hydrocarbons

predict how the alkane and alkene will react with the bromine water. Students should note that cyclohexane and cyclohexene are flammable, giving yellow flames. A yellow flame indicates incomplete combustion, occurring because insufficient oxygen reacts with the hydrocarbon to allow complete combustion. The cyclohexene flame is smokier as soot formation is favoured in the presence of unsaturated compounds. Further details of soot formation can be found here: https://eic.rsc.org/feature/clearing-the-air-around-smoke-formation/2500478.article

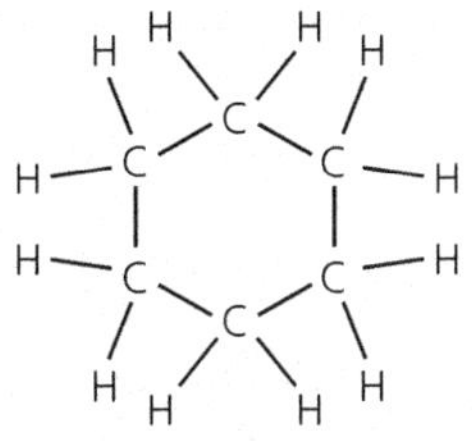

Figure 10.2 Structural formula of cyclohexane

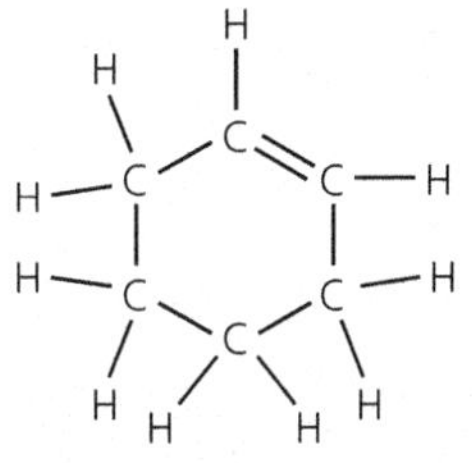

Figure 10.3 Structural formula of cyclohexene

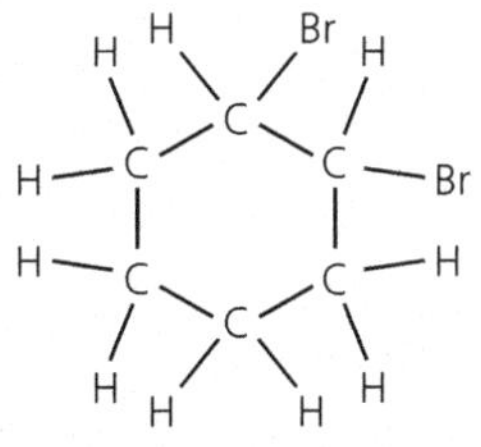

Figure 10.4 Structural formula of 1,2-dibromocyclohexane – a member of the homologous series of haloalkanes.

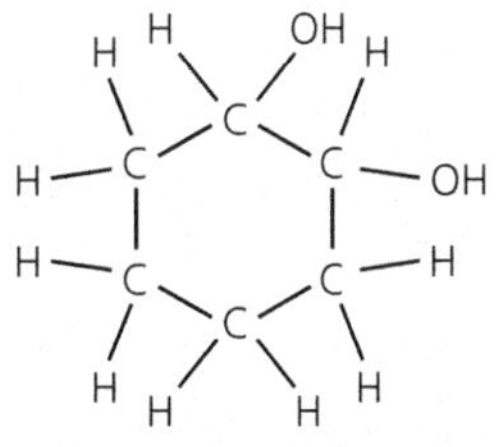

Figure 10.5 Structural formula of cyclohexan-1,2-diol – a member of the homologous series of alcohols

The alkane – cyclohexane – does not readily react with the bromine water (although in bright light, for example, sunlight, a reaction would occur). Cyclohexene readily reacts with bromine water to form 1,2-dibromocyclohexane (Figure 10.4). This is an addition reaction as a bromine molecule, Br_2, is added across the C=C double bond. The double bond is broken and two C–Br single bonds are formed. The product, 1,2-dibromocyclohexane, is colourless, explaining why the orange-brown bromine water decolourises. Bromine is soluble in organic solvents such as cyclohexane, explaining why cyclohexane turns orange-brown when bromine water is added. A general equation for addition of bromine to an alkene is

$$C_nH_{2n} + Br_2 \rightarrow C_nH_{2n}Br_2$$

Cyclohexene is oxidised by acidified potassium manganate(VII) to form cyclohexan-1,2,-diol (Figure 10.5). The strongly coloured purple manganate(VII) ion (MnO_4^-) is reduced in the reaction to the weakly coloured pink manganese(II) ion (Mn^{2+}), accounting for the decolourisation.

10 Organic chemistry

Table 10.3 Observations from tests on cyclohexane and cyclohexene

Chemical	Burning	Acidified potassium manganate (VII) solution	Bromine water
Cyclohexane	Yellow flame	No change	No change. Note bromine is soluble in cyclohexane so colour may remain in the non-aqueous layer.
Cyclohexene	Yellow flame – more smoky than cyclohexane	Decolourises from purple to colourless	Decolourises from orange-brown to clear and colourless.

Questions to ask
- What does the type of flame tell you about the extent of combustion?
- Which observations show one hydrocarbon is more reactive than the other?
- Predict the structure of the addition product of cyclohexene with bromine.
- Write a general equation for the addition of bromine to an alkene.

The tests in this practical are general characteristic tests for unsaturated hydrocarbons.

Using structural formulae and 3-D molecular models will help students recognise that alkanes have only single covalent bonds between all atoms, creating a *saturated* structure which means that no further atoms can be added to alkane molecules. This explains why no reaction is observed with bromine water.

Students may confuse distinct descriptions of solutions such as colour (coloured/colourless) and clarity (clear/opaque). Bromine water is a clear, orange-brown solution that turns clear *and* colourless when added to an unsaturated compound. Encourage students to refer to before and after observations of chemicals when discussing and noting observations to ensure the *change* is emphasised.

Differences in the burning test are harder to distinguish. Alkanes and alkenes combust readily. Alkenes burn with a smoky flame because the ratio of carbon to hydrogen is lower than that of alkanes.

Fractional distillation

In general, crude oil extracted from oil fields worldwide comprises a mixture of alkanes. For crude oil to be useful, components must be separated into portions called fractions. Fractions are separated by fractional distillation. This process vaporises almost all the crude oil fed into a fractional distillation tower. The tower can vary in height from 25 to 100 m and contains an arrangement of condensers. As vapourised gases rise up the tower, they condense along the temperature gradient in fractions related to the size of the molecules. This creates fractions that are piped off to make gasoline, kerosene and diesel, among others.

10.1 Hydrocarbons

KEY ACTIVITY

Fractional distillation of crude oil

Genuine crude oil cannot be used in school laboratories due to the presence of carcinogens such as benzene in the mixture. A crude oil substitute can be made http://science.cleapss.org.uk/Resource-Info/RB032-Crude-oil-substitute.aspx This mixture can be distilled using a fractionating column and Liebig condenser, or, on a smaller scale, by students using a side-arm boiling tube (see https://edu.rsc.org/resources/the-fractional-distillation-of-crude-oil/754.article).

'Light' fractions (for example, gasoline) contain hydrocarbons with shorter carbon chains. 'Heavy' fractions, used to make products such as diesel and bitumen, contain longer-chain hydrocarbons. Crude oil contains more heavy than light hydrocarbon fractions, the opposite of the requirements of industry and human society. The process known as *cracking* is used to make short-chain hydrocarbons from long chain molecules in heavy fractions.

Enrichment

Fractional distillation and cracking are used on an enormous scale at huge industrial sites worldwide. Locations are generally coastal, such as Grangemouth in Scotland, Fawley in southern England, Jamnagar in India, and the Paraguaná Peninsula in Venezuela. Further details can be found at www.essentialchemicalindustry.org/processes/distillation.html. The petrochemical industry is fundamental to the modern world, processing 36 billion barrels of oil (5700 billion litres) in 2018, worth $2.2 trillion in 2018 prices. Crude oil, or more correctly petroleum (from the Latin petra – rocks and oleum – oil) is the major source of hydrocarbon-based raw materials for fuels, polymers, pharmaceuticals and industrial chemicals. The Essential Chemical Industry website includes a description of the formation and processing of crude oil: https://www.essentialchemicalindustry.org/processes/extracting-oil-and-natural-gas-fracking.html

Cracking

In cracking, large chain molecules are heated, vaporised and passed over a catalyst to break the large molecule into smaller ones. Cracking involves passing vaporised alkanes such as decane (ten carbon atoms) over a hot catalyst. The heat breaks the decane molecules into smaller alkane and alkene molecules, an example of thermal decomposition. These are separated by further distillation. Changing conditions and catalysts leads to different products. For example, high temperature steam cracking produces large quantities of alkenes, while using alumina/silica catalysts forms branched and cyclic alkanes.

10 Organic chemistry

Cracking is a highly complex and energy-intensive process, requiring multiple stages, temperatures ranging from 727–877 °C and pressures from near vacuum to 100 atmospheres. The fundamental chemical process has not changed in decades, but progress is continually being made on improving energy efficiency and safety.

KEY ACTIVITY

Cracking hydrocarbons

Practical details for a large-scale teacher demonstration of cracking are at https://edu.rsc.org/resources/cracking-hydrocarbons/681.article In this demonstration, liquid paraffin is vaporised and passed over heated broken pieces of ceramic crucible, acting as a catalyst. The gaseous products are alkenes, which are collected over water. These are tested with bromine water and acidified potassium manganate(VII) solution to demonstrate unsaturation. Alkenes present in the mixture will decolourise bromine water (orange-brown to clear and colourless) and potassium manganate (VII) solution (pink-purple to clear and colourless).

A microscale practical can be carried out by students. Details can be found at http://science.cleapss.org.uk/Resource-Info/PP061-Microscale-Cracking-of-Liquid-Paraffin.aspx A glass pipette is used as the reaction vessel, with paraffin absorbed onto mineral wool at one end of the pipette, with aluminium oxide powder catalyst. The catalyst is heated with a spirit burner. Gaseous product passes directly into test tubes of bromine water and acidified potassium manganate (VII) solution to demonstrate unsaturation. Both should decolourise due to the presence of alkenes such as ethene.

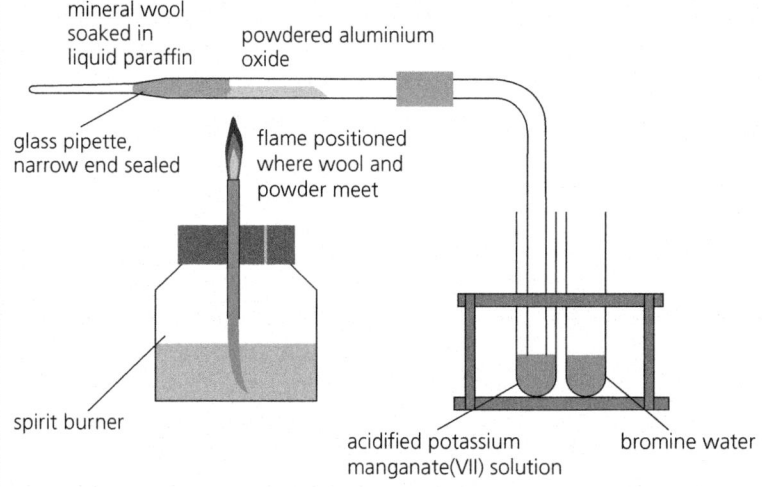

Figure 10.6 Cracking paraffin

> A catalyst increases the rate of reaction by providing a surface on which long hydrocarbon molecules split into smaller molecules. Paraffin molecules heat up as they move across the hot catalyst. The heat provides activation energy to break bonds in the paraffin molecules. Some parts of the paraffin molecules split up into carbon particles (soot). These are black and stick to the catalyst. This may mean that the catalyst eventually stops working.
>
> **Questions to ask**
> - Why do alkenes decolourise bromine water and acidified potassium manganate (VII) solution?
> - Why does the catalyst turn black?
> - Why does the paraffin need to pass over a hot catalyst?
> - Write a balanced equation to model the cracking process, starting with dodecane ($C_{12}H_{26}$) and forming two different alkenes and one alkane.
>
> Molymod® kits and other modelling systems (see above) may be used to help visualise the cracking process. Note that there is no right answer to what is formed in cracking, as this varies.

10.2 Natural products

Natural products are substances produced by living organisms. These substances range from simple structures such as ethanol, a product of fermentation of sugars by yeasts, to highly complex structures such as deoxyribonucleic acid (DNA), which contains the genetic code for all living organisms. Methane and ethene are natural products as these are produced by living organisms. Their chemistry is discussed above. This section discusses homologous series of alcohols, carboxylic acids, esters and includes amino acids and DNA.

Alcohols

Table 10.4 shows the first five members of the alcohol homologous series. Methanol (CH_3OH), the first member, is highly toxic to humans. The second member is ethanol (CH_3CH_2OH), commonly known as alcohol. The general formula for alcohols is $C_nH_{2n+1}OH$. The characteristic functional group of the alcohols is –OH, called the *hydroxyl* group.

10 Organic chemistry

Table 10.4 Names, molecular formulae, boiling temperatures and structural formulae of some alcohols

Number of carbon atoms	Prefix	Suffix	Molecular formula	Structural formula(e)	Boiling temperature/ °C
1	Meth-	-anol	CH_3OH	H–C(H)(H)–O–H	65
2	Eth-	-anol	C_2H_5OH	H–C(H)(H)–C(H)(H)–O–H	78
3	Prop-	-anol	C_3H_7OH	H–C(H)(H)–C(H)(H)–C(H)(H)–O–H	97
4	But-	-anol	C_4H_9OH	H–C(H)(H)–C(H)(H)–C(H)(H)–C(H)(H)–O–H	118
5	Pent-	-anol	$C_5H_{11}OH$	H–C(H)(H)–C(H)(H)–C(H)(H)–C(H)(H)–C(H)(H)–O–H	138

Ethanol is made by fermentation of sugars, such as glucose, by yeast in the following reaction:

$$C_6H_{12}O_6(aq) \rightarrow 2CH_3CH_2OH(aq) + 2CO_2(g)$$

The source of sugar varies. Making beer by brewing cereal grains such as barley, wheat, rice or corn is one of the oldest known procedures common to many human societies. A culture of yeast is added to grain, while flavour is produced by adding other ingredients. In the UK, hops, which give beer a malty taste, are common. The malt-flavoured spreads Marmite® and Vegemite are brewing by-products. Grape juice is fermented to wine, using natural yeasts on grape skins. Allowing fermentation to stop naturally produces a dry wine, with low sugar content. Stopping fermentation at an earlier stage produces a sweeter wine, as more sugar remains in the product.

Ethanol mixes readily with water. This property is applied in measuring the strength of a drink containing ethanol to give the 'per cent proof' value. The proof value is twice the volume per cent of ethanol. Thus, spirits, which are around 80% proof, contain 40% by volume of ethanol. The strength of an ethanol–water mixture can be increased by distillation.

Enrichment

Products of fermentation of sugars are a complex mixture, including ethanol in aqueous (water-based) solution. Ethanol concentration varies from about 0.5% to about 15% depending on fermentation conditions. This percentage value shows the volume (in cm³) of ethanol in 100 cm³ of the mixture. Example percentages in commercial drinks are 3.5–6.5% (beer) and 10–15% (wine). Higher ethanol concentrations cannot be achieved through fermentation, as ethanol concentrations above about 15% kill the yeast. To produce alcoholic drinks with higher concentrations, such as spirits like whisky and vodka, fermented mixtures undergo distillation.

KEY ACTIVITY

Simple distillation of an ethanol–water mixture

Students can undertake simple distillation to separate a mixture of ethanol and water. Anti-bumping granules are added to ensure steady boiling and prevent the mixture from becoming superheated and potentially flash-boiling. When the mixture is heated, ethanol boils at 78 °C and passes into the condenser, cooling back to liquid which is collected in a receiver, usually a beaker. The water, which boils at 100 °C, remains in the flask. The separated fraction contains less water than the original mixture, so has a higher concentration of ethanol. The presence of ethanol can be tested using acidified potassium dichromate(VI) solution, which changes colour from orange to green.

10 Organic chemistry

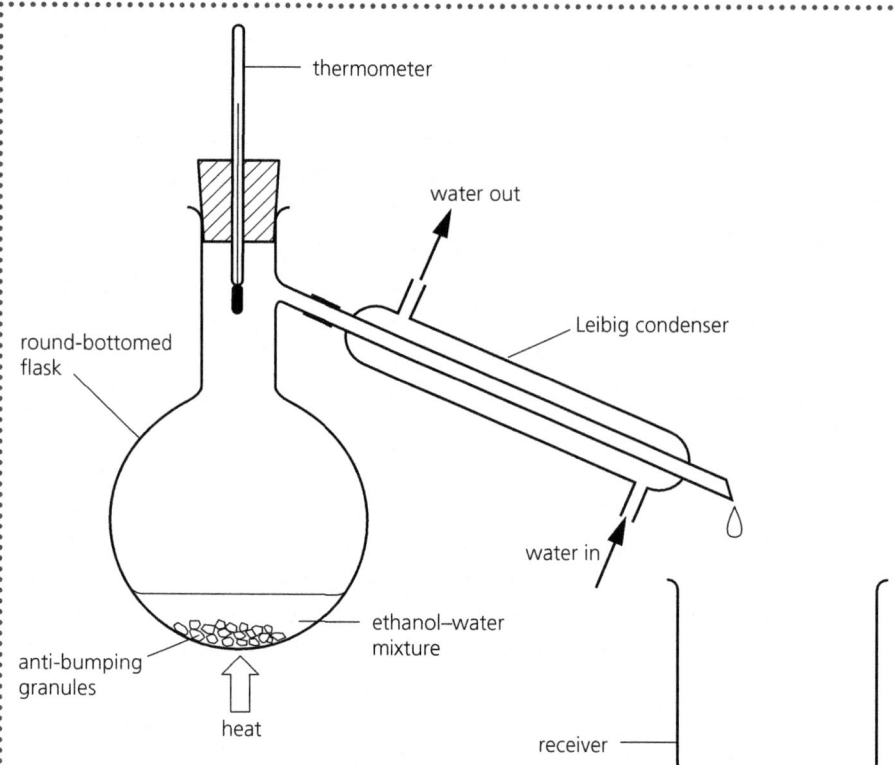

Figure 10.7 Apparatus required for distillation of an ethanol–water mixture

Questions to ask
- Why are anti-bumping granules added?
- What is the temperature of the vapour going into the condenser? Why is the vapour at this temperature?
- What remains in the flask? How can we determine this?
- How can we test the distillate? Should the liquid in the flask be tested as well? And the mixture beforehand?

Simple distillation does not allow complete purification of ethanol from an aqueous solution, as some water evaporates at the boiling temperature of ethanol. Further purification of ethanol is possible using fractional distillation, which achieves approximately 95% pure ethanol. Purification beyond this requires addition of substances such as cyclohexane or toluene, or molecular filters such as zeolites.

Producing ethanol-based fuel from fermentation of crops is an interesting context for considering the social, economic and environmental impact of chemistry. The USA and Brazil are the world's largest producers of bioethanol, mainly through fermentation of corn (USA) and sugar cane (Brazil). Using agricultural land for growing fuel crops rather than food crops has consequences. Students can investigate the history of these fuels, the chemistry involved in producing bioethanol and social, economic and environmental impacts. To provide practical context, students can ferment sugars with yeast: https://edu.rsc.org/resources/fermentation-of-glucose-using-yeast/470.article

Carboxylic acids

Carboxylic acids are common in nature. For example, the simplest carboxylic acid is methanoic (or formic) acid (formula: HCOOH), which is released when red ants bite. Formic acid is also found in bee stings. Other common carboxylic acids are:

- ethanoic acid (CH_3COOH), found in vinegar
- propanoic acid (or propionic acid, formula CH_3CH_2COOH) which gives Swiss cheese its flavour
- butanoic acid (or butyric acid, formula $CH_3CH_2CH_2COOH$) is the taste of rancid butter
- citric acid (or 2-hydroxypropane-1,2,3-tricarboxylic acid $HOOCCH_2C(COOH)(OH)CH_2COOH$) is found in fruit juices
- ethanedioic acid or oxalic acid (HOCOCOOH) is toxic and found in edible plants such as rhubarb, spinach and sorrel.

About 4 kg of spinach contains enough oxalic acid to poison a person. In rhubarb, the oxalic acid is mainly found in the leaves but a human would need to eat huge quantities to cause any problem.

Carboxylic acids form a homologous series. The general formula for carboxylic acids is $C_nH_{2n+1}COOH$. The characteristic functional group of carboxylic acids is the carboxyl group, formula –COOH. This group has the structure shown in Figure 10.8.

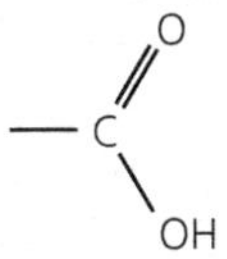

Figure 10.8 The carboxyl group

These are weak acids, meaning that only a small proportion of the acid molecules are ionised in solution:

$$CH_3COOH(aq) \rightleftharpoons CH_3COO^-(aq) + H^+(aq)$$

Solutions of carboxylic acids have higher pH values than equimolar solutions of strong acids such as hydrochloric acid. For example, a 0.1 mol dm^{-3} solution of ethanoic acid at 25 °C has a pH of 2.9, while a 0.1 mol dm^{-3} solution of hydrochloric acid at 25 °C has a pH of value of 1.0.

10 Organic chemistry

KEY ACTIVITY

The chemical properties of carboxylic acids

The carboxyl group gives these compounds acidic properties (see Chapter 7). These can be demonstrated in a class practical investigating the reactions of ethanoic acid with bases and metals. Full instructions are available at https://edu.rsc.org/resources/the-acidic-reactions-of-ethanoic-acid/462.article The reactions of carboxylic acids are summarised in Table 10.5. Test results show that ethanoic acid reacts as any mineral acid, such as hydrochloric, sulfuric and nitric acids. However, as a weak acid, fewer hydrogen ions are present in solutions of ethanoic acid than in solutions of mineral acids. Other members of the series, such as propanoic acid, would react in a similar way to ethanoic acid.

Table 10.5 Reactions of carboxylic acids

Test	What to do	Result	Explanation
With Universal Indicator	Add a few drops of Universal Indicator solution. Note the colour. Keep this as a reference tube for Tests 2 and 3.	A red colour forms in the test tube.	The solution is acidic due to the presence of hydrogen ions.
With sodium hydroxide solution.	Add a few drops of Universal Indicator solution. Note the colour. Then add sodium hydroxide solution dropwise, counting the drops and noting any colour changes. Continue additions until no further colour change occurs.	An alkaline solution forms within a few drops.	Sodium hydroxide solution is a strong alkali. This is a neutralisation reaction forming a salt (sodium ethanoate) and water. $CH_3COOH(aq) + NaOH(aq) \rightarrow CH_3COONa(aq) + H_2O(l)$
With sodium hydrogencarbonate solution.	Add a few drops of Universal Indicator solution. Note the colour. Add sodium hydrogencarbonate solution dropwise, counting drops and noting any colour changes. Continue additions until no further colour change occurs.	An alkaline solution forms but more sodium hydrogencarbonate solution is needed than sodium hydroxide solution to produce the same colour.	Sodium hydrogencarbonate is a weak alkali. This is a neutralisation reaction forming a salt (sodium ethanoate), water and carbon dioxide. $CH_3COOH(aq) + NaHCO_3(aq) \rightarrow CH_3COONa(aq) + H_2O(l) + CO_2(g)$
With magnesium ribbon.	Add a small piece of magnesium ribbon to the ethanoic acid. Note what happens. Test the gas produced.	Fizzing (effervescence) occurs. The gas produced will pop when tested with a lighted splint.	The acid and magnesium react to produce a salt (magnesium ethanoate), and hydrogen gas. $2CH_3COOH(aq) + Mg(s) \rightarrow (CH_3COO)_2Mg(aq) + H_2(g)$

> **Questions to ask**
> - How do these reactions compare with acids such as hydrochloric acid and nitric acid?
> - Predict the reactions of propanoic acid with the reagents shown here.
> - Write equations for reactions observed in this experiment.

Esters

Carboxylic acids react with alcohols to produce sweet-smelling compounds called esters. These are found in ripe fruit such as pears, bananas, melons and pineapples. The smells of esters are distinctive and often very pleasant, so they are valued in perfumes. Specialist chemical companies, such as Sigma-Aldrich, produce esters for perfumes. Esters are non-toxic, so make good artificial flavourings. Ethyl ethanoate smells and tastes of pears, used in making pear drop sweets. Other esters are used as honey, rum or parsnip flavouring in foods. The ester, methyl salicylate, is responsible for the minty scent found in perfumes and scented candles, and the flavouring in chewing gum and mints. Ester combinations yield a variety of scents and tastes. Chemists can 'fine tune' an ester mixture to specific requirements. For example, strawberry yoghurts include an ester mixture to enhance flavour.

Reactions producing esters can be slow. In nature, fruit takes time to ripen as ester concentrations increase gradually. When making an ester in the laboratory from an alcohol and a carboxylic acid, it can take a week for the equilibrium mixture to form at room temperature. A catalyst, hydrogen ions, usually from concentrated sulfuric acid, is added to lower the activation energy and increase the rate of reaction (see Chapter 6).

An example of an ester is ethyl ethanoate, produced in the reaction between ethanol and ethanoic acid:

$$CH_3CH_2OH(l) + CH_3COOH(l) \rightarrow CH_3COOCH_2CH_3(l) + H_2O(l)$$

ethanol + ethanoic acid → ethyl ethanoate + water

This reaction is known as esterification. Water is always a product. The water molecule is made from a hydrogen atom from the alcohol and a hydroxyl group from the carboxylic acid. The functional group is the ester group, which is –COO–.

Ester names comprise two parts. The first part derives from the alcohol, for example, *ethyl* from ethanol. The second part derives from the acid, for example *ethanoate* from ethanoic acid. However, perhaps confusingly, ester *formulae* are written the opposite way. So, in an ester formula, the acid-derived section comes first, the functional group –COO– is in the centre and the alkyl group from the alcohol is

10 Organic chemistry

on the right-hand side. Hence, butan-1-ol, formula $CH_3CH_2CH_2CH_2OH$ and ethanoic acid, formula CH_3COOH react to form the ester butyl ethanoate, formula $CH_3COOCH_2CH_2CH_2CH_3$. Butyl ethanoate has a sweet smell of apple or banana.

Esterification reactions are reversible reactions (see Chapter 6). The forward reaction is shown above. In the reverse reaction, hydrolysis, the ester reacts with water to form the original alcohol and carboxylic acid. At equilibrium, the two reactions occur at the same rate. If esters are needed in large quantities, chemists extract the ester as it is made. This increases overall yield by favouring the forward reaction (see Le Chatelier's principle, Section 6.3).

KEY ACTIVITY

Making an ester

Students can make esters using small quantities of alcohols and acids. Full details are available at http://science.cleapss.org.uk/resource/tl006-making-esters.pdf The practical can be extended by substituting the alcohol with 2-methyl-propan-1-ol to make other esters. Molecular models can be used to familiarise students with ester structures and changes occurring during the reaction. In esterification, the –H from the alcohol and –OH from the carboxylic acid combine to form water; and the O– from the alcohol bonds with the C=O from the ethanoic acid to form the ester linkage. Use molecular models to show how the products form from the reagents.

Ester synthesis is catalysed by concentrated sulfuric(VI) acid, which is a known hazard. Risk is minimised by pre-dispensing one drop into reaction test tubes. Alternatively, an acidic ion-exchange resin can be used as a catalyst (for example, Amberjet 1200H).

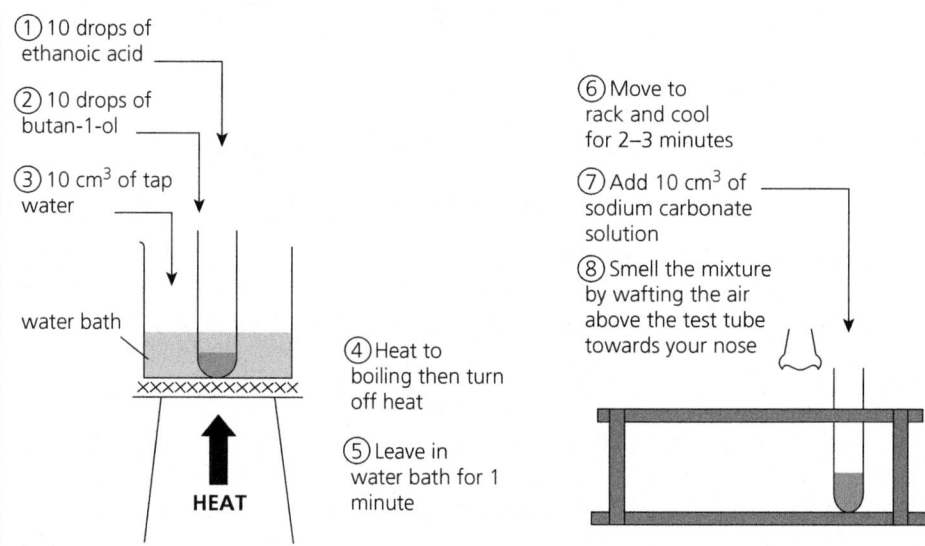

Figure 10.9 Making an ester

> **Questions to ask**
> - What is the name of the organic product?
> - Describe the smell of the product.
> - Where/when can similar chemicals to this one be found?
> - Why is the acid needed (it does not appear in the equation for the reaction)?

Enrichment

In humans, chemicals known as *steroids* include sex hormones (progesterone, oestrogen and testosterone) and stress hormones such as cortisol. Today, we take access to a wide range of steroids for granted, using them for birth control, in body-building and as treatments for many illnesses including arthritis, multiple sclerosis, asthma and eczema. Steroids were first recognised as a group of chemicals in the 1930s, when chemists found they have similar structures. Their potential as treatments was immediately understood, but steroids are produced naturally in very small amounts, are very difficult to extract from animal tissue and have complex structures. So, once their functions and potential as medicines were known, chemists in many countries wanted to synthesise them. Plant material provided a possible route.

German chemists found the steroid *stigmasterol* could be obtained from soybeans. Making stigmasterol in large quantities then using it to synthesise human steroids was the challenge. Step in African–American chemist Percy Julian (1899–1975). Percy was born in Montgomery, Alabama, the grandson of enslaved people. He left school at 14, as at that time black people were not entitled to education beyond this age. Percy's parents encouraged him to continue studying, and he was eventually accepted at a university in the US state of Indiana. On his first day, a white student welcomed him with a handshake. Although he faced many challenges due to his race, Percy said, 'In the shake of a hand my life was changed. I soon learned to smile and act like I believed they all liked me, whether they wanted to or not.' Percy progressed to Vienna to complete a doctoral degree studying medicinal plants. In 1936 he was hired by a Chicago company working on soybean products. One day he was told water had leaked into a tank of purified soybean oil, forming a solid white mass. Percy realised the white mass was stigmasterol. In 1940 he developed the first novel, large-scale process producing progesterone from stigmasterol. His first batch of about 0.5 kg was valued at around $60 000 (about $500 000 today) and was transported to the buyer in an armoured car.

Amino acids

Amino acids are not a homologous series but comprise a group of twenty compounds. Amino acids each have one central carbon atom in which three of four available bonds are bonded in the same way. These bonds are to an amino group, a carboxyl group and a hydrogen atom. The fourth bond is to what is known as an R-group, which is different in each amino acid. In glycine this is simply a hydrogen atom –H; in alanine, the R-group is a methyl group –CH_3; in serine, the R-group is a primary alcohol functional group –CH_2OH.

Amino acids bond together to form polymers, and these polymers are proteins. Amino acid sequences are controlled by the genetic code, that is, the order of bases in DNA. Protein structures vary enormously, enabling the existence of a huge variety of functions and structures. Many proteins have catalytic functions in living organisms. These have an *active site* that forms temporary bonds with small molecules such as glucose and adenosine triphosphate (ATP), facilitating a reaction that helps to maintain life.

> **KEY ACTIVITY**
>
> ### Making models of amino acids
>
> Students can make models of amino acids, then investigate the importance of amino acid side chains. The structure of the first four alcohols (Table 10.4) is a good template: replace hydroxyl groups (–OH) with an amino group (–NH_2). Handling and observing models and discussing observations and ideas with peers is valuable. Students may work in pairs or individually to make some or all molecules, pooling their models across the whole class. Each student or student pair requires a model kit with at least 5 carbon, 2 nitrogen, 4 oxygen and 11 hydrogen atoms.
>
> **Figure 10.10** Structure of glycine and alanine

 ## Polymers

Polymers are substances comprising molecules made up of one or more repeating units bonded together. The repeating unit in a polymer molecule is called a *monomer*. The prefix *poly-* means 'many', so a polymer molecule may be thousands of atoms long. Naturally occurring polymers include carbohydrates (sugars) such as starch and cellulose which are found in plants; proteins, such as keratin, present in hair and nails, and collagen, responsible for the elasticity of skin; and deoxyribonucleic acid (DNA), present in all plant and animal cells (with a few exceptions, such as mature red blood cells). Numerous synthetic polymers exist, many made from derivatives of crude oil, particularly ethene. Others, such as nylons, polyesters and proteins, form from two or more monomers. These are called co-polymers.

KEY ACTIVITY

Investigating natural polymers

To investigate natural polymers, students can be given one natural polymer each and invited to carry out desk-based research leading to a report or presentation. The task should be focused on analysis of polymer structure, bonding and function. This activity will develop research skills and provide information on a range of polymers that allows students to realise similarities and differences between them, recognising properties and structures of homologous groups.

Addition polymers

Addition polymers include those based on ethene. The simplest is poly(ethene), which is the formal, systematic name for polythene. It forms when ethene molecules bond together forming long chains. Poly(ethene) (abbreviated to PE) was first made by accident in 1898 when a German chemist, Hans von Pechmann, was heating a substance called diazomethane under pressure. His assistants found a white waxy substance in the reaction vessel. They found out the substances comprised chains containing $-CH_2-$ only. They did not capitalise on their discovery. In 1933, two British chemists, Eric Fawcett and Reginald Gibson, working at ICI (Imperial Chemical Industries) in Cheshire, made the same white waxy substance accidentally by heating ethene with another compound, benzaldehyde, under high pressure. Chemists eventually worked out how to produce the white solid (known at the time as polyethylene) on demand, and how to use it. During the Second World War poly(ethene) was used secretly to insulate cables. Poly(ethene) manufacture has developed since then and, since it is an extremely versatile material, it is now probably the world's most commonly made polymer. Poly(ethene) is measured by density and is often referred to as low-density (LDPE) or high-density (HDPE).

10 Organic chemistry

An equation for the addition polymerisation of ethane is shown in Figure 10.11.

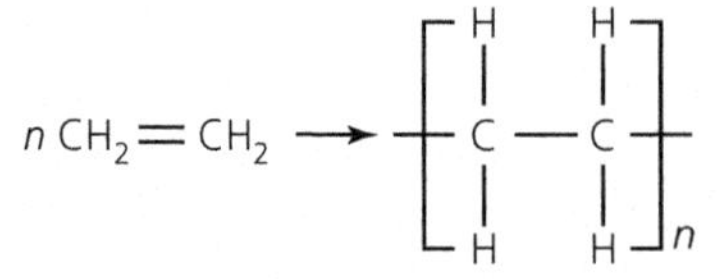

$$n\,CH_2=CH_2 \longrightarrow \left[\begin{array}{cc} H & H \\ | & | \\ C-C \\ | & | \\ H & H \end{array}\right]_n$$

Figure 10.11 A reaction equation for the polymerisation of ethene

Two activities based on poly(ethene) are described. These are: modelling an addition polymerisation reaction and comparing the properties and structures of polymers based on the ethene monomer.

> **KEY ACTIVITY**
>
> ## Modelling addition polymerisation
>
> Students can model addition polymerisation reactions using Molymod sets or similar modelling kits. Students begin by building a model of an ethene molecule from two black carbon atoms, four white hydrogen atoms, two flexible grey bonds and four rigid grey bonds. Line the students up shoulder to shoulder, holding their model molecules. Ask the first student in the line to initiate polymerisation by breaking the double bond in their ethene molecule. The next student breaks their double bond, joining this to the free bond in the first student's molecule. This process repeats, until multiple ethene molecules have polymerised into a long chain poly(ethene). Polymerisation terminates when no further monomer molecules are present. The polymer produced will have as many repeating units as there are students in the group. This is likely to be a fraction of the actual number in a real molecule. In addition polymerisation, atom economy is 100% as all atoms in the monomer molecules are present in the polymer molecule.
>
> Ensure students understand limitations of this model of the process. Bonds are not physical objects. The model suggests the addition of one monomer to the next requires two steps: breaking open the double bond and then forming a bond between the two monomers. In practice, these steps occur simultaneously in a fraction of a second. Polymerisation rate increases at higher temperature and pressure due to more frequent successful collisions between monomer molecules. Addition of a catalyst increases reaction rate. For example, peroxides enhance the rate of formation of low density polyethene (LDPE), and Ziegler–Natta catalysts increase the rate of formation of high density polyethene (HDPE).
>
> All monomers based on ethene have a carbon–carbon double bond. The monomers vary by the chemical groups attached to the carbon atoms in the double bond. These include hydrogen atoms, methyl groups and halogen atoms.

10.3 Polymers

> **Questions to ask**
> - What type of polymerisation reaction is modelled when making poly(ethene)?
> - What is the atom economy of the reaction? Explain why this happens.
> - How and why would the monomer and polymer react with bromine water?
> - What reaction conditions increase the rate of polymerisation? Why?
> - How does polymerisation end?

In a further activity, students can investigate properties of poly(ethene) by carrying out simple tests and comparing properties and structures of ethene and poly(ethene).

KEY ACTIVITY

Investigating everyday polymers

Well-known polymers made from monomers based on ethene include LDPE and HDPE; poly(propene) (PP); poly(chloroethene), known as poly(vinylchloride) or PVC; poly(styrene) (PS); and poly(tetrafluoroethene) (PTFE). Chemists write polymer names formally with brackets around the name of the monomer.

Another common polymer is poly(ethylene terephthalate) (abbreviated to PET or PETE). This is a co-polymer made from two monomers. One monomer is a dicarboxylic acid and the second is a diol, an alcohol with two –OH groups. These react in a condensation reaction (see below), which forms ester linkages (–COO–) between monomers so, technically, PET is a polyester. PET is often transparent, with glass-like clarity. It is used to package products which manufacturers want consumers to see, for example, shower gels, water, juices and chocolates.

Students can compare and contrast items made from polymers, relating these uses to polymer structure and properties. Polymers are identified by recycling numbers. Each polymer type has a recycling number from 1 to 7 imprinted in a small triangle, usually on the object's underside. On some items, the plastic abbreviation, for example, LDPE, PP or PVC may be present instead of the number. Polymer recycling numbers, properties and formulae are in shown in Table 10.6.

Create an information sheet for each plastic that includes the plastic recycling number, monomer formula and name, plastic formula and name, T_m and T_g values (Table 10.6). Make an 'Other' sheet for plastics with recycling number 7, which indicates a less common plastic or a combination of plastics. Add a model of the monomer to help clarify the differences between the polymers.

The terms 'polymer' and 'plastic' have different meanings. Plastics are materials produced from polymers, with additives that alter properties. Polymers are long chain molecules made from one or more monomers bonded together.

10 Organic chemistry

Table 10.6 Common polymers listed by recycling numbers

Recycling number	1	2	3	4	5	6
Common name	Polyethylene terephthalate	High-density polyethylene/ polythene	Polyvinylchloride	Low-density polyethylene/ polythene	Polypropylene	Polystyrene
Abbreviation	PET or PETE	HDPE	PVC or V	LDPE	PP	PS
Formal name	poly(ethenebenzene-1,4-dicarboxylate)	High-density poly(ethene)	Poly(chloroethene)	Low-density poly(ethene)	Poly(propene)	Poly(phenylethene)
Monomer formula and structure	Co-polymer of $HO_2CC_6H_4CO_2H$ and $HOCH_2CH_2OH$	C_2H_4	CH_2CHCl	C_2H_4	CH_2CHCH_3	$CH_2CHC_6H_5$

10.3 Polymers

Recycling number	1	2	3	4	5	6
Polymer formula and structure	(PET aromatic ester structure)$_n$	$\begin{bmatrix} H & H \\ -C-C- \\ H & H \end{bmatrix}_n$	$\begin{bmatrix} H & H \\ -C-C- \\ H & Cl \end{bmatrix}_n$	$\begin{bmatrix} H & H \\ -C-C- \\ H & H \end{bmatrix}_n$	$\begin{bmatrix} H & H \\ -C-C- \\ H & CH_3 \end{bmatrix}_n$	$\begin{bmatrix} H & H \\ -C-C- \\ H & C_6H_5 \end{bmatrix}_n$
T_m values /°C	~126	~130	~147	~110	~165	~150
T_g values /°C	67	27	81	−125	−10	100

Notes

The T_m value is the melting temperature. Polymers normally melt over a temperature range. This is because polymers contain long molecules of varying length, so there is no precise temperature at which the change to liquid occurs. The value given is mid-range. The symbol (~) that precedes the T_m figure in Table 10.6 represents that the value is not precise.

The T_g value is the glass transition temperature. At or below this temperature the polymer structure becomes glassy, crystalline and brittle. Below T_g, a plastic may shatter more easily, but has higher tensile strength.

T_m and T_g values enable estimation of conditions under which polymers can be used. For example, HDPE is a hard, rigid plastic at room temperature and pressure, while polypropene (PP) is often more flexible.

Make a display to reinforce polymer properties and molecular structure by asking students to bring clean samples of polymers to the lesson. Place information sheets numbered 1–7 for each polymer in a row along a bench or desk. Add a model of the monomer to help students connect the structure and function. Students sort polymer samples using the recycling numbers on each example, placing each next to the corresponding information sheet. Any samples without recycling numbers should be placed in 'Other' (number 7). Students can tabulate how items are used, noting monomer and polymer formulae, T_m and T_g. Extend this by developing research skills in which students explore polymer prices and other uses.

Questions to ask

- Describe similarities and differences between the monomers.
- How do physical properties vary within each individual type and between types of polymer?
- What are polymers used for? Which polymers have two or more distinctly different uses?

Further information can be found at www.chem1.com/acad/webtext/states/polymers.html

Enrichment

The properties that make polymers useful, such as stability in a wide range of conditions and temperatures and their adaptability to many thousands of uses are precisely those that make them challenging to recycle. In Europe, mechanical recycling is the main route, involving grinding, washing, separating, drying and re-granulating polymers into *recyclates* that can be made into new products. Some plastics are recycled using chemical methods, including burning (pyrolysis), catalysis, adding solvents and reacting with steam. All these are energy-rich processes that are expensive to run and are imperfect, as the amount of plastic products we use is greater than recycling capacity. To solve this problem, chemists have taken a step towards the dream of zero-waste plastics by developing new polymers that can be easily recycled into the original monomers. In 2019, a team working at the Lawrence Berkeley National Laboratory in California, US created a group of new polymers called polydiketoenamines or PDKs. Monomers in PDKs are held together by covalent bonds. Although these bonds are very strong, they can be broken by adding small amounts of concentrated acids to the polymer at room temperature, reforming the original monomers. This is a much more straightforward recycling process than that for a conventional polymer. In theory, PDK can be made, used, recycled and remade into new forms an infinite number of times, without changing the polymer's properties. At the time of writing, PDKs are very expensive to make so it may be a while before we see 'PDK' on items used in everyday life.

Condensation polymers

Condensation polymers are made when two or more monomers react successively to make a long molecule, releasing (condensing) a small molecule, often water (H_2O), whenever the two monomers combine. Well-known condensation polymers include nylon and polyesters (see above).

KEY ACTIVITY

Making nylon, a condensation polymer

A popular activity is making nylon, a condensation polymer, via the 'nylon rope trick'. Full details are available at https://edu.rsc.org/resources/making-nylon-the-nylon-rope-trick/755.article

In this demonstration, nylon (6,10) forms at the boundary between the layers of two chemicals: diamine (1,6-diaminohexane, $C_6H_{16}N_2$) and sebacoyl chloride (decanedioyl dichloride, $C_{10}H_{16}Cl_2O_2$ dissolved in cyclohexane). The numbers in brackets, (6,10), correspond to the numbers of carbon atoms in the diamine monomer and sebacoyl chloride respectively.

When setting up, take care to layer the two liquids correctly, placing the aqueous diamine solution in the beaker first. The sebacoyl chloride is placed on top to form the upper layer.

The monomers polymerise, forming nylon and a small molecule, hydrogen chloride. Small quantities will produce several metres of nylon. The polymer is drawn out as a long thread, the rope. The polymer can be washed in water and its properties observed.

Writing the formula for polymers is complicated because the molecules are large, so chemists employ abbreviations to focus on functional groups. Substituting symbols such as ■ and ♦ for carbon chains in formulae of organic substances help students focus on functional groups and how these change during a reaction. These symbols minimise distractions created by large numbers of atoms between groups. The reaction can then be shown as:

$$H_2N-■-NH_2 \text{ and } ClOC-♦-COCl \rightarrow [-OC-♦-CONH-■-NH-]_n + 2nHCl$$

Questions to ask

- What small molecule is produced in this polymerisation reaction?
- Write an equation for the formation of the polymer. Use the pictorial formulae $H_2N-■-NH_2$ and $ClOC-♦-COCl$ to represent the diamine and sebacoyl chloride respectively.
- Why did the two solutions form layers? Explain the relative positioning of the two layers.
- Is this an efficient method for industrial preparation of nylon? Explain your answer.

10 Organic chemistry

> **Maths**
>
> To strengthen mathematical skills, students can calculate the numbers of molecules of each monomer required for this nylon chain and the numbers of carbon atoms in one molecule.
>
> Numbers of molecules of each monomer
>
> - The repeat unit of nylon (6,10) is $(CO-(CH_2)_8-CONH-(CH_2)_6-NH)_n$, M_r value = 282.
> - Assume one nylon (6,10) chain has an M_r value of 130 000.
> - The number of molecules of each monomer is 130 000 ÷ 282 = 461.
>
> Number of carbon atoms in one molecule of nylon (6,10)
>
> - Each repeating unit includes 16 carbon atoms.
> - There are 461 repeating units in one molecule.
> - Therefore there are 16 × 461 carbon atoms in one molecule = 7376.

In industry, large quantities of nylon are made quickly and cheaply from diamines and dicarboxylic acids in aqueous solution at about 230 °C. The temperature is raised to about 270 °C and steam is removed to maintain the reaction pressure. The pressure is reduced and the polymer extrudes as a lace. At the time of writing, a video outlining the industrial process is available here: https://www.youtube.com/watch?v=4GxeSO7DyaE

> **Enrichment**
>
> Recycling plastics is a multi-million pound, highly technical industry. Some countries, including China, have stopped accepting imported plastic waste. At the time of writing, a Sky News documentary is freely available, which can be a useful starting point for discussions and research, at https://www.youtube.com/watch?v=oRQLilXLAIU

Deoxyribonucleic acid (DNA)

The structure of DNA was first proposed by scientists James Watson and Francis Crick in 1953, using X-ray diffraction photographs of DNA molecules taken by another scientist, Rosalind Franklin, working with Maurice Wilkins at King's College, London. DNA encodes the genetic information of all living organisms. DNA (deoxyribonucleic acid) is a condensation polymer with a complex structure. The polymer comprises two chains that curl together in a double helix. Each chain has three types of molecule: on the outside, a carbohydrate called ribose, which has five carbon atoms (compared to six in glucose) alternating with phosphate (PO_4^{3-}) groups; on the inside, four nucleotides (called thymine, cytosine, guanine and adenine) lie flat in horizontal pairs across the two ribose–phosphate chains. Thymine always pairs with adenine and cytosine with guanine. One chain is the coding sequence that dictates the order of amino acids in proteins.

Proteins

Proteins are condensation polymers of amino acids (see above). The arrangement of amino acids in a protein is determined by the genetic code: each gene provides the information required for one protein. Proteins perform an enormous variety of functions in living organisms. They may be structural, such as keratin which forms hair and nails. Functional proteins have specific roles in metabolism, such as myoglobin, the protein that carries oxygen in muscle, and haemoglobin, which transports oxygen around the body in red blood cells. Other functional proteins include enzymes, which are biological catalysts, and antibodies, which form as part of our natural defence mechanism when a disease-causing agent, such as a virus or bacterium, enters the body.

Enrichment

Stephanie Kwolek (1923–2014) was an American chemist. In 1946 she took what she expected to be a short-term, post-graduate job with the US chemical company DuPont. Opportunities for women were (unusually) available, as many men were still serving in the armed forces overseas after the Second World War. DuPont was researching new materials for lightweight body armour to improve the protection for soldiers, sailors and aircrew. Nylon (see above) was a recent success, so testing new polymers seemed a good option. By 1964 Stephanie was running her own polymer research group. She decided to try polymerising two substances (poly-p-phenylene terephthalate and polybenzamide) at a temperature of 0–40 °C. Polymer solutions are usually clear, colourless and thick. This time, the solution was thin, opaque and looked like buttermilk. Stephanie considered throwing it away but decided to try spinning a fibre from it. To her surprise, the new fibre proved to be stronger than nylon and five times stronger than steel. The polymer was patented in 1965 as Kevlar. Kevlar is best known for its use as body armour, but also has many other applications including ropes, cables, tennis rackets and aeroplanes. Her invention led to numerous awards. Stephanie Kwolek retired in 1986, 40 years after starting her temporary job.

10 Organic chemistry

> **Careers**
>
> The petrochemical industry is a vast international operation, with many opportunities for careers within the UK and worldwide. Chemists are employed at all stages of production, from testing oil drawn from new sites, through separation and refinery, to formulation and quality control. Chemical engineering combines an understanding of chemistry with industrial processes for handling millions of tonnes of product per year. British Petroleum (BP) provides a useful careers website: https://www.bp.com/en/global/bp-careers/professionals.html
>
> Synthesis of materials can be studied in university degree programmes such as Materials Science. New materials include polymers that can be easily returned to their starting monomers (https://science.sciencemag.org/content/360/6387/398); 'wood sponge' that separates spilled oils (https://pubs.acs.org/doi/abs/10.1021/acsnano.8b05763); and a gold–platinum alloy 100 times more durable than high-strength steel (https://interestingengineering.com/scientists-develop-the-most-wear-resistant-metal-alloy-in-the-world).
>
> Pharmaceutical chemists synthesise novel substances to tackle human diseases and illnesses. These include various types of cancer, heart disease and neurological conditions such as Parkinson's and Alzheimer's diseases. When the biochemistry of the disease is known, chemical compounds can be developed to target the biomolecules involved. For example, gene mutations cause production of malformed enzymes for which chemical inhibitors can be developed. Pharmaceutical chemists work in multi-disciplinary teams that include biochemists, microbiologists and clinicians. They carry out multi-step syntheses, purify compounds and investigate their products with spectroscopy. RSC Careers has more information: www.rsc.org/careers/future/medicinal-chemist
>
> The study of biochemistry is based on an understanding of organic chemistry, particularly focusing on reactions that occur in living organisms. Biochemists investigate metabolic processes, genetics, proteins and protein structures, the functions of organs of the body, the production and action of hormones, cell structures and biosynthesis. The Biochemistry Society offers information about careers in biochemistry https://www.biochemistry.org/education/careers
>
> Analytical chemists use separation and analytical instruments on a daily basis in industries as diverse as food production, water quality, sports and health services. They use their understanding of the molecular structure and properties of organic compounds to determine the presence of compounds in samples, for example, detecting illegal performance enhancing drugs in athletes at major competitions and toxins in hospital patients. See RSC Careers: www.rsc.org/careers/future/assistant-analyst-drug-control-centre

10.4 Resources

References and further reading

Atkins, P. (2003) *Molecules* (second edition), Cambridge University Press.

Brock, W.H. (2011) *The case of the poisonous socks: Tales from chemistry*, Royal Society of Chemistry.

Brunning, A. (2016) *Why does asparagus make your wee smell?*, Orion.

Clarke. C. (2012) *The science of ice cream*, Royal Society of Chemistry.

Hanson, J. (2011) *Chemistry in the kitchen garden*, Royal Society of Chemistry.

Hartings, M. (2016) *Chemistry in your kitchen*, Royal Society of Chemistry.

Hutchings, K. (2000) *Classic chemistry experiments*, Royal Society of Chemistry.

Kind, V. (2004) *Contemporary chemistry for schools and colleges*, Royal Society of Chemistry.

Lister, T. (1995) *Classic chemistry demonstrations*, Royal Society of Chemistry.

Nicholson, J.W. (2011) *The chemistry of polymers*, London: Royal Society of Chemistry.

Selinger, B. and Barrow, R. (2017) *Chemistry in the marketplace*, CSIRO Publishing.

Websites

Kind, V. (2004) *Beyond appearances: Students' misconceptions about basic chemical ideas.* Available at https://edu.rsc.org/resources/beyond-appearances/2202.article

Authoritative information on chemical industries: www.essentialchemicalindustry.org/index.php

Practical and safety support and guidance: http://science.cleapss.org.uk/ and https://www.sserc.org.uk/

BEST evidence in science teaching project at the University of York Science Education Group: www.york.ac.uk/education/research/uyseg/research-projects/bestevidencescienceteaching/

http://molview.org/ – a free website that allows common molecules to be displayed and rotated in a range of models.

RSC Teach Chemistry: https://edu.rsc.org/ – a wide range of resources for teaching and learning chemistry, including worksheets, simulations, practicals and demonstrations.

RSC online training module on organic chemistry: www.rsc.org/careers/cpd/teachers/about-courses/course-listings

Clark, J (2017) *Basic organic chemistry*: www.chemguide.co.uk/orgmenu.html

CLEAPSS (L195) Safer chemicals, safer reactions: http://science.cleapss.org.uk/Resource-Info/L195-Safer-chemicals-safer-reactions.aspx

The RSC carried out a global practical measuring vitamin C in food, which is an interesting investigation for students: www.rsc.org/learn-chemistry/resource/res00001280/measuring-vitamin-c-in-food-a-global-experiment

Other resources

Molymod® – molymods.com – available from school equipment suppliers such as Better Equipped, Philip Harris, SciChem and TimStar.

Standard laboratory equipment is required for practical activities. Microscale practical activities may require equipment sometimes not found in schools. These can usually be purchased from school equipment suppliers or made in-house. CLEAPSS offers guidance at http://science.cleapss.org.uk/Resource-Info/TL018-Microscale-chemistry-poster.aspx

Quickfit glassware is excellent for the distillation activity – see Section 10.2.

Earth science

Stuart Jones and Christopher Saville

Introduction

What is Earth science?

Earth science is the study of the planet we live upon, focusing on the materials that make up the Earth: that is, rock, water and atmosphere. Earth science includes various branches of science. It is an exciting subject with many interesting and practical applications. Some Earth scientists use their knowledge of the Earth to locate and develop energy and mineral resources. Others study the impact of human activity on Earth's environment, and design methods to protect the planet. Frequently, Earth scientists use knowledge about Earth processes such as volcanoes, earthquakes, floods and tsunamis to plan towns and cities that will not directly expose people to these dangerous events.

Understanding how the Earth works as a system and how humans interact with the Earth is critical at the present time. Understanding the causes and potential societal consequences of natural Earth processes and the production, availability, and potential depletion of natural resources (for example, water, minerals, and energy) is essential because these impact our economy, our security and the safety and sustainability of our environment. The context for understanding modern Earth processes lies in deciphering records of Earth's past. Investigating these records, as well as modern human interactions with Earth processes and resources, is therefore critical to the wellbeing of humanity and the planet.

The study of the chemistry of the Earth, known as geochemistry, is an important marriage between geology and chemistry that, over the last 50 or 60 years, has grown to touch every aspect of Earth science. Much of this progress has come from innovation in analytical techniques and the ability to assess the nature of natural materials with increasing precision using finer scales. It is difficult to do justice to geochemistry in a single chapter. Nonetheless, core concepts and ideas are included as a starting point. This chapter provides some background subject knowledge, suggested teaching sequences, practical exercises and best approaches to deliver geochemistry in school.

A world of rock

Our world is made from rocks and their component minerals. This fact can be difficult to comprehend, since much of the Earth's surface is covered by vegetation, soil, concrete or water, but if you dig down deep enough, solid rock will always be found. The importance of chemistry (geochemistry) to Earth science lies with the chemical elements that act as building blocks for minerals, rocks and fossils and understanding how the elements become distributed throughout the Earth.

About 4.54 billion years ago, our solar system formed from a cloud of gas and dust (see Chapter 1). This slowly contracted under the mutual gravity of all of its particles. Early Earth, and other planets that formed, chemically reflected the abundances of particles in this cosmic dust. The early Earth was very hot due to factors such as gravity, which forced particles together, impacts between particles, and radioactive decay (which was much more intense than observed today). The early Earth was probably partially or largely molten. Denser metallic liquids sank to the centre of the Earth while less dense silicate liquids rose to the top, like oil rises to the surface of water. In this way, the Earth differentiated into a metallic, mostly iron *core* with a silicate *mantle* and later, a thin *crust*.

The most common elements that displaced into the mantle during the formation of the Earth's iron core were oxygen and silicon (Table 11.1 and Figure 11.1).

Table 11.1 Chemical composition of rocks in the Earth's mantle and crust

Element	Symbol	% by weight	
		Mantle	Crust
Oxygen	O	44.8	46.6
Silicon	Si	21.5	27.7
Magnesium	Mg	22.8	1.5
Iron	Fe	5.8	5.0
Aluminium	Al	2.2	8.1
Calcium	Ca	2.3	3.6
Sodium	Na	0.3	2.8
Potassium	K	0.03	2.6
Other		0.3	2.1

The crust is a thin shell on the outside of Earth that solidified from cooling of the mantle. The crust accounts for less than 1% of the Earth's volume. It is broken into *tectonic plates* that move, allowing heat to escape from the interior of the Earth. The plates move due to convection currents occurring in the molten layer underneath them. During the formation of the crust, compounds known as minerals formed from elements found in the mantle. Minerals are inorganic building blocks (both elements and compounds) from which rocks are formed. 5650 minerals are recognised by the International Mineral Association and about 2000 of these are naturally occurring Earth materials. Some are rare and precious, such as elements gold

(Au) and diamond (C). The rarest elements (lower shaded area of Figure 11.1) are not the heaviest but are the *siderophile* (iron-loving) elements. Processes relocating them deep into the Earth's core cause depletion of available sources. Interestingly, only around twenty minerals are common. These include quartz (silicon dioxide, SiO_2) and feldspar (a complex mixture of aluminium oxide, Al_2O_3, and silicon dioxide). The vast majority of rocks form from combinations of a few common minerals, referred to as *rock-forming minerals*. The rock-forming minerals are: feldspars, quartz, amphiboles, micas, olivine, garnet, calcite and pyroxenes.

Elements present in rock-forming minerals are shown in the upper-left section of Figure 11.1.

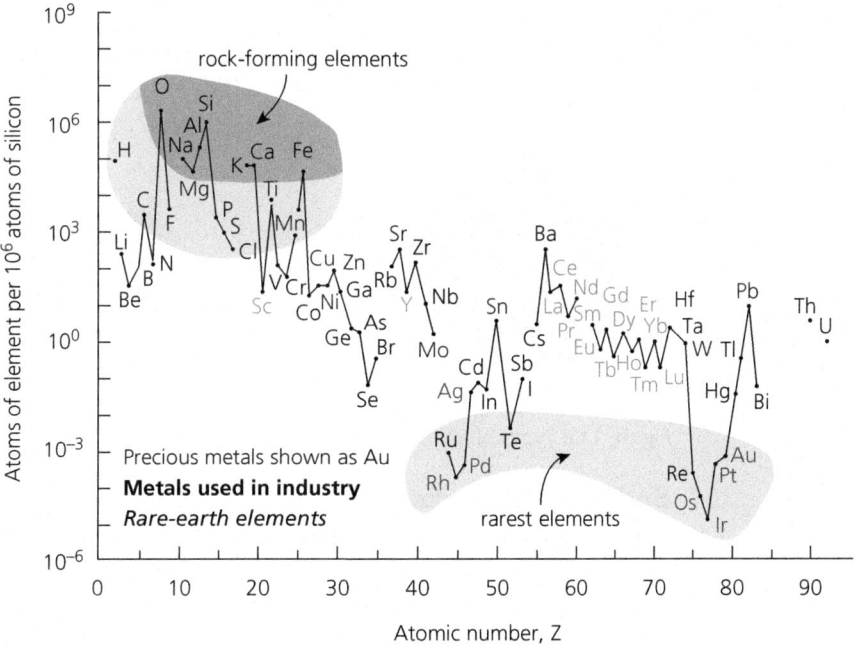

Figure 11.1 Abundance of chemical elements in Earth's upper continental crust as a function of atomic number

There are three types of rocks: igneous, sedimentary and metamorphic.

Igneous rocks form by cooling and crystallisation of a silicate melt. Molten rock material from which igneous rocks form is called *magma*. Magma is molten silicate material and may include already formed crystals and dissolved gases. The name magma applies to silicate melts within the Earth's crust; when magmas reach the surface, they are referred to as *lava*. The principal constituents of magma are oxygen, silicon, aluminium, calcium, sodium, potassium, iron and magnesium. Silica (silicon dioxide, SiO_2) is the most abundant component and ranges in abundance from 35 to 75%. The viscosity of magma is largely controlled by the silica content, while its melting point is controlled by the water content.

The relatively recent discovery of plate tectonics provides a grand unifying theory that helps explain how rocks form. Recent evidence shows how different igneous rocks form by partial melting of the crust at different depths. For example, basalts form in the upper mantle, granites form in the deep mantle and andesites form in subduction zones (areas of seabed that are being dragged downwards by plate movements). Types of rock can be identified by mineral composition and texture. Igneous rocks formed from lava tend to cool quickly, forming small mineral crystals or even no crystals. Rocks formed from magma deeper in the crust cool down more slowly than those formed from lava, forming larger crystals. Chemical composition and the cooling temperature of the original magma determine when minerals crystallise and hence which type of igneous rock forms (Figures 11.2 a and b).

a Igneous rock: obsidian (volcanic glass)   **b** Igneous rock: granite

Figure 11.2 Types of Igneous rocks

Sedimentary rocks form from pre-existing rocks or pieces of once-living organisms. The starting materials for sedimentary rocks are rock outcrops. Processes of physical and chemical weathering break down these source materials into small fragments of rock such as gravel, sand or silt. Sedimentary rocks form when these initial sediments solidify by cementation and compaction (Figures 11.3a and b). The most significant feature of the majority of sedimentary rocks is that they form layers or strata in chronological order. Consequently, sediments can provide a preserved record of former climates and landscapes. Frequently, fossil remains of animals and plants that lived during these time periods are found preserved in their respective sediment units. Since some 70% of rocks on the Earth's surface are sedimentary, and sediments are of great economic importance, there is a very good chance that we encounter sedimentary rocks or an associated sedimentary process at some point in our daily lives.

11 Earth science

a Sedimentary rock: oolitic limestone   b Sedimentary rock: sandstone

Figure 11.3 Types of Sedimentary rocks

Metamorphic rocks form when the original texture, composition and mineralogy of igneous, sedimentary and existing metamorphic rocks are changed by conditions of high pressure and extreme high temperatures (Figures 11.4a and b). Metamorphosis often occurs in the environment, most commonly at the centre of mountain belts, such as the Alps and the Andes, where tectonic plates meet. Metamorphic rocks and associated igneous rocks formed deep in the crust make up about 85% of the continental crust. The process of metamorphism does not melt rocks, but transforms them into denser, more compact material. New minerals are created by rearrangement of mineral components or reactions with fluids that enter the rocks. Pressure or temperature can change previously metamorphosed rocks into new types. Metamorphic rocks are often squished, smeared out, and/or folded. Despite these uncomfortable conditions, metamorphic rocks do not get hot enough to melt, or they would become igneous rocks.

a Metamorphic rock: garnet mica schist   b Metamorphic rock: a banded gneiss

Figure 11.4 Types of Metamorphic rocks

Students' prior knowledge and misconceptions

Studies of students' prior understanding of geology show that many do not understand that rocks are actually materials from the Earth's crust. Indeed, some students are only likely to identify large, jagged samples as rocks, describing smaller ones as stones or pebbles. Rock material is not regarded as 'natural' material. Students sometimes think of cut and polished samples as man-made while rock is a material from a quarry. There are also frequent misunderstandings about how natural cut stone can be used for building, and how this differs from bricks and other manufactured building products. Many man-made products such as limestone and gypsum are sourced from rocks for use in the building industry. Furthermore, students do not understand that rock material comprises compounds called minerals. They frequently use rock and mineral interchangeably.

Undoubtedly, one of the biggest challenges is to explain geological (deep) time to students. The mathematics curriculum includes understanding of numbers in the hundreds of thousands and millions. However, students are unlikely to have a good grasp of the magnitude of the ages involved in geology. A student is likely to just accept that a rock or fossil is just 'really old' without knowing what this means. Like historians, geologists strive to establish the sequence of events that created an array of geological features such as rocks, preserved fossils and key changes in Earth's history and landscapes. When possible, they provide a date on which each event happened. Geologists specify the age of one geological feature or rock with respect to another in a sequence as *relative age* and the age of a feature given in years as its *absolute age*.

Choosing a route

A sensible starting point to help elucidate the story of the changing Earth at secondary level is outlining important key events that have occurred since the Earth formed. Over geological timescales, in millions of years, the Earth's climate has changed many times, with many defining moments. These include the slow drift of continents which has forced readjustment of ocean circulation patterns; the rise and fall of mountain chains; giant volcanic eruptions; impacts by asteroids; and the emergence of life on Earth. All of these factors, and many more, have played a role in shaping the Earth we live upon. Starting by placing key events in geological order mirrors the way geologists work to understand processes on Earth and how these potentially interact.

KEY ACTIVITY

Geological deep time

This practical activity involves sorting and hanging pictures representing important evolutionary events on a 'washing line of time'. The requirements for each group of students (or whole class) are: a washing line (rope/string 4.6 m long); 20 clothes pegs; and a set of pictures of the items listed in the Event column in Table 11.2.

The washing line represents the 4600 million (4.6 billion) years since the Earth formed. Tell students which end represents the origin of the Earth, and which is the present day. On a 4.6 m washing line, 1 million years = 1 mm.

11 Earth science

- Ask students to order the events in Table 11.2 by time, from the one that occurred furthest in the past to the most recent. It may be helpful to place a marker midway to represent 2.3 billion years since the formation of the Earth.
- Invite students to peg events on the line in the position of the time that they think the events happened and record their answers.

Table 11.2 Key geological events through Earth's history

Event	Time of occurrence/ millions of years ago	Distance from present day/cm
First human (genus *Homo*)	2	0.2
First grasses	55	5.5
The mass extinction of 80% of animal and plant species at the start of the Tertiary period (K–T extinction)	65	6.5
First flowering plants (for example, *Archaefructus*)	130	13
First birds (for example, *Iberomesornis romerali*)	150	15
First mammals (for example, *Morganucodon*)	220	22
First dinosaurs (for example, *Thecodontosaurus*)	225	22.5
The mass extinction of 83% of all species on Earth at the start of the Triassic period (P–Tr extinction)	251	25.1
First reptiles (for example, *Hylonomus*)	325	32.5
First plants with seeds	360	36
First amphibians (for example, *Ichthyostega*)	360	36
First plants and animals on land	420	42
First animals with hard parts (for example, a trilobite)	545	54.5
First multicellular organisms	1200	120
First eukaryotes	2000	200
First bacteria	3500	350
The origin of the Earth	4600	460

There are other items that could be used to introduce the concept of deep time (the geological timescale), including, for example, using marked up rolls of wallpaper, 24 hours of the day, or even a toilet roll. The 'unit' (wallpaper roll, 24 hours, toilet roll) represents 4600 million years. Divide the unit into 1-million-year sections, then mark events on the unit at the point these occurred.

Questions to ask

- Ask students to explain their reasoning for the order they produce. Compare their order with the correct one (see Table 11.2). Discuss any differences.
- Which events were difficult to place on the timeline?
- What can they say about the order in which the events occurred?
- Humans have existed for 2 million years, while bacteria have been around for 3500 million years. Will either still be around 3500 million years from now?

> **Enrichment**
>
> Two scientists, Louis Agassiz (1807–1873) and Charles Lyell (1797–1875), proposed different theories about how the Earth's features formed. Agassiz proposed the catastrophe theory, which suggested that features formed in a series of major events, particularly glaciation, which caused sudden changes. He modified this to the daring hypothesis that glaciers and ice sheets covered much of the northern hemisphere from the North Pole to the Mediterranean and Caspian Seas, proposing a global Ice Age that was the last major event to blight the Earth. Lyell proposed the uniformitarian theory, which proposed that changes were gradual, and caused by events such as volcanic eruptions as well as glaciation. He also suggested changes were still occurring. Over time, evidence was collected that proved Agassiz's catastrophe theory to be false. Lyell became famous as an influential geologist whose book *Principles of Geology* helped to establish geology as a science. See Section 11.7 for an activity relating to these scientists.

11.1 Understanding how rocks are formed

In primary school, students are likely to have seen rock samples as small pieces of different rocks in the classroom, or when taking part in a school science trip to a beach, forest, hillside, or even in a museum. They are likely to have sorted rocks by making observations and using results of simple tests, such as scratching rock with a nail. Understanding how rocks form is more helpful. This approach is engaging and likely to help students understand processes involved in the formation of the dynamic Earth.

How does sediment stick together to form a rock?

The simplest starting point for students to appreciate formation of rocks is to investigate the process of sediment hardening into a sedimentary rock, known as *lithification*. Physical compaction (squeezing together of grains) of sediment alone is not sufficient to create a sedimentary rock. Progressive burial of sediment leads to grains becoming increasingly closely packed due to dewatering and compaction. Mudrocks may have up to a 90% reduction in thickness during burial. Compaction of between 40–50% occurs in coarser-grained sediments (Figure 11.5). If compaction continues, due to even greater thicknesses of overlying sediment and/or tectonic pressure, the faces of grains may dissolve where they are squeezed against neighbouring grains. This is called *pressure dissolution*. A sedimentary rock that continues to be deeply buried will experience ongoing alteration in its chemical and physical nature (known as *diagenesis*). This is a result of chemical reactions between the rock and new fluids passing through the rock as well as increasing pressure and temperature. These reactions may dissolve previously formed cements and/or form new cements and minerals (Figure 11.6). Commonly, changes in pressure and temperature lead to mineral transformations that alter the mineralogy of the sedimentary rock. To demonstrate that high pressures and temperatures are needed to make sedimentary rocks, ask a student to squeeze a handful of sand as hard as possible to see if it is possible to make a rock. They will find that they cannot do it.

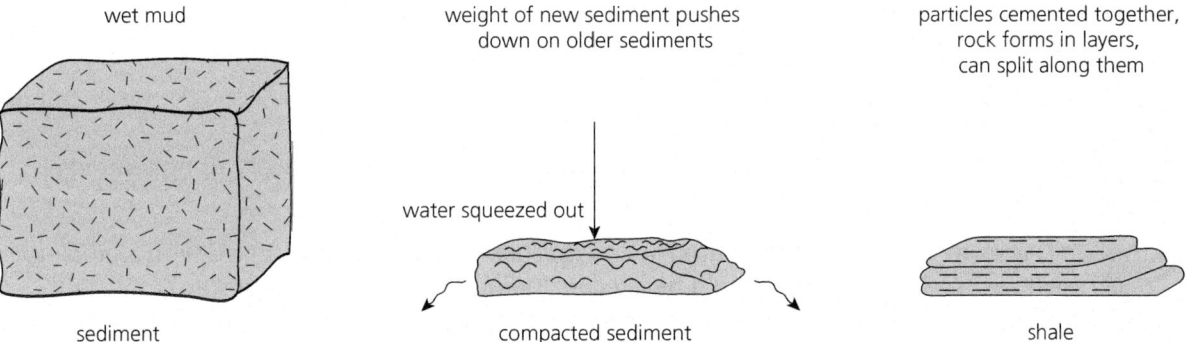

Figure 11.5: At the time of deposition muds can be as much as 90% water and undergo rapid dewatering and eventually leading to a mudrock

Minerals that form the necessary cement to 'stick' (or bond) the grains together primarily come from groundwater, which is present in most sediments. In most cases, common cementing minerals are quartz (SiO_2), calcite ($CaCO_3$), various compounds of iron, and clay minerals. Naturally occurring rocks tend to be more strongly cemented than those which can be made in school. Also worth exploring is the way that some sedimentary rocks are created from direct precipitation out of solution, without passing through the loose-sediment stage, but they will undergo compaction and cementation.

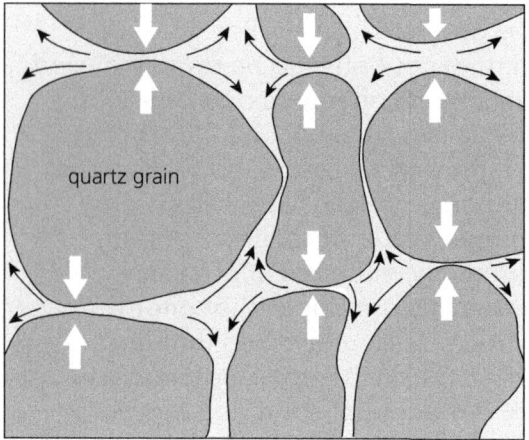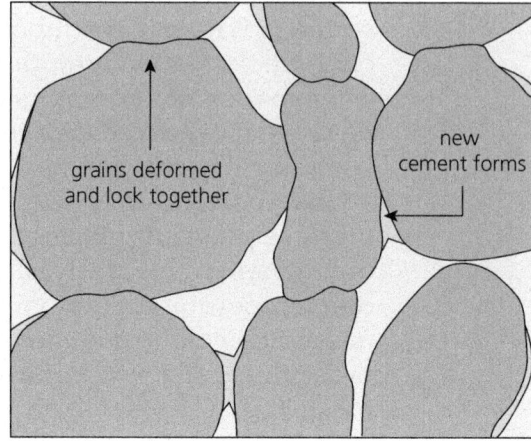

Figure 11.6: As a sediment is progressively buried under new layers, it starts to experience chemical and physical changes. Commonly, grains experience greater concentrations of stress at points of contact, leading to pressure dissolution. New cements that bind grains together occur in all sedimentary rocks.

Activity: Making sedimentary rocks

KEY ACTIVITY

The idea of grains being bound together can be tested in the laboratory using a variety of binding agents (cements).

In this experiment, students make rock pellets using sand and sand mixed with a binding agent. Several binding agents can be used including plaster of Paris (gypsum), clay, salt and sugar. Changing proportions of cement and sand permits investigation of how this affects the rock. The pellets are dried before testing. Full instructions are available via the Geological Society website at: https://www.geolsoc.org.uk/ks3/webdav/site/GSL/shared/pdfs/education%20and%20careers/RockCycle/Making%20Rocks%20Experiment.pdf

The equipment required is simple but includes 20 cm^3 syringes with their nozzles removed. These are used to create identical-sized pellets. If these are unavailable, pellets can be made in the base of small plastic or paper cups. About 3–5 cups will be needed for each student group, allowing for each group to test 2–3 sand/cement mixes and to adjust the sand to cement ratio.

Before carrying out each part of the activity, ask students to predict how strong they think each of their rocks will be when it has dried.

When the rocks have become hard (which may take a day or so), students can devise a fair test to compare their strength, to see which were made with the strongest cements. If possible, give them a piece of real sandstone to test as well. If they plan a series of tests, they should try the least destructive test first!

> **Discussion points**
> Many sedimentary rocks were once loose sediments. Natural cements deposited by fluids flowing through the spaces between the grains fixed them in place. Compacting sand into a plastic pot replicates sedimentary grains pushing together during burial. Fluids can be expelled with continuous new layers of sediment being laid down onto each other. The process of taking a loose unconsolidated sediment and the changes into a sedimentary rock can be complex.

An alternative version of the make your own rock activity is available at https://www.earthlearningidea.com/PDF/Make_your_own_rock.pdf. A larger scale, more complex version of the rock pellet task is available at https://www.earthlearningidea.com/PDF/341_Make_sed_structures.pdf. A possible extension activity involves comparing and contrasting sediments and rocks on Mars and Earth. The NASA Mars website has some great exercises and many fascinating 'raw' images from the surface of Mars (https://mars.nasa.gov). To get to the images, go to The Red Planet, then Geology, Mars Facts and Find out more. The website www.earthlearningidea.com has many teaching strategies for geology and Earth science.

11.2 Identifying rocks and their characteristics

In daily life, students experience small rocks, or pieces of exposed rock. This does not help them develop awareness of a continuous rock crust making up the surface of the Earth. In everyday speech the words rock and mineral are used imprecisely, whereas they have precise meanings for geologists. Students must understand that rock refers to material that makes up that part of the Earth beneath our feet. Soil, sand or water form a shallow covering over underlying rock. Opportunities to look at cliffs, quarries and caves to see exposed rock faces will help. Good photographs and videos can be a substitute for first-hand experience.

Collect good-sized hand samples or borrow a 'box of rocks' from a local university geology department. The virtual rock kit found at www.earthscienceeducation.com/virtual_rock_kit/index.htm may also be helpful.

The Geological Society has a useful resource designed for students aged 11–14 showing local scenery created by different rock types around Britain (https://www.geolsoc.org.uk/KS3Resources). This connects very well with discussing the rock cycle.

11 Earth science

KEY ACTIVITY

Making fake shelly limestone

This activity helps students consider that rock types, including sedimentary rocks, could contain fossils. They may know that vertebrate fossil bones of dinosaurs have been found in limestone at various sites worldwide. Limestone describes a large range of carbonate-rich ($CaCO_3$) sedimentary rocks that form in a variety of settings, regardless of origin. For example, limestones can form from direct organic precipitation from seawater or even from geothermal hot springs. Limestones frequently contain fossils from a marine environment. Shelly limestone comprises shells and skeletons of invertebrates that live in the sea. The shells are broken up by waves, so are not usually whole. Shells are typically made of carbonate, the same material as sedimentary rock. Splitting open a sample of shelly limestone may reveal beautifully preserved seashells (Figure 11.7).

Students may appreciate the challenge of adapting the make a rock activity above to make a fake shelly limestone using modern seashells and plaster of Paris mixed in a plastic container. Adding cold black coffee and/or yellow food colouring makes the mixture more realistic. The result can be compared with real shelly limestone (Figure 11.7).

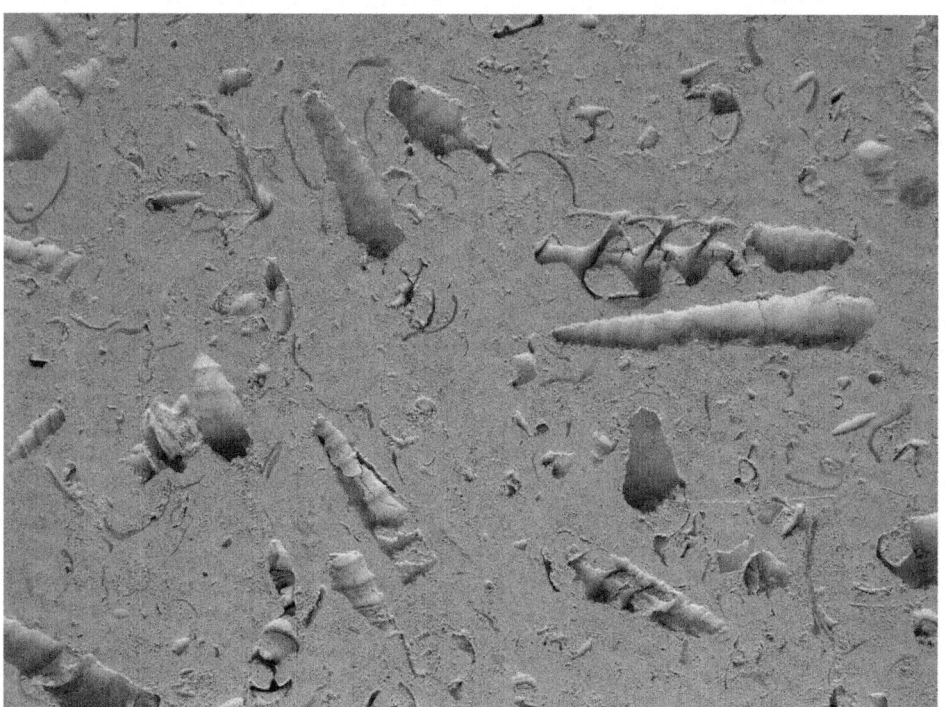

Figure 11.7 Gastropods and bivalves in the Upper Jurassic Portland limestone, Purbeck quarry, southern England. Scale of the large gastropod is about 5 cm

11.2 Identifying rocks and their characteristics

Minerals in granite

Ask students to examine a sample of granite with a hand lens (see Figure 11.2b). They will find lots of little pieces of different-coloured interlocking crystals called grains. Each type of grain is a different mineral and they are usually about 5 mm across. Most granites (an igneous rock) comprise four minerals, coloured grey, pink, white and black. Granites have more light-coloured crystals than dark. These are usually white or pink feldspar, glassy quartz and shiny mica.

Students can complete a table, stating the colour, lustre and size of grains in samples of minerals. Students should estimate the amount of each mineral in the sample. Discuss how to do this: for example, measuring a fixed area and counting the numbers of grains of each type present. Discuss the range and variety, including which are most and least common. Relate this to the abundance of chemical elements present in the minerals (see Table 11.1).

Granite is widely used in everyday lives for kitchen worktops, monuments and in commercial buildings. Students should discuss why granite is a good material for this. Answers include that granite is durable and will withstand weathering; it can be polished to shine and reflect light, which makes it look attractive; it can be shaped to form curved structures and pillars; the stone is available in variety of colours to suit different purposes; and it can be engraved.

KEY ACTIVITY

Comparing crystalline and granular rocks

To help students appreciate that the Earth's crust is rock, encourage them to become aware of rocks in their locality and to identify them. Collect examples of different types of rock from your local area and further afield. This investigation sequence uses clues in the rocks to find out how they formed.

Begin with two rocks, one made from sediment with obvious grains (a sedimentary rock such as sandstone, or chalk, which is a type of limestone), and the other a crystalline igneous rock with big crystals (such as a granite).

Organise students in groups of three. Two students will each choose one of the two rocks and describe it carefully. The third person will record key words and phrases used. These descriptions are reported to the class. Pooling responses should identify key properties of rocks, namely: their colour; that they are made of grains or crystals; have rough surfaces; feel cold, heavy and dense. Note that colour can be misleading as many minerals have the same colour and one mineral can have many different colours.

Ask students to predict what happens to the two rocks when placed in water. Often they will say just that rocks will sink. So watch very carefully when both rocks are placed into water together, then left for about a minute. Bubbles of air emerge from the sandstone, but many fewer from the granite.

Questions to ask
- What do you notice about where the bubbles come from?
- Why do they come from here?
- What does this tell you about the rock?
- Why is the sandstone different from the granite?
- What effect will being submerged have on the mass (weight) of each rock?

The answers are that most of the bubbles rise from the top of the rock, because air in the spaces (known as *pores*) in the rock rises, allowing water to flow into the base. This shows the rock is porous and that the spaces interconnect, meaning that the rock is permeable. The granite has no connected spaces, so air and water cannot flow through. For more information, visit the Rock detective section of www.earthlearningidea.com (https://www.earthlearningidea.com/PDF/Rock_detective.pdf).

KEY ACTIVITY

Identifying rocks

This activity involves using a key for identifying rocks. A set of large hand specimens is required. If a set of samples is not available, a virtual rock kit can be found at https://www.earthscienceeducation.com/virtual_rock_kit/index.htm.

To start, take photographs of one good example of an igneous, a sedimentary and a metamorphic rock, and label these (see Figures 11.2–11.4). Students can match rock specimens to the photographs and describe the distinctive features of each.

Next, progress to using a key to identify unknown rock specimens. In primary school, students will have used a dichotomous key (a Yes/No decision tree), to distinguish between objects and classify them. They may have used this technique with rocks. Classification using observational skills and simple tests to place items in categories is an important scientific skill. Figure 11.8 is a suitable key for rock identification. Keys work best when hand specimens are available which match the descriptions accurately.

11.2 Identifying rocks and their characteristics

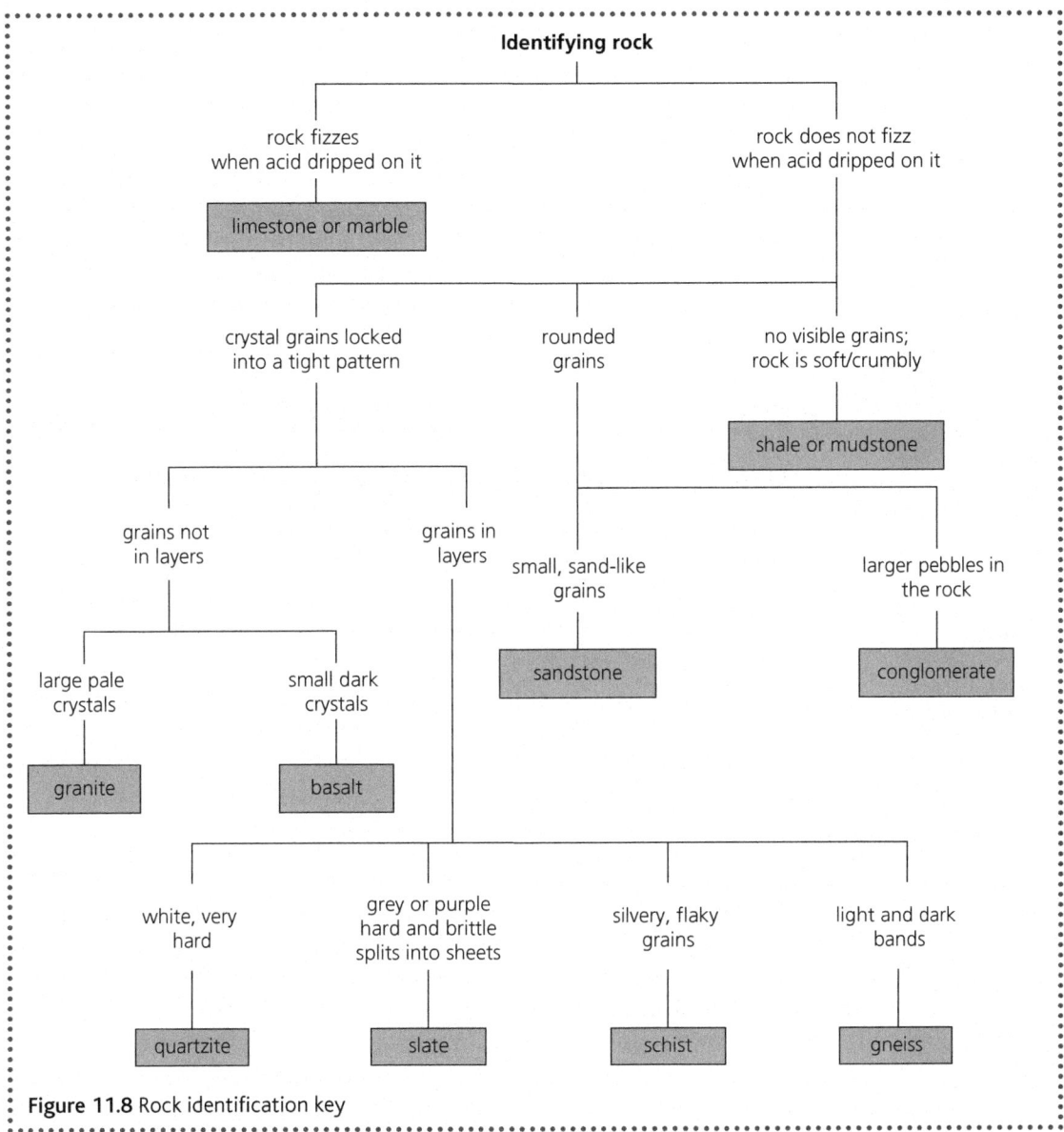

Figure 11.8 Rock identification key

Cross-disciplinary

Carbon is one of the most abundant chemical elements on Earth. Of all its forms, diamond – crystalline carbon – is the most valuable, with diamonds symbolising love and permanence in many societies. The most prolific source of diamonds worldwide is an igneous rock known as kimberlite pipe, which forms deep within the mantle when magma flows into deep fractures in the Earth. On cooling, the rock adopts the shape of huge carrots filling these fractures. The name kimberlite comes from the town

Kimberley in South Africa. In 1869 a diamond, named the Star of South Africa and weighing around 17 g (83.5 carats), was found near Kimberley, initiating a diamond rush – people flocked to the town seeking their fortunes. Over about forty years, miners dug a huge open-pit mine, the Big Hole, near the town. The Big Hole is 214 m deep, with a perimeter of 1.6 km and a surface area of 7 hectares, and yielded about 15 million diamonds. The largest diamond ever found is the Cullinan, found at the Premier No. 2 mine, Cullinan, South Africa, in 1905. Uncut, it weighed 621.35 g (3106.75 carats) and had an estimated value of about 2 billion US dollars.

11.3 Physical, chemical and biological weathering and erosion

Weathering and erosion are often confused. Processes that cause small pieces to break from rocks are *weathering*. *Erosion* is a separate process involving moving water, air or ice transporting weathered rock fragments away from the parent rock. *Physical weathering* occurs when rocks are exposed to the effects of the weather, such as extreme changes in temperature, causing small pieces to break from the surface. *Chemical weathering* causes removal of rock material by chemical interactions, for example, acidic rain reacting with limestone. Plants and animals living in or around rocks may cause *biological weathering*.

To understand erosion, students can observe images of sediments in a desert, noting that these have been transported long distances, and that changes in size, shape and composition of the mixture of fragments have taken place on the journey. If possible, introduce students to a laboratory-based stream table (videos are available online, for example, https://www.youtube.com/watch?v=q52hEv1tnWY). These mimic in miniature how streams transport rock debris, deposit this where the stream enters the sea. Weathered rock falls under gravity, then is carried away in the stream. This is erosion.

Sedimentary rocks can only form from rocks that already exist and are exposed to weathering at the Earth's surface. Weathered rock fragments are transported from the parent rock (igneous, metamorphic or sedimentary) then deposited as sediments on the seabed. These are compacted over time (see Figures 11.5 and 11.6) to form 'new' sedimentary rocks. The timescale involved in weathering, erosion and sedimentation is challenging to understand, as these processes occur over many hundreds and thousands of years.

11.3 Physical, chemical and biological weathering and erosion

Ornamental rocks on shop fronts, in graveyards and as building stones for churches and cathedrals are chosen to resist weathering. A general belief is that natural building stone is highly durable and weather-resistant. This is not always the case. All building stone is prone to physical weathering caused by freeze-thaw processes, salt weathering and wet–dry cycling. Prompt students to discuss the qualities of a good building stone and the processes that could cause deterioration over time.

Chemical weathering occurs where mineral constituents in the rock react with water, carbon dioxide and oxygen from the air. Chemical weathering is often more subtle, taking place at the sub-microscopic level. Exposed rock surfaces that are porous or fractured offer greater potential for chemical reactions to take place causing breakdown of the building material (Figure 11.9).

Figure 11.9 Weathered Carboniferous sandstone next to newly replaced cut sandstone from Durham Cathedral, UK.

An activity illustrating physical 'freeze-thaw' weathering is available at https://www.earthlearningidea.com/PDF/337_Freeze_thaw.pdf. This involves preparing a set of rock samples that have been frozen in a domestic freezer then allowed to thaw repeatedly. These steps must be completed before the lesson. The frozen and thawed samples are compared with a control set of the same rocks. Students should observe that sandstone (a sedimentary rock) generates more small particles over this short time period than other rock types. Relate this to the structure of the rock and how permeable to water it is (see the Comparing crystalline and granular rocks activity above). When frozen, water expands, exerting pressure within the rock. Over several freeze-thaw phases this pressure weakens the structure, causing grains to fall off. Less porous rocks are not as prone to freeze-thaw weathering. Images of physical weathering are available online (see Websites, Section 11.7 below).

Chemical weathering of limestone

Loose limestone (largely calcium carbonate) fragments are often used as the surface of paths, driveways and car parks, as well as in making roads. The fragments are susceptible to chemical weathering. Atmospheric carbon dioxide dissolves in rainwater forming carbonic acid (a weak acid, see Chapter 7):

$$CO_2 + H_2O \rightarrow H_2CO_3$$

Carbonic acid reacts with limestone:

$$CaCO_3 + H_2CO_3 \rightarrow Ca(HCO_3)_2$$

In words, limestone and carbonic acid combine to form calcium hydrogencarbonate. This is soluble in water, so is washed away by rain. Students can test this by making a solution of carbonic acid by blowing exhaled air into water, then adding limestone powder. If Universal Indicator is added, the colour changes from green to yellow/orange/pink as carbon dioxide dissolves, then back to green/blue when the limestone is added and reacts with the carbonic acid solution. The activity is described in full at www.earthlearningidea.com/PDF/214_Weathering_limestone.pdf

Good explanations of chemical weathering on limestone landscapes are available on the British Geological Survey webpages at https://www.bgs.ac.uk/discovering-geology/geological-processes/weathering/

Biological weathering

Chemical and physical weathering cause alterations and modifications to mineralogical components of rocks. When these processes are assisted by biological action, the process is called biological weathering. Organisms may alter rock by mechanical and chemical actions. Plant roots may expand and penetrate cracks, fractures, pores and other discontinuities, eventually causing rocks to rupture and disaggregate. Lichen bodies (thalli) exhibit similar behaviour to roots, expanding up to 3900% due to their high gelatine content. Figures 11.10 to 11.12 show examples of rocks being weathered by a plant or an animal. For each situation shown, ask students to:

→ identify the environment being weathered
→ identify the organism doing the weathering
→ explain how the organism causes the weathering
→ explain what happens to the products of the weathered rock.

11.3 Physical, chemical and biological weathering and erosion

Figure 11.10 Tree roots taking advantage of natural fractures in a rock.

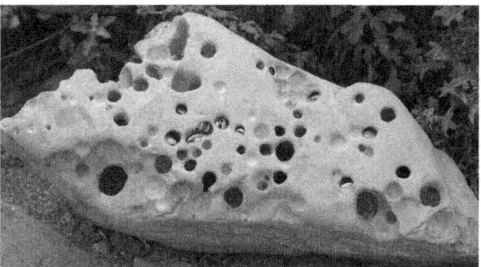

Figure 11.11 Piddock clam are rock-burrowing molluscs with ridged shells that they use to grind holes in solid rock, where they can live for up to 8 years. When they die, they leave empty holes that can be used by crabs, sea anemones and other animals.

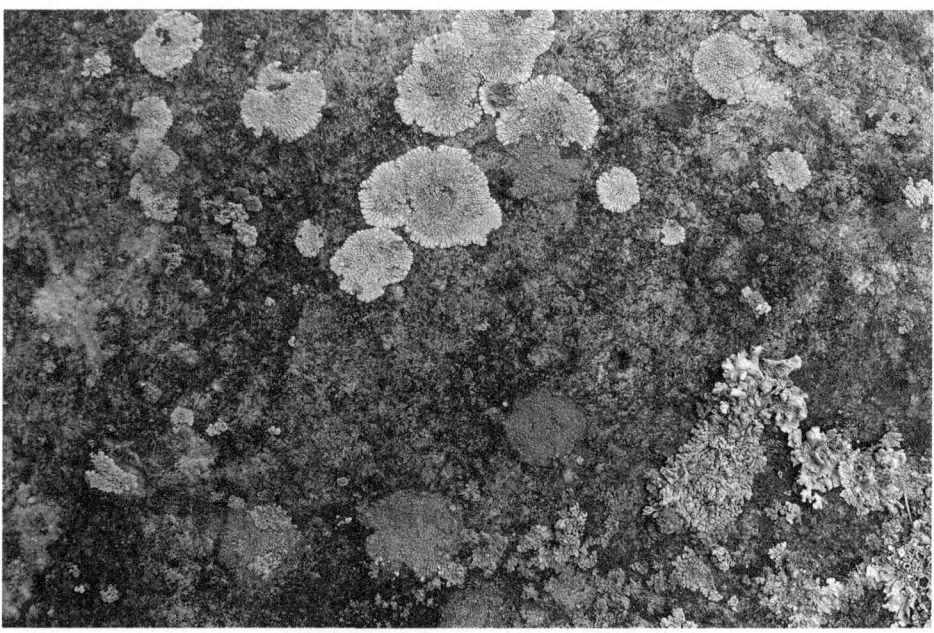

Figure 11.12 Lichens are among the first living organisms to establish on barren rock. Lichens comprise algae and fungi, which live together symbiotically. Lichens produce a dilute, acidic solution that slowly causes some minerals in rock to decompose.

11.4 The rock cycle

Understanding the rock cycle requires knowledge of rock formation, weathering and erosion. To introduce this, begin with sedimentary rocks, then expand the model to include metamorphic and igneous rocks.

The sedimentary rock cycle: a simple model for rock formation

Having introduced students to the formation of sedimentary rocks and weathering, the next logical step is to link sediments, sedimentary products and weathering processes into a cycle. A helpful way of doing this involves students working in pairs to carry out a card sort, linking images with terms associated with rock formation. Put together a series of cards like those shown in Figures 11.13 to 11.17.

Figure 11.13 Sedimentary rock

Figure 11.14 Weathered sedimentary rocks

11.4 The rock cycle

Figure 11.15 Mobile sediments with ripples

Figure 11.16 Layers of sedimentary rocks

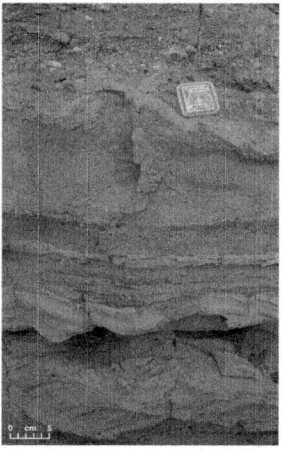

Figure 11.17 Sediments

The terms are:

- Weathering
- Erosion
- Compaction and cementation
- Burial

Ask students to arrange the rock cards in order and link them to the associated processes. Note that the terms may be used more than once.

11 Earth science

KEY ACTIVITY

To consolidate learning, these activities present individual processes.

Why are all samples of sand the same?
This short (150 second) video explains how sand forms, how it ends up at the beach, and why sand across the world tends to be the same https://sites.usask.ca/geol109/230/
The video revises sedimentary processes with links to minerals and where sediments originate.

Transport of sediment by wind
Wind is a very powerful force that can move large quantities of sediment produced by weathering. The Dust bowl activity available at https://www.earthlearningidea.com/PDF/61_Dust_bowl.pdf invites students to model wind erosion by making a mini-desert. Note students with asthma or an allergy to dust need additional protection to be able to participate in this activity.

Students make a mixture of fine rock grains with diameters of 0.125–4 mm. The mixture is placed on a tray, or in a box. Students blow an air current via a straw/empty pen barrel over the mixture. They examine which grains move, where the grains fall, and the shapes formed.

Typically, the air current from a straw carries sediment of fine sand (< 1 mm) and smaller. This is because air is a fluid with low density and viscosity that cannot exert sufficient pressure to move larger grains. This models movement of wind-blown very fine-grained dust entering the Earth's upper atmosphere and moving across the globe. Dust from the Sahara desert in Africa has been transported into Europe. In certain weather situations, Saharan dust can affect air pollution and pollution levels. Wind can erode topsoil from farmland, causing dust clouds; and form sand dunes of various shapes.

Sorting and rounding of grains
Water, ice, wind and gravity transport sediments over considerable distances. During transportation, weathering processes continue, causing sediment to change character. Commonly, rock fragments and grains hit and scrape into each other as they move. This removes corners and angular edges in a process called *rounding*. Sediments close to their sources are less well rounded than those transported long distances (Figure 11.18). Sediment grains of different sizes are selected and separated by transportation a process called *sorting*. Glaciers transport sediment of all sizes. Glacial sediments are poorly sorted, as various sized grains are deposited along glaciated valleys. In comparison, wind-blown sediments of a sand dune are well sorted, as grains in a dune are nearly all the same size (Figure 11.18). This activity illustrates this principle in water: www.earthlearningidea.com/PDF/230_Sediment_transport.pdf

11.4 The rock cycle

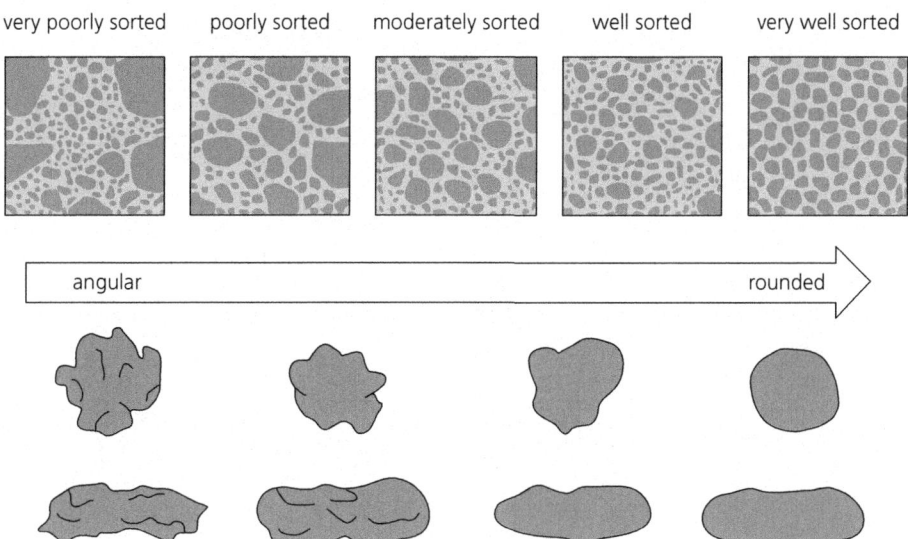

Figure 11.18 Rounding and sorting of grains allows a sedimentologist to infer how far grains have travelled from source areas and suggest how grains were transported

Questions to ask

→ Why do you think the grains move as they have done? Look closely at where the grains have fallen.
→ Compare the sizes of the grains with the distance they have been blown. What do you notice?
→ Explain how sand from the Sahara can land in northern Europe.
→ How does wind erosion affect our local area?
→ Consider the difference between rounding and sorting of grains. Well-rounded grains are not always found in well-sorted sediments. Why is this? What does this tell you about how far grains may have travelled?

Introducing the rock cycle

The next step is to widen the cyclical process of rock formation, weathering and erosion to include igneous and metamorphic rocks. Diagrams of the cycle are included in many school textbooks. For students aged 11–14 this Geological Society resource www.geolsoc.org.uk/ks3/gsl/education/resources/rockcycle.html may be helpful, together with www.earthlearningidea.com/PDF/52_Rock_cycle.pdf Other resources are recommended at the end of the chapter.

Ask students to label rock types in Figure 11.19, using images from Figures 11.2–11.4.

Even better, use actual rock samples from a rock box or available specimen collection.

11 Earth science

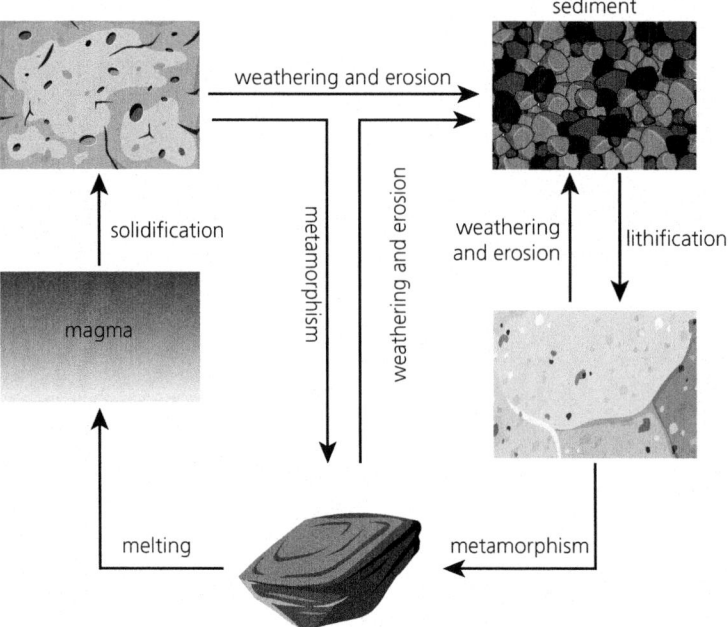

Figure 11.19 The rock cycle illustrating the processes that form each rock type

11.5 The Earth's atmosphere

Previous knowledge and experience

A big challenge for students is understanding the controversy surrounding debates about global warming. While there is considerable scientific evidence available, students may well be confused by strong messages being made by substantial numbers of non-scientists who argue that global warming is not an issue or is unimportant. The key disputed point surrounds causes of increased global average air temperature, especially since the mid-twentieth century. Debates focus on whether this warming trend is unprecedented, and/or is within normal climatic variations, and whether humankind has contributed significantly to an increase. To support students' understanding, develop knowledge about the Earth's early atmosphere and how this has changed over time. Then discuss greenhouse gas emissions and atmospheric pollutants (Chapter 11 in Biology).

Formation of the Earth's atmosphere

There are many uncertainties about the Earth's early atmosphere, in part because the Earth is 4.6 billion years old. The understanding we have at the moment suggests that when the Earth formed from a hot

11.5 The Earth's atmosphere

mix of gases and solids, it was very different from our Earth today. Its gravitational pull held onto hydrogen and helium molecules that had been swirling in the protoplanetary disc (https://phys.org/news/2015-01-planets.html). The surface of the early Earth was probably covered in a molten magma ocean with abundant volcanic activity yielding new gases that formed the early atmosphere (Figure 11.20). The first atmosphere survived only a short time, as heat from the Sun caused lightweight atoms to zip about so rapidly that they eventually achieved escape velocity from the gravitational pull and zoomed into space. Atoms that did not escape on their own were carried away by intense solar winds. While the Earth's primary atmosphere disappeared, volcanic activity on the Earth's surface yielded new gases. Also, the early Earth experienced frequent impacts from asteroids, meteorites and comets. Although volcanoes are thought to be the major contributors to the Earth's early atmosphere by volume, comets are regarded as an important source of more complex organic molecules.

Once the Earth had developed a magnetic field capable of deflecting the solar wind, these gases began to accumulate to form a second atmosphere from a series of ongoing complex reactions. Large quantities of volatile elements are bonded to solid minerals inside the Earth. When the minerals melt, these volatile gaseous elements separate from the minerals and dissolve in magma. Near the Earth's surface, where pressure decreases, these gases come out of solution and form bubbles that burst out of lava and into the atmosphere. Volcanic gases comprise 70% to 90% water with small amounts of carbon dioxide (CO_2), sulfur dioxide (SO_2) and traces of nitrogen (N_2) and ammonia (NH_3). So, the Earth's second atmosphere consisted of these gases plus other gases such as methane (CH_4) and carbon monoxide (CO).

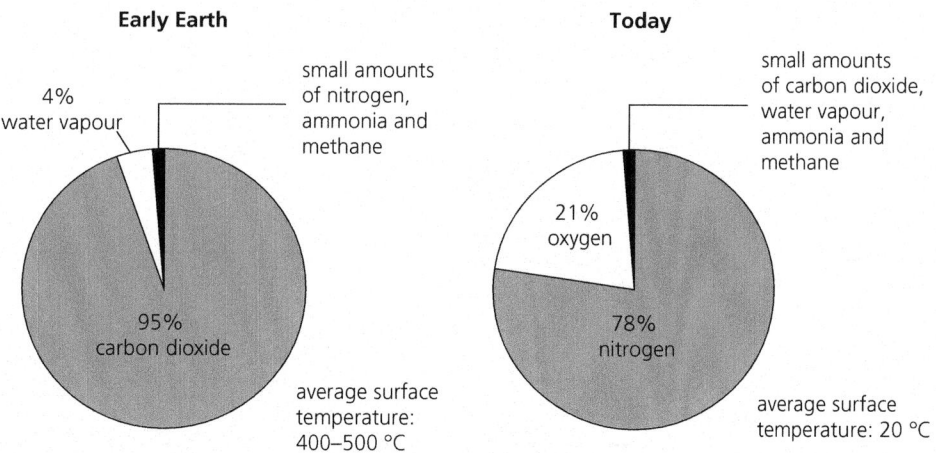

Figure 11.20 Pie charts showing the approximate percentages of gases in the Earth's early atmosphere and in the Earth's atmosphere today

The Earth's atmosphere has evolved over geological time. When the Earth's surface cooled sufficiently water vapour condensed into rain that filled the oceans, lakes and streams, or sank into the shallow crust to become groundwater. Permanent oceans only appeared around 3.85 billion years ago.

The first photosynthetic organisms, cyanobacteria (also known as blue-green algae), appeared on Earth between 3.8 and 3.5 billion years ago, enabling significant amounts of oxygen (O_2) to be produced. By about 2.4 billion years ago the Earth's atmosphere contained about 1% of present-day oxygen levels. Earth scientists refer to the transition from an oxygen-free atmosphere to an oxygen-containing atmosphere as the *Great Oxygen Event*. However, atmospheric oxygen levels became substantial only about 600 million years ago (0.6 billion years), when the Earth's modern atmosphere dominated by a mixture of oxygen (O_2) and nitrogen (N_2) gases was created.

Oxygen levels have varied over the last 600 million years reflecting changes in life on Earth. The evolution of the first plants around 470 million years ago led to a peak oxygen level of 35% around 300 million years ago, during the Carboniferous (coal-bearing) geological period. This period has the highest recorded oxygen levels. Extensive coal deposits from that time represent carbon that was not incorporated into carbon dioxide as lignin-rich flora decayed, a process that reduces atmospheric oxygen. Since then, oxygen levels declined to the present value of 21%.

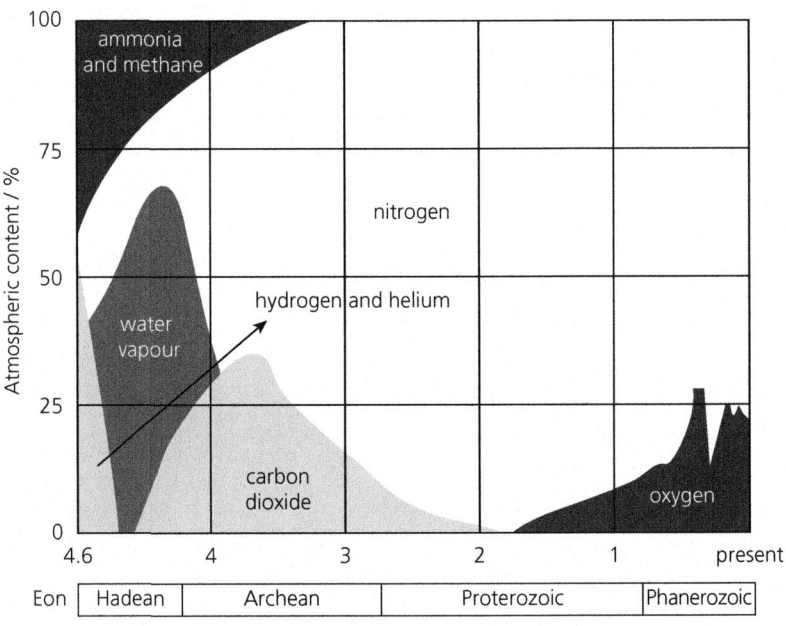

Figure 11.21 Composition of Earth's atmosphere through geological time

The evolving atmosphere: oxygen increasing and carbon dioxide decreasing

The accumulation of surface water in the Earth's oceans, lakes and as groundwater led to a radical drop in atmospheric carbon dioxide concentration. This occurred for two reasons:

1. Carbon dioxide dissolves in water to form carbonate ions which react with calcium ions to form solid carbonate minerals such as chalk and limestone. This process is very important for marine organisms that make their shells from calcium carbonate, becoming important components of many limestones, such as shelly limestone (see activity below). The reverse reaction occurs in chemical weathering (see p.364 Chemical weathering of limestone, in section 11.3).
2. During chemical weathering, carbon dioxide reacts with silicate rocks in the presence of water to produce carbonate ions. This forms carbonate minerals, which trap carbon dioxide in rock, removing it from the atmosphere.

A useful point to make is that huge amounts of carbon dioxide are present in limestone. Some of this stored carbon dioxide is released when limestone is converted into building materials such as concrete. This needs to be considered as an additional source of greenhouse gases.

KEY ACTIVITY

Limewater breath experiment

Bubbling carbon dioxide into limewater is a straightforward test for the presence of the gas (see Chapter 9). Limewater is a dilute, aqueous solution of calcium hydroxide. Exhaled carbon dioxide gas produces a fine precipitate of calcium carbonate (chalk) in reaction with limewater:

carbon dioxide + calcium hydroxide → calcium carbonate + water

                       limewater   limestone

$$CO_2 + Ca(OH)_2 \rightarrow CaCO_3 + H_2O$$

The experiment could be extended to demonstrate that carbon dioxide dissolves in water forming carbonic acid, lowering the pH of the solution. To track this, bubble exhaled air into water via a straw, measuring the pH of water beforehand and afterwards. The pH should drop to around pH 4–5. This process is replicated on a global scale as carbon dioxide dissolves into the Earth's oceans. Acidification of global oceans is a major effect of global warming, occurring as a direct consequence of climate change. This activity therefore provides a good link to greenhouse gases and global warming.

As the early atmosphere of Earth evolved, water and carbon dioxide were gradually removed and oxygen levels increased when the first photosynthesising cells (cyanobacteria)

11 Earth science

evolved around 3.5 billion years ago. Discuss with students whether the first appearance of photosynthetic organisms generated significant volumes of oxygen. Oxygen levels steadily increased to higher levels oxygen about 600 million years ago (see above).

Questions to ask

- → Name four gases in the Earth's atmosphere.
- → Give the proportions of gases in the Earth's atmosphere.
- → Why is evidence for the Earth's early atmosphere limited?
- → What gases are believed to have been released from volcanoes during the first billion years of the Earth's existence?
- → How did the oceans form?
- → Why did the amount of carbon dioxide in the earlier atmosphere decrease?
- → What gas in the current day atmosphere was not present in the atmosphere 4.6 billion years ago?
- → What has happened to most of the water vapour in Earth's early atmosphere?

Enrichment

A stromatolite is a layered rock left behind by a colony of bacteria, specifically, cyanobacteria. Cyanobacteria live in shallow water and are sometimes referred to as blue-green algae. Stromatolites tell us when photosynthesising organisms first evolved and proliferated on Earth. The oldest found so far is about 3.45 billion years old. A stromatolite has distinct layers. These formed when sediment clung to the sticky surface of a group of cyanobacteria so the bacteria could not receive sunlight to photosynthesise. The bacteria moved upwards over the sediment to get to the sun, leaving the sediment forming a layer underneath. More sediment fell, repeating the process, building layer upon layer to create the structure that ultimately fossilised.

Stromatolites live in a few locations worldwide, usually in water too salty for other animals to inhabit, so no predators can snack on them. One location is the Hamelin Pool Marine Nature Reserve at Shark Bay, off the western coast of Australia. Here stromatolites are considered living fossils, since they continue to grow. See www.sharkbay.org for excellent images of stromatolites. Meanwhile, https://stromatolitesgl2019.wordpress.com provides details about their anatomy and biology. Modern stromatolites in small streams deep in the Atacama Desert of the Chilean Andes thrive in the complete absence of oxygen. These high-altitude stromatolites use two different forms of arsenic to perform photosynthesis and respiration https://theconversation.com/ancient-microbial-life-used-arsenic-to-thrive-in-a-world-without-oxygen-146533

Greenhouse gases

The small amount of carbon dioxide that remained in the Earth's atmosphere after the oceans formed plays an essential role in keeping the Earth habitable. Carbon dioxide lets solar radiation (light) from space pass through it but prevents IR radiation (heat) rising from the Earth's surface from escaping back into space. When the amounts are small, this regulates atmospheric temperature, keeping the Earth's surface from becoming cold enough to freeze. However, when present in large amounts, carbon dioxide and other greenhouse gases including methane (CH_4), water (H_2O) and nitrous oxide (N_2O), trap more heat in the atmosphere than is necessary to keep the Earth temperature constant and comfortably habitable. The overall impact is that the Earth's atmosphere is gradually heating up, a process known as *global warming*.

Why should we be concerned about the greenhouse effect and climate change?

During the last century, global population has grown rapidly, increasing the level of intensive use of the Earth's resources. Emissions of gases such as carbon dioxide and methane from burning fossil fuels (oil, gas and coal), agriculture, cement production and deforestation have greatly increased. Evidence from the geological record is consistent with calculations showing that adding large amounts of carbon dioxide to the atmosphere warms the world. Warming causes the Earth's polar icecaps to melt, leading to higher sea levels and flooding of low-lying coasts. Weather patterns are changing, leading to significant changes in rainfall patterns, creating droughts in some countries and severe storms in others. Also, the acidity of the oceans is increasing as these absorb more carbon dioxide. At the same time, oxygen levels in seawater are decreasing, leading to significant changes in marine environments. Ultimately this means that the quality of life that humans and other living organisms experience on the Earth will diminish significantly over time unless there are global agreements that enable reversal of these changes.

The Paris Agreement, signed in 2015 by 196 countries, was the first recognised global attempt to combat climate change. The agreement aims to keep a global temperature rise well below 2 °C above pre-industrial levels. To achieve this, countries agreed to make efforts to limit the temperature increase to 1.5 °C. To enact the agreement, power will need to be generated from sustainable rather than carbon-based sources, and emissions from burning carbon-based fuels will need be reduced in many nations to zero.

There is international recognition that this requires investment in new power generation methods, building well-insulated homes with carbon-free heating, and changing transportation. The agreement also promotes aid to developing countries to enable them to achieve its aims.

To promote discussion of climate change, global warming and arguments against these, a *Café Scientifique* (Scientific Café) discussion with students may be a useful approach. A Café Scientifique discussion is always held outside a traditional academic environment, so in school this could be a dining hall, sports hall, arts room/theatre. The website cafescientifique.org gives more information. In school, ensure that students are ready to discuss the issues, for example, supporting continued use of fossil fuels and extending application of green renewable energy resources.

One point students may have come across is the notion that volcanoes emit more carbon dioxide than human activities. This point is discussed next.

Do the Earth's volcanoes emit more carbon dioxide than human activities?

A simple answer is 'No'. All studies to date of global volcanic carbon dioxide emissions indicate that present-day subaerial (land) and submarine (seabed) volcanoes release less than 1% of carbon dioxide released currently by human activities. For example, the eruption of Eyjafjallajökull in Iceland in 2010 released 150 000 tonnes of carbon dioxide per day. The eruption curtailed global air travel, which generated a net reduction of atmospheric carbon dioxide by between 1.3 and 2.8 million tonnes. Intense volcanic release of carbon dioxide in the deep geological past did cause global warming, and possibly mass extinctions, but even this is debated by scientists at present.

Published scientific estimates of global carbon dioxide emission rates for all land and submarine volcanoes range from 0.13 to 0.44 gigatonnes per year. In contrast, in 2010, human activity was responsible for 35 gigatonnes of carbon dioxide emissions. This is 80–270 times larger than annual global volcanic carbon dioxide emissions.

11.5 The Earth's atmosphere

Figure 11.22 Gases commonly produced during volcanic eruptions

Very large volcanic eruptions inject significant amounts of carbon dioxide into the atmosphere. The 1980 eruption of Mount St. Helens in Washington state, north-west US vented approximately 10 million tonnes of carbon dioxide into the atmosphere in only nine hours. However, humanity requires only 2.5 hours to produce the same amount. Large explosive eruptions occur globally roughly once every 10 years. Meanwhile, humanity's carbon dioxide emissions are ceaseless and increase every year. There is little doubt among Earth scientists that carbon dioxide emissions due to human activity dwarf global volcanic carbon dioxide emissions.

Websites provide activities related to greenhouse gases and climate change, for example https://scied.ucar.edu/games-sims-weather-climate-atmosphere A simulation of the greenhouse effect is available at https://gizmos.explorelearning.com/index.cfm?method=cResource.dspDetail&ResourceID=372

Other common atmospheric pollutants

Substances that would not be in the atmosphere, or are present at higher concentrations than would be observed due to natural phenomena alone, are *pollutants*. Human activity, especially energy production, manufacturing and transportation, has changed atmospheric concentrations of trace gases and added new gases and aerosols (tiny particles or droplets) to the Earth's atmosphere. Table 11.3 shows common atmospheric pollutants.

Table 11.3 Common atmospheric pollutants and their sources

Pollutant	Source of pollutant
Carbon monoxide (CO)	Incomplete combustion of any fuel containing carbon atoms
Nitrogen oxides (NO$_x$)	Oxidation of atmospheric nitrogen inside combustion engines in cars, buses, lorries, etc.
Sulfur dioxide (SO$_2$)	Combustion of a fossil fuel which contains sulfur impurities
Particulate carbon, (C, soot)	Incomplete combustion of any fuel containing carbon atoms

Atmospheric pollution has increased since the late eighteenth-century Western European industrial revolution. Burning coal and fossil fuels for power generation, heavy industry and home heating in large cities created clouds of soot (carbon particulates or aerosols) which mix with fog to produce *smog*. Soot alone can cause breathing problems and blackens the external surfaces of buildings. Today, burning coal, charcoal or wood for heating and cooking is much less common in developed nations, reducing this source of smog formation. However, major pollutants in towns and cities are emitted from vehicle engines and exhausts. Nitrogen is not present in fuels: the high temperature and pressures inside vehicle engines cause atmospheric nitrogen and oxygen to react, forming nitrogen oxides, including nitrogen monoxide (NO) and nitrogen dioxide (NO$_2$). These two pollutants are referred to by the common formula NO$_x$, as the proportions in which they are formed varies considerably. NO$_x$ gases are soluble, so cause acid rain which is responsible for causing environmental damage to buildings and green spaces. They also react in the atmosphere with other pollutants to make *photochemical smog*, visible as a haze on a warm day over cities worldwide.

An excellent website showing up-to-date real-time sources of air pollution in the UK can be found at https://uk-air.defra.gov.uk/air-pollution/. An equivalent site by the US Environmental Protection Agency is https://www.epa.gov/climate-indicators. The US site includes data on the weather, climate data and ocean acidity. Real-time estimates of air quality are available at https://waqi.info

> ### Enrichment
>
> Gravestones provide a useful means of observing and measuring weathering of different types of rock over 100 to 300 years (or longer). This is a relatively easy time span to imagine. In the UK, gravestones are made of local and imported natural rocks. Common gravestone rock types include limestone and sandstone (sedimentary), marble (metamorphic) and granite (igneous). Gravestones provide information about how these

rocks behave under changing climatic conditions and intensities of atmospheric pollution. This information helps assess the durability of natural building materials (see Weathering, above).

Weathering of gravestones in town centre cemeteries is probably dominated by chemical weathering brought about by atmospheric pollution (see above). Vehicle exhaust fumes produce sulfur and nitrous oxides, which form acid rain, an agent involved in chemical weathering. An excellent activity, *Will my gravestone last?* is available at https://www.earthlearningidea.com/PDF/135_Gravestones.pdf The task suggests visiting a graveyard and considering atmospheric pollutants causing weathering, as well as other activities.

Most of us would find it fascinating and no doubt convenient to see the future by looking into a crystal ball. A vision of the Earth's climatic future would be especially helpful. At present, the best conclusion from scientific evidence points to the climate warming and continuing to warm as greenhouse gases are added to the Earth's atmosphere. There is a high probability that the average global temperature will ultimately increase by 2–4 °C. This will cause widespread environmental changes.

There are prior examples of large natural temperature changes. the huge eruption of the Indonesian volcano Krakatoa in 1883 led to a 0.5 °C drop in average global temperature. The largest recorded volcanic eruption in human history, the 1815 eruption of Tambora (also in Indonesia) was followed by a 'year without a summer'. The massive eruption of the Philippines volcano Pinatubo in June 1991 introduced a vast quantity of fine ash and sulfurous gas into the stratosphere which quickly spread into the northern and southern hemispheres. The dust and gas reflected incoming solar radiation, causing average surface temperatures to fall by about 0.5 °C. Meanwhile, there was a substantial increase in the level of ozone depletion.

Volcanic effects may reverse temporary upward trends in average global temperature attributed to continued emission of greenhouse gases for a short time period – when and if large eruptions take place. Nonetheless, the overwhelming prospect is that the Earth will become warmer. Scientists investigate this using past intervals of rapid environmental change found in the geological record. Reconstructions of warm intervals in the Earth's past history provide insights into physical and biological responses to global warming.

Additional activities relating to the Earth's atmosphere are listed at the end of the chapter.

11.6 Earth system science: climate change

In the mid-1980s, NASA began a systems approach to cataloguing elements in the Earth's environment. This resulted in *Earth system science* or, more simply, study of the *Earth system*. Earth system models acknowledge important synergies and constant interaction between the atmosphere (air), hydrosphere (water, including oceans, rivers, lakes), cryosphere (ice), biosphere (life), lithosphere or geosphere (solid rock) and human influences (the anthroposphere). Consequently, the Earth system is often represented by interlinking and interacting spheres of processes and phenomena.

Figure 11.23 Components of the Earth system

The atmosphere, hydrosphere, biosphere, cryosphere and geosphere form the simplest collection, with increasing input from the anthroposphere due to human activity. At the time of writing, the latest thinking concludes that reducing the system to a series of discrete spheres is unhelpful, because in reality no part of the Earth system can be considered in isolation from any other.

Earth system science embraces chemistry, physics, biology, mathematics and applied sciences, transcending disciplinary boundaries to treat the Earth as an integrated system. It seeks deeper understanding of physical, chemical, biological and human interactions that determine the past, current and future states of the Earth. Earth system science provides a physical basis for understanding the world in which we live and upon which humankind seeks to achieve sustainability.

At the time of writing, the most recent research evidence from NASA and the UK Natural Environment Research Council (NERC) is based on the premise that Earth is the only known planet to sustain life. Collective evidence so far suggests very strongly that life on Earth is critically dependent on the abundance of water as liquid, vapour and ice. Furthermore, carbon forms the very basis of living organisms and exists in multiple forms. Atmospheric carbon is fully oxidised as carbon dioxide. Methane and particulate carbon (soot) are present in the atmosphere contribute to the greenhouse effect, helping to make the Earth habitable. Strong evidence also suggests that the Earth's atmosphere and electromagnetic field protect the planet from harmful radiation, while allowing useful radiation to reach the surface and sustain life. Finally, the Earth exists within the Sun's zone of habitation and, with the Moon, maintains the precise orbital inclination needed to produce our seasons.

> These remarkable factors have contributed to Earth maintaining a temperature range conducive to the evolution of life for billions of years. The great circulation systems of Earth – water, carbon and the nutrients – replenish what life needs and help regulate the climate system. Earth is a dynamic planet; the continents, atmosphere, oceans, ice and life ever changing, ever interacting in myriad ways. These complex and interconnected processes comprise the Earth system, which forms the basis of the scientific research and space observation that we refer to as Earth system science (NASA, 2011).

KEY ACTIVITY

Researching the impact of climate change

Use Google Earth® to investigate how climate change could affect the Earth. Ask students to create presentations showing examples of the impact of global warming around the world.

To use Google Earth, go to start Google Earth then look at the Layers panel (bottom-left corner). Layers show information placed on top of the Google Earth satellite images. Examples are borders of countries, names of cities, and buttons linking to images of places.

Layers can be switched on and off by clicking on relevant boxes to tick or untick them. Experiment with different Layers to see what happens to the Google Earth view.

Use the panel to find out more about climate change: clicking the plus (+) icon next to Global Awareness in Layers will show more layers. Find the UNEP (United Nations Environment Programme) icon. Click the tick box to switch this layer on. The UNEP *Atlas*

11 Earth science

> *of Our Changing Environment* shows places around the world where the environment is changing. Locate these places by looking for the square blue and black UNEP icons. Reasons for change differ from place to place. For example, urban areas become larger with increasing population size; other places may be impacted by climate change. Alternatively use https://na.unep.net/atlas/googleEarth.php

Additional activities relating to climate change are suggested in Section 11.7 below.

Careers

Studying Earth sciences leads to a wide range of interesting and satisfying careers, many of which require knowledge of specialist areas, although many are interlinked. Knowledge about the structure of the Earth and how landforms were made leads to work as a geologist. Geologists study rocks and rock formations and natural processes that shaped them. They try to understand the Earth's natural history and the substances that comprise the Earth. This knowledge is useful in understanding dramatic events such as earthquakes, mudslides and volcanic eruptions. Geologists work in mining, engineering and the energy supply industry. Some geologists may become mineralogists, specialising in the study of minerals, the crystalline substances found in the Earth's crust, particularly those of economic value, such as sources of metals (copper, iron, aluminium). Mineralogists work worldwide in a wide variety of settings, in the field, including remote locations and the laboratory. Geophysicists also study the Earth, focusing on a range of magnetic, gravitational and seismic techniques. They work extensively in mining, energy and construction providing computer modelling and data analysis valuable in mineral exploration and civil engineering settings. Specialist seismologists study earthquakes and their effects on nearby and distant landforms, human activity. Soil scientists study the chemical, physical and biological properties of soils. This knowledge is essential to the construction industry, where detailed knowledge of the ground underlying proposed buildings, bridges, dams and foundations is vital to success of the project.

Earth science also leads to work in hydrology. Hydrologists develop knowledge about how water circulates on the Earth, particularly on the surface in rivers and lakes, and underground in aquifers and caves. Their understanding of how water accumulates, penetrates and moves through the Earth's surface is essential to water companies, flood-control agencies and in management of global water resources. Oceanographers also work with water, specialising in marine environments. Their knowledge about the conditions in the Earth's oceans, the ocean floor and current circulation patterns is applied in designing harbours, in exploration and developing drilling platforms.

Volcanology is a specialist area of geology devoted to studying volcanoes. Volcanologists' expert knowledge of eruptions, lava flows, particle and gas emissions are extremely important when these occur. They also continuously monitor living volcanoes, checking for the likelihood of an eruption and providing advice as required.

Earth scientists with an interest in the geological history of the Earth may become palaeontologists, studying the fossil record. Palaeontologists study existing fossils, try to find new specimens in the field, and may act as curators of museum fossil collections. Dinosaurs are probably the most well-known fossilised creatures, but palaeontologists study all fossilised life, using the information to understand in greater detail what the Earth was like in the past and how environmental events may have impacted life on Earth.

Atmospheric scientists are Earth scientists who specialise in understanding the Earth's atmosphere. They may play a role in predicting weather conditions, using information from weather satellites and knowledge of ocean currents, providing information about the causes and effects of the weather we experience. Atmospheric scientists may also work in public health, monitoring and advising on air quality and methods for controlling atmospheric pollution. They may use existing records about the weather, climate patterns and air quality to provide advice to make predictions about weather systems and the impact these may have on life on Earth. Climate scientists and meteorologists are working on improving models of the Earth's weather systems. Supercomputers are used to handle very large amounts of data to forecast weather patterns. These can be used to mitigate effects of climate change on the planet.

11.7 Resources

References and further reading

Branney, M.J. and Zalasiewicz, J. (2020) *Volcanoes: a short introduction*, Oxford University Press.

Houghton, J.T. (2015) *Global warming: The complete briefing* (5th edition), Cambridge University Press.

Jones, S.J. (2015) *Introducing sedimentology*, Dunedin Academic Press.

Kind, V. and Kind, P.M. (2008) *Teaching secondary science: How science works*, Hodder Education.

Maslin, M. (2021) *Climate change: a very short introduction*, Oxford University Press.

Park, G. (2018) *Introducing geology: A guide to the world of rocks*, Dunedin Academic Press.

Reiss, M.J. and Winterbottom, M. (editors) (2021) *Teaching secondary biology* (3rd edition), Hodder Education.

Zalasiewicz, J. (2018) *Geology: A very short introduction*, Oxford University Press.

Websites

Geological time

www.enchantedlearning.com/subjects/Geologictime.html.

www.ucmp.berkeley.edu/

Rocks and rock formation

www.earthlearningidea.com

www.earthscienceeducation.com

www.geolsoc.org.uk/ks3/gsl/education/resources/rockcycle.html

Plate tectonics

http://volcanoes.usgs.gov/about/edu/dynamicplanet/index.php

http://mineralsciences.si.edu/tdpmap/

https://www.geolsoc.org.uk/Plate-Tectonics

The Earth's early atmosphere and life in extreme environments

https://forces.si.edu/atmosphere/02_02_01.html

www.bbc.co.uk/bitesize/guides/zwc6w6f/revision/1

https://astrobiology.nasa.gov/classroom-materials/

Climate change

www.carbonfootprint.com

https://climate.nasa.gov/earth-apps/

www.esrl.noaa.gov/gmd/ccgg/trends

http://serc.carleton.edu/NAGTWorkshops/climatechange/recommended.html

www.teachclimatechange.org

http://co2degrees.com/

https://climatekids.nasa.gov/carbon/

https://interactive-atlas.ipcc.ch

https://climatehub.uk/

Carbon Capture and Storage (CCS)

www.globalccsinstitute.com/

www.ccsassociation.org/

https://www.iea.org/fuels-and-technologies/carbon-capture-utilisation-and-storage

www.co2crc.com.au/

https://zeroemissionsplatform.eu/

www.co2captureproject.org/a/digital_in_depth_tool.html

www.ccsbrowser.com/#

www.zeroco2.no/

Videos

GeoBus Numerous short 1-minute videos explaining geological concepts: https://www.youtube.com/channel/UC0Sry1dA-hGuThgTKu4iGNg/videos

Zero Emissions Platform The hard facts behind carbon capture and storage: https://youtu.be/aHtbDmzjYgg

Zero Emissions Platform Safe storage: Closing the carbon loop – CO_2 capture and storage: https://www.youtube.com/watch?v=GglSLuWP5cM

New Gen Coal Carbon capture and storage: https://vimeo.com/16307994

Crash Course The hydrologic and carbon cycle: https://www.youtube.com/watch?v=2D7hZpIYlCA

Shell Take a journey more than 2000 metres underground with Shell's carbon capture and storage project: https://www.youtube.com/watch?v=sqkXYKRFkFc

Earth Institute Carbon capture sequestration and reuse, Webisode 1: The basics: https://www.youtube.com/watch?v=3dm1esCpzR0

Other resources

Virtual Rock kit: https://www.earthscienceeducation.com/virtual_rock_kit/index.htm

Making bracelets from UV beads and testing these in 'normal' and UV light to enhance understanding of the electromagnetic spectrum: https://www.carolina.com/pdf/boothdemos/uv_bead.pdf

UV beads: available at https://www.amazon.co.uk/rosenice-1000-UV-Coloured-Plastic-Jewellery/dp/B075K5YZDR/ref=sr_1_5?dchild=1&keywords=uv+beads&qid=1605276868&sr=8-5

A hypothesis that was proved wrong – how the Earth's features were formed. Chapter 4, Experiments and Evidence, Kind and Kind (2008) (see References).

Climate change resources including greenhouse effect are available at the British Geological Survey: www.bgs.ac.uk/discovering-geology/climate-change/

Images of retreating glaciers to support discussion of reasons for this see: https://earthobservatory.nasa.gov/images/7679/glacial-retreat

Tasks to support calculation of personal carbon footprints and possible changes to reduce this: https://co2analysis.com/

Index

acetylsalicylic acid (aspirin) 12
acid rain 221, 377, 378, 379
acids and alkalis 196
   acid-metal reactions, salt and hydrogen production 208–09
   careers 221–22
   learning progression 197–98
   making salts 209–14
   models of 206–08
   neutralisation and titration 214–20
   pH scale 200–05
   properties of 199–200
   resources 222–24
   strength of acids and bases 215–16
   students' prior knowledge 196–97
   teaching sequence overview 198–99
   titration 217–20
activation energy
   and catalysts 170, 181–82
   collision theory 172–73
   endothermic reaction 163
   exothermic reaction 162
   thermite reaction 159
addition polymers 335–40
Agassiz, Louis (1807–1873), catastrophe theory 354
air, gases in 81, 273, 373–74
alcohols 325–29
   breathalysers 301–02, 303
   enthalpy change of combustion 166–68
   fuel crops 328–29
   reaction with carboxylic acids 331–32
   simple distillation 327–28
alkali metals (Group 1) 104–05
   reactivity of 105–07, 117–18
   resources 121
alkalis see acids and alkalis
alkanes 316–25
   chemical properties of 320–22
   physical properties of 318, 319
alkenes 316–25
   chemical properties 320–22
   physical properties of 318, 319
allotropes, carbon 100
aluminium (Al) 90

cation test, inconclusive results 301
displacement reaction 248
extraction by electrolysis 248, 261–62
reaction with iodine 35, 36, 51, 160–61
thermite reaction 159–60, 242
amino acids 334, 343
ammonia
   diffusion activity 84–85
   fountain experiment 76–77
   Haber process 186, 192
   reactions producing 290
   red litmus test 291
ammonium chloride
   endothermic reaction 159
   reversible reaction 185–86
ammonium cyanate, forming urea 313
ammonium nitrate, endothermic reaction 158–59
analogies 132, 134
   demonstrating dynamic equilibrium 188–89
analytical chemistry see chemical analysis
analytical chemists 344
anions (negative ions) 287, 288
   beach objects, test to identify 308
   confusion with anode 251
   and electrolysis 254
   identifying in salts 294, 296, 299–300, 307
   ionic compounds 237–38
   salts giving inconclusive results 301
antacid powders, comparing 217
anthropomorphism 119, 125, 132
argon (Ar) 82, 90, 95
Arrhenius, Svante (1859–1927), alkalis 207, 208
aspirin, history of 11–12
asteroid mining 246
atmosphere
   evolution of gases in 373–74
   formation of 370–72
   greenhouse gases and climate change 375–77
   pollution of 377–79

atmospheric scientists 383
atomic mass (A_r) 49, 53, 274
atomic number (Z) 113
   and electron configurations 116–17
   periodic table 10, 90, 93–95, 96
   used to define an element 99
atomic structure 113–14
atomism see particle theory
Avogadro, Amadeo (1776–1854) 48–49
Avogadro's number 48–49, 52, 54
balancing equations 44–46
   writing balanced half-equations, redox 235–38
Bartlett, Neil (1932–2008), noble gas compound 112
bases 208
   neutralisation reactions, salts 209–10
   pH of 201
   strong and weak 215–16
bath bomb making activity 213–14
battery technology 266
beach objects, identifying 307–08
Bentley, Wilson (1865–1931), snowflake uniqueness 146
Berthollet, Claude Louis (1748–1822), reversible reactions 186
big ideas in chemistry 5–6
   chemical reactions 9
   periodic table 10
   small particles, all matter made of 6–8
billiard-ball model of atomic structure, Dalton 67, 68
biochemistry 344
bioethanol 328–29
biological weathering 364–65
Bohr, Neils (1885–1962), atomic model 68, 114–15, 148
boiling point/temperature 77–79
   alcohols 326
   alkanes and alkenes 318, 319
   of gases 79, 80, 81
   identify a substance by 284
   increased by impurities 282
bond-breaking
   chemical vs physical changes 39
   endothermic reaction 161

Index

students' misconceptions about 156
bond energies/enthalpies 161–64
bonding *see* chemical bonding
breathalysers 301–02
bromine (Br) 90, 108, 109
   displacement reactions 110
bromine water, alkenes decolourising 320–22, 324–25
Brønsted, Johannes Nicolaus (1879–1947) 207
Brønsted–Lowry acid–base model 207, 208, 214
Bunsen, Robert (1811–1899) 305
burning *see* combustion reactions
Café Scientifique discussions, climate change 376
calcium (Ca) 90, 348
   extraction by electrolysis 248
   flame emission spectrum 304
   flame test for cation 288, 301, 303, 308
   reactivity 240
calcium carbonate (limestone) 183, 248, 293, 358, 364, 373
candle wax
   burning 35, 37
   melting point of 73
carbon (C) 90 *see also* organic chemistry
   covalent bonding 141, 143
   electrodes 258–59
   and metal extraction 246–47, 249
   not used to extract aluminium 261–62
   and reactivity series 240
carbon–carbon (C–C) bonds 143, 313–14
carbon–carbon (C=C) double bond 141, 317, 319–20, 336
carbon dioxide
   atmospheric 81, 371, 373–74, 375
   and carbonated (fizzy) drinks 81, 189–91
   emitted by volcanoes 376–77
   in limestone 373
   limewater tests for 291, 292–93, 373
   reactions producing 290
carbonates, reaction with acids 212–14
carbonic acid, chemical weathering 364

carboxylic acids 329–31
   making esters with 186, 331–32
careers in chemistry
   acids and alkalis 221–22
   chemical analysis 310
   chemical bonding 149
   chemical change 56–57
   Earth science 382–83
   electrochemical processes 266
   engineering 193
   inorganic chemistry 119
   organic chemistry 344
   particle theory 86
   polymers 344
catalysts
   in addition polymerisation 336
   cracking of paraffin 323–25
   in ester synthesis 332
   increasing rate of reaction 181–83
   and position of equilibrium 192
   student misunderstandings 170
cations (positive ions) 287, 288
   beach objects, test to identify 308
   flame test colours given by 303
   identifying in salts 294–95, 297–98, 307
   metallic lattices 135
   salts giving inconclusive results 301
chemical analysis 270
   careers 310
   case studies 305–09
   chemical reactions to identify substances 286–88
   handling inconclusive results 300–02
   identifying gases from reactions 289–94
   identifying ions in salts 294–300
   learning progression 272
   physical properties to identify substances 281–86
   pure substances vs mixtures 273–81
   resources 310–12
   spectroscopy 302–05
   students' prior knowledge 271–72
   teaching sequence overview 273
chemical bonding 127
   comparison table 146–47
   covalent 140–43
   as holding power 127–29
   hydrogen bonds 145, 148

ionic 137–39
   metallic 134–36
   modelling scientific ideas for 131–34
   nature of 130–31
   quantum theory 130–31
   representational challenges 148–49
   resources 149–51
   teaching a scientific view 127–31
   van der Waals forces 144–45
chemical change 26–27, 29–30 *see also* energetics
   chemical reactions, five examples 35–37
   equations and formulae 42–46
   kalium and plumbate activity 31–34
   learning progression 28–29
   Maillard reactions 42
   measuring 48–57
   modelling 46–48
   resources 57–59
   students' prior knowledge/ misconceptions 27–28
   vs physical change 38–42
chemical elements
   defining 98–102, 274–75
   finding in the periodic table 96–97
   origins of 2–3
   resources 121
   variants of periodic table 98
   vs compounds 102–04
chemical engineers 57, 193, 344
chemical reactions 9 *see also* chemical analysis; chemical change; energetics
   extent of 153, 185–93
   identifying substances using 286–88
chemical weathering 363, 364, 373, 379
chemistry 2
   applications of, aspirin 11–12
   big ideas in 5–10
   ethical issues 14–16
   history of 2–5
   practical 13–14
   scientific principles of 10–11
chemistry teaching 1
   conceptual change, achieving 16–17

levels of observation, Johnstone's triangle 17–18
mathematical requirements 21–23
resources 24–25
use of models 18–21
chlorine (Cl) 90, 108
blue litmus test 291
displacement reactions 110
electrolysis 250–52, 255–57
microscale diffusion of 85
reaction with hydrogen 165
reaction with magnesium 234–35
reaction with sodium 127
chlorofluorocarbons (CFCs) 14–16
chocolate 74
chocolate and egg experiment 40–41
chromatography
careers using 310
paper chromatography 204–05
separating mixtures using 279–81
cinnabar, mercury from 70
citric acid 215
making bath bomb activity 213–14
making sherbert activity 199–200
climate change 375–76, 380–83
ocean acidification 373
over geological timescales 352
promoting discussion of 376
researching impact of 381–82
'clock' reactions 180–81
cocoa beans 74
collision theory 172–73
combustion reactions
alkanes and alkenes 320–22
bond energies in molecules 163
burning elements in oxygen 206
candle wax 35
energy released when fuels burn 166–68
exothermic reactions 162, 164–65
of methane 164
pollutants from burning fuels 378
students' misconceptions 156
compounds 99–100, 102–04 see also organic chemistry
ionic, splitting 237–38
and metal extraction 248–49
concentration
of hydrogen ions, pH scale 200–01
and rate of reaction 176–81
students' misconceptions 187
concept maps 253–54
condensation polymers 341

nylon 341–42
copper (Cu) 90, 117
displacement reactions 227, 242
electrodes 259
extraction technique 247
flame colour of cation 303
history and uses 228
purifying by electrolysis 259
reactivity 240
soluble salt, making 210
copper(II) oxide and sulfuric acid 212
copper(II) sulfate
anhydrous vs hydrated, reversible reaction 187
displacement reaction 158, 227–28
electrolysis reaction 258–60
covalent bonding 140–42
intramolecular 144
students' vs scientific ideas 126
teaching challenges 142–43
cracking of hydrocarbons 323–25
crude oil, fractional distillation of 322–23
Curie, Marie 101
cyclohexane and cyclohexene 320–22
Dalton, John (1766–1844), atomic model 66–67, 68
Davy, Humphrey, Sir (1778–1829) 207, 248
decomposition reactions
ammonium chloride 185–86
hydrated copper(II) sulfate 187
hydrogen peroxide 35, 37, 181–83
mercury calx 66
water 166, 258, 263–64
deep time 352–53
delocalised electrons, metallic bonding 134–36
Democritus (460–370 BCE) 6, 65
density
energy density 167
float–sink test, plastic density 285–86
identifying a substance using 284–85
of noble gases 111
poly(ethene) 335–36
deoxyribonucleic acid (DNA) 342–43
diamonds 361–62
diffusion 84–85

displacement reactions
chlorine, bromine and iodine 109–10
copper(II) sulfate solution and magnesium 158, 227
of hydrogen from an acid 208–09
of metals in redox reactions 242–46
thermite reaction 159–60
dissolving 85–86
distillation
ethanol-water mixture 327–28
fractional, crude oil 322–23
DNA (deoxyribonucleic acid) 342–43
double (covalent) bonds 141
hydrocarbons 317, 319–20, 336
dust bowl activity 368
dust explosions 185
dynamic equilibrium 185–93
Earth's atmosphere 370
formation of 370–72
greenhouse gases 375–77
oxygen and carbon dioxide levels, changes in 373–74
pollutants 377–79
Earth science 347–51
careers 382–83
climate change 380–82
diamonds, igneous rock 361–62
key geological events 352–53
resources 383–86
rock characteristics 357–59
rock cycle 366–70
rock formation 354–57
rock identification 360–61
students' prior understanding 351–52
weathering and erosion 362–65
Earth scientists 383
electrical conductivity
and metallic bonding 136
noble gases 111
electrolysis 225, 228
aluminium extraction 261–62
of common salts, summary table 260
concept map for 253–54
copper(II) sulfate solution 258–59
electroplating 266–67
extraction of metals 247, 248
and hydrogen fuel cells 262–66
language of 253
sodium chloride solution 255–57

Index

visual representations 257
zinc chloride 250–52
electron configurations 114–17
  and periodic table position 117–19
electrons 7, 8
  atomic structure 113
  bonding 130–31
  redox reactions 226–28, 231–32
electroplating 266–67
electrostatic attractions, chemical
  bonding 129, 130, 147
  covalent bonds 126
  intermolecular forces 144
  ionic bonds 126, 137, 138, 139
  neutral molecules, metallic
    bonds 131, 135
  quantum theory explanation
    130–31
  scientific view 126–27
elements see chemical elements
endothermic reactions 154
  calculations 166
  decomposition of water 166
  demonstrations of 158–59
  ice-pack example of 159
  investigation of 156–58
  using bond enthalpies to
    explain 161–64
energetics 152 see also rate of
  reaction
  bond enthalpies 161–68
  demonstrations of energy
    change 158–61
  energy in chemical reactions
    154–55
  exothermic and endothermic
    reactions 156–58, 161–64
  learning progression 152–53
  students' misconceptions 155–56
  students' prior knowledge 152
  teaching sequence overview
    153–54
energy see also activation energy
  chemical reactions 9
  and electrolysis 255
  and electron configurations
    114–15
  and evaporation 79–80
  in making and breaking bonds
    20–21
  and reactivity 117–18, 241
  released in burning a fuel 166–68
  representing in equations 44

energy density, calculating 167
enthalpy (H) 154
  bond enthalpies 161–64
  change of combustion 166–68
environmental scientists 221
equations, chemical 42–46
  balancing 44–46
  ionic, writing 235–38
  moles 55–56
  notation for 46
  reacting mass ratios 50, 51
equilibrium, dynamic 153, 181–93
equivalence point, neutralisation 214,
  216, 217, 218
erosion 362, 368, 370
esters 331–33
ethanoid acid 329, 330
  reaction with ethanol 331–32
  reaction with magnesium 209, 330
ethanol 78, 325–28
  bioethanol 328–29
  breathalysers detecting 301–02
  and chemical bonding 128
  enthalpy of combustion 168
  esterification reaction 331
  purification of 328
ethical issues 14–16
evaporation 79–80
exothermic reactions 152–53
  activity investigating 156–58
  energy level diagrams 161–62
  hydrogen-chlorine reaction 165
  measuring temperature change
    in 154–55
  nitrogen-hydrogen reaction 165
  students' misconceptions
    about 169
Farman, Joe, Antarctic ozone hole
  15–16
fast reactions 169, 171
fermentation
  of crops, fuel production 328–29
  of sugars, alcoholic drinks 326–27
fizzy drinks, dynamic
  equilibrium 189–91
flame emission spectrometry 303–05
flame tests, cation identification 288,
  294–95, 301, 302–03
  beach objects 308
  composition of rocks 309
flavour scientists 57
flavourings, artificial 331
float/sink density test 285–86

fluorine (F) 90, 108, 145
food
  chemical reactions in 193
  flavour scientists 57
food scientists 221–22, 267
forensic scientists 310
formulae, chemical 42–43
formulation chemists 278–79, 310
formulations of everyday household
  products, investigating 279
fossil fuels
  effect of burning 202, 375–76, 378
  hydrogen fuel as alternative
    262–66
fractional distillation 322–23
freeze-thaw weathering 363
fuels
  burning, energy released from 162,
    166–68
  effect of burning fossil fuels 202,
    375–76, 378
  ethanol-based 328–29
  hydrogen fuel cells 262–66
  students' misconceptions
    155–56
  students' prior knowledge 314
gases 80–82
  boiling and condensing 74–82
  noble gases 111–12
  particle models 82–84
  reactions that produce 290
  students' misconceptions 61
  tests identifying 289–94
geological (deep) time 352–53
geologists 149, 352, 382
geology see rocks
Gerhardt, Charles Frédéric, salicylic
  acid 12
global warming 370, 373, 375–77,
  379
  researching impact of 381–82
gold (Au) 90, 101, 247
  density of identifying counterfeit
    coinage 285
  history 3, 285
  plating 241, 250, 267
  reactivity 240
  Goldschmidt, Hans, thermite
    reaction 242
granite, minerals in 359
granular vs crystalline rocks 359–60
gravestones, weathering of 378–79
greenhouse gases 373, 375–77, 379

groups, periodic tables 89–90, 94, 97, 115, 119
  Group 0: noble gases 111–12
  Group 1 alkali metals 104–07
  Group 7: halogens 108–10
  and reactivity 117–18
Grove, William, hydrogen fuel cells 262
Haber process, ammonia production 186, 191, 192
half-equations, redox reactions 226–27, 232–33
  reactions of metals 238–39, 241
  writing 235–38
halogens (Group 7) 108–10
  reactivity pattern 118
  resources 121–22
helium (He) 82, 90, 95, 111, 112
history of chemistry 2–5
  aspirin 11–12
  particle theory 65–70
  periodic table 93–96
Hodgkin, Dorothy (1910–1994) 18
Hoffman, Felix, side effects of salicin 12
Hoffmann voltameter 262–64
holding power (chemical bonding) 127–29
homologous series, organic compounds 314
  alcohols 325–29
  carboxylic acids 329–31
  esters 331–33
  hydrocarbons: alkenes and alkanes 316–25
hydrocarbons 316–17
  chemical properties of alkanes and alkenes 320–22
  cracking 323–25
  Fractional distillation 322–23
  making models of 319–20
  physical properties of alkanes and alkenes 317–19
hydrochloric acid 214–15
  change in concentration, effect on reaction rate 176–78
  reaction with magnesium 174–76, 208–09
hydrogen bonds 145, 148
hydrogen chloride 165, 185–86
hydrogen (H) 90
  alkali metal reactions producing 107
  atomic structure 113, 116
  bonding 145, 148
  displaced in metal–acid reactions 208–09
  fuel cells 262, 264–66
  and oxygen 35, 36–37
  and oxygen, balancing equations 44–45
  and oxygen, reacting mass ratios 49–50
  reaction with chlorine 165
  reaction with nitrogen 165
  reactions producing 290
  squeaky pop test 291
  testing for 292, 293
  water decomposition 166
'hydrogen monoxide', states of 38
hydrogen peroxide, decomposition 35, 37
  effect of adding a catalyst 181–82
hydrologists 382
igneous rocks 349–50
  comparing with sedimentary 359–60
  granite 359
  identifying 360–61
inconclusive results, handling 300–01
indicators 198
  alkali metal test 106
  carboxylic acid test 330
  and the pH scale 202–06
  for tracking a neutralisation reaction 216–17
inert electrodes, carbon rods 253, 258
inert gases see noble gases
intermolecular bonds/forces 144
  comparing with covalent, ionic and metallic 147
  students' ideas vs scientific view 126
  vs intramolecular 144
interstellar medium (ISM) 140
invisible objects, investigating 114
iodine (I) 90, 108, 109
  clock demonstration 180–81
  displacement reactions 110
  reaction with aluminium 160–61
ionic bonding 137–39
  comparing with covalent, intermolecular and metallic 147
  student vs scientific view 126
  students' misunderstandings 137, 138, 148
  students' vs scientific ideas 126
  vs covalent 143, 145, 147
  vs metallic 147
ionic compounds see also salts
  electrolysis of molten 250–52
  'splitting' into constituent ions 237–38
ionic equations/half-equations, writing 235–38
ionic lattices 137–39
ions in salts, identifying 294–300
  inconclusive results, handling 300–01
iron (Fe) 90, 117
  displacement reaction 159–60, 242
  extraction from a match head 249
  extraction technique 246, 247
  Haber process catalyst 192
  reaction with sulfur 35, 36, 55–56, 103–04
iron(III) oxide 159–60, 242, 246, 249
isotopes 274
Johnstone's triangle 17–18, 98–99
Julian, Percy (1899–1975), stigmaterol 333
kalium activity, chemical change 31–34
Kevlar, discovery of 343
key (geological) events through Earth's history 352–53
kimberlite 361–62
kinetic particle theory simulator 71
Kirchhoff, Gustav (1824–1887), spectrometer 305
Kroto, Harry, carbon–60 model 18–19
krypton (Kr) 82, 90, 111
Kwolek, Stephanie (1923–2014), Kevlar 343
Lavoisier, Antoine and Marie-Anne 11, 65–66
law of conservation of mass 43–44, 66, 67
law of definite proportions, Proust 66, 67
law of multiple proportions (Dalton's law) 66
Le Chatelier's principle 191–93
Lewis, Gilbert (1875–1946), acid model 19, 198, 207, 208

Index

limestone (calcium carbonate) 248, 308, 351
  activity to make 358
  carbon dioxide present in 373
  chemical weathering of 364
  fossils found in 358
  identifying 361
limewater tests for carbon dioxide 291, 292, 373
liquids (solutions), particle models 82–84
lithium (Li) 90, 105
  flame colour of cation 288, 303
  physical properties 106
  reaction with water 107
  reactivity 240
Lovelock, James (1919–), detection of CFCs 15
Lowry, Thomas Martin (1874–1936) 207
Lyell, Charles (1797–1875), uniformitarian theory 354
macro scale 17–18
macroscopic view
  acids and alkalis 196, 199–200
  electrolysis 256, 258, 259, 263
  elements 99, 101–02
  sensory information 60–61
  symbol vs word equations, teaching 44
magnesium (Mg) 90, 135
  reaction with in oxygen 30, 206, 230–34
  reaction with dilute acid 174–76, 208–09, 290
  redox reactions 227–28, 230–34
magnets, use of
  as chemical bonding analogy 134
  to test for iron 103, 104, 278
Maillard reactions 42
marble chips (calcium carbonate) 183, 293
materials
  development of new 4–5
  identifying 306–07
  mixtures of 276–77
  understanding of term 274, 275
materials scientists 149, 344
maths 4, 21–23, 164–66, 220, 342
matter (stuff) 2, 274–75 *see also* particle theory
  big idea: all matter is made of very small particles 6–8

and chemical bonding 131
and chemical reactions 9
in the interstellar medium (ISM) 140
medicine
  chemists working in 56, 149, 310
  history of aspirin 11–12
  iodine solution 109
  use of noble gases 112
melting/melting point point 72–74, 282–84
Mendeleev, Dimitri (1834–1907), periodic table 10, 93–95, 96
mercury 70
mercury calx 65–66
metal extraction 246–49
  aluminium 261–62
  reduction reactions 226–28
metal reactions
  with acids 209
  displacement 242
  reactivity series 240–41
  redox 239, 243–46
metallic bonding 134–36
  comparing with covalent, ionic and intermolecular 147
  students' ideas vs scientific view 126
metallic lattices 134–36
metamorphic rocks 351
  formation of 369–70
methane 81
  combustion 164
  greenhouse gas 375, 381
  properties 318
  reaction with oxygen 19–20, 47
Midgley Jr., Thomas (1889–1944), synthesis of first CFC 15
mineralogists 382
minerals 348–49 *see also* rocks
mixtures 276–78
  formulations 278–79
  separating using chromatography 279–80
  understanding of term 274, 275
  vs compounds 99–100
  vs pure substances 273
mnemonics 231
modelling
  addition polymerisation 336–37
  chemical changes 46–48
  scientific ideas for chemical bonding 131–34

models in chemistry 18–21
  amino acid 334
  for gas, liquid and solid 82–84
  particle theory: atomic structure 67–70
molecular mass (M_r) 49, 145
molecular models 18–19
  esters 332
  limitations of 148–49
  to represent bonds 132–34, 322
  to understand chemical reactions 46–48
moles (mol) 48–56
Molina, Mario, ozone depletion 15
Molymod® kits 47, 48
monomers 334–35
  formula and structure 338
  nylon made from 341–42
  polymerisation model 336–37
  polymers made from 337, 338
  zero-waste plastics 340
Moseley, Henry (1887–1915), X-ray spectroscopy 96
Nagaoka, Hantaro (1865–1950), Saturnian model of the atom 68
nanoscale activity 64–65
nanotechnology 86
natural polymers 335
natural products 316
  alcohols 325–29
  amino acids 334
  carboxylic acids 329–31
  esters 331–33
  prior knowledge of 314
  steroids 333
neon (Ne) 82, 90, 116
neutralisation 207, 208
  acids and carbonates 212–13
  activities demonstrating 199, 205, 217
  equivalence point 214, 216, 217, 218
  explanations of 207, 214
  of hydroxide ions 210–11
  making salts by 209–10
  misconceptions about 197, 198
  of oxide ions 211–12
  reactions, choosing an indicator to track 216–17
  soap-making (saponification) 211
  strong and weak acids and bases 215–16
  and titration 217–20

neutrons 43–44, 49, 113
nitric acid 209, 215
   anion tests 299, 300, 301
nitrogen (N) 81, 82, 90
   boiling point 79
   bonding 141, 142, 145
   in Earth's atmosphere 371, 372
   oxides 289, 378
   reaction with hydrogen 165, 186
noble gases (Group 0) 111–12
   periodic table position and reactivity 118, 119
   resources 122
non-inert electrodes, copper rods 253, 259
nylon 341–42
octet rule 127, 142
oganesson (Og) 90, 111, 113
OILRIG mnemonic 231
opacity change and reaction rate 179–80
organic chemistry 313–14
   hydrocarbons 316–25
   learning progression 315–16
   natural products 325–34
   polymers 334–44
   students' prior knowledge 314–15
   teaching sequence overview 316
Ostwald, Wilhelm (1853–1932), the mole 48
oxidation *see* redox reactions
oxides
   and air pollution 378
   of metals 238–40
   reactions with acids 211–12
   in water, pH values of 206
oxygen (O) 16, 90
   in the atmosphere 372–74
   burning elements in 206
   glowing splint test 291
   reacting with hydrogen 35, 36–37, 44–46, 49–50
   reaction with methane 19–20, 47
   reaction with Mg 30, 206, 230–34
   reactions of metals with 238–39, 240
   reactions producing 290
ozone 100
ozone depletion, CFCs 14–16
palaeontologists 383
paper chromatography 204–05
paraffin, cracking 324–25

Paris Agreement (2015), climate change 375–76
particle theory 60
   careers 86
   diffusion and dissolving 84–86
   gases 74–82
   historical perspective 65–70
   learning progression 62–63
   models of atomic structure 67–70
   models for solid, liquid and gas 82–84
   resources 87–88
   simple particle model, substances 63–65, 71–74
   students' knowledge/misconceptions 60–62
particles, big idea in chemistry 6–8
particulate carbon (soot) 378, 381
perfume
   diffusion 84
   esters 331
periodic table 10, 89–90
   atomic structure 112–14
   careers 119
   electron configurations 114–17
   elements 98–102
   elements vs compounds 102–04
   Group 0: noble gases 111–12
   Group 1: alkali metals 104–07
   Group 7: halogens 108–10
   historical overview 93–96
   learning progression 92
   relating properties of elements to 117–19
   resources 120–22
   structure of 96–98
   students' knowledge and misconceptions 91
petrochemical industry 323, 344
pH scale/values 200–01
   carboxylic acids 329
   and food science 221
   indicators 202–06
   and neutralisation reaction 216–17
   strong and weak acids and bases 216
   water in the environment, effect on nature 202
pharmaceutical chemists 56–57, 344
phlogiston theory 11, 16
physical properties
   alkanes and alkenes 317–19
   Group 1 (alkali metals) 106

   identifying substances using 281–86
physical vs chemical changes 38–42
   confusion between 27
physical weathering, 'freeze-thaw' 363
plastics *see also* polymers
   recycling of 285–86, 340, 342
   students' prior knowledge of 314–15
plumbate activity, chemical change 31–34
pollution
   atmospheric 377–79
   environmental scientists 221
poly(ethene)/polythene 335–37, 338
   intermolecular bonding 144
polymers 334–35
   addition polymerisation 335–37
   amino acids 334
   condensation 341–43
   DNA 342–43
   formula and structure 339
   investigating everyday 337–40
   natural 335
   nylon, making 341–42
   proteins 343
   recycling challenges 340
   vs plastics 337
potassium (K) 90, 105, 348
   extraction by electrolysis 248
   flame colour of cation 303
   reaction with water 118–19
   reactivity 107, 240
   salts of, displacement reactions 109–10
   salts of, identifying 307
practical chemistry 13–14
precious metals 3, 101, 285, 348–49
prefixes 317
   alcohols 326
   alkanes and alkenes 318
   'cyclo' 320–21
   'poly' 334
pressure
   of a gas, Le Chatelier's principle 191
   metamorphic rock formation 351
   and sedimentary rock formation 355, 356
Priestley, Joseph, oxygen 16
proteins 334, 343
protons 8, 49, 99, 113

Proust, Joseph (1754–1826) 66
purity of substances 273–75
   confusion about 'pure' 28, 40
   introduction of concept 62
   and the law of definite proportion 66
   melting point determining 282–84
   mixtures compared to 276–78
qualitative analysis *see* chemical analysis
quantum theory 130–31
rainbow reaction 205
Ramsay, Sir William (1852–1916), noble gases 82
rare elements 3, 246, 348–49
rate of reaction 168–69
   and catalysts 181–83
   and collision theory 172–73
   concentration affecting 176–81
   factors affecting 173–84
   fast and slow reactions, discussion of 171
   learning progression 153
   measuring 171–72
   students' misconceptions about 169–70
   and surface area 183–84
   teaching sequence 153–54, 170–71
   and temperature changes 174–76
Rayleigh, Lord, discovery of argon 82
reacting mass ratios 49–51, 55–56
reactivity series
   and metal extraction 247–48
   metals 240–46
reactivity trends
   alkali metals 105–07, 117–18
   halogens 109–10, 118
recycling
   of copper 259
   of plastics 285–86, 340, 342
recycling numbers, polymer identification 337–39
red cabbage indicator, making 204–05
redox reactions 225–26 *see also* electrolysis
   careers 266–67
   equations and half-equations 235–38
   fuel cells 262–66
   learning progression 230
   magnesium and chlorine reaction 234–35
   magnesium and oxygen 230–34
   in reactions of metals 226–28, 238–40, 243–46
   resources 267–69
   students' prior knowledge 228–29
   teaching sequence overview 226
reduction *see* redox reactions
reversible reactions 185–86
   hydrated copper(II) sulfate example 187
   misconceptions about 187
rocks 347–51
   formation of 354–57
   identifying 308–09, 357–62
   rock cycle 366–70
   students' misconceptions 351–52
   weathering and erosion 362–65
Rowland, F. Sherwood, ozone depletion 15
Rutherford, Ernest (1871–1937), atomic structure 67, 68, 113
salicin and salicylic acid 12
salts 40
   giving inconclusive test results 301
   identifying 307
   identifying ions in 294–300
   making by neutralisation 209–14
   produced by acid-metal reactions 208–09
   reaction with halogens 108
sand, formation of 368–69
sandstone 351, 361, 363
Scheele, Carl Wilhelm, lactic acid 313
Schönbein, Christian Friedrich, hydrogen fuel cells 262
Schrodinger, Erwin (1887–1961), atomic structure 69
scientific ideas, mnemonics for remembering 231
scientific inquiry, practice of 13–14
scientific principles 10–11
scientific views of chemical bonding
   modelling 131–34
   teaching 127–31
   vs students' conceptions 125–27
sediment transport 368
sedimentary rocks 350–51
   making 356–57
seismologists 382
separation of mixtures 277–78
   by chromatography 279–81
   by distillation 322–23, 327–28
   evaporation 79–80
   plastic recycling 285–86
sherbet-making activity 199–200
silica (silicon dioxide) 140
   in magma 349–50
   sand 368–69
silicon (Si) 90, 97
   in the Earth's mantle and crust 348
simple particle model 63–65, 71–74
sink-float test, plastic recycling 285–86
slow reactions 169–70, 171
smell
   diffusion activity 84
   of esters 331, 332
smelting, metal extraction 246
snowflakes, uniqueness of 146
soap-making 211
sodium chloride (NaCl) 237, 274, 276
   dissolving 39, 85
   electrolysis 255–57, 260
   from carbonate-acid reaction 213
   from hydroxide-acid reaction 210–11
   identifying ions in 294
   melting point of 73–74
sodium hydroxide
   cation identification test 295–96, 297, 298, 301
   and hydrochloric acid 210–11
   reaction with carboxylic acids 330
sodium (Na) 90, 104
   atom and ion 137
   electron configuration 115, 116
   flame colour of cation 303
   flame emission spectrum 304
   metallic lattice 135
   reactivity 105, 106, 107
sodium thiosulfate reactions 179–80
solids
   bonding 63–64, 144, 145
   dissolving in water 61–62, 85, 158
   particle models 82–84
   reversible reaction between 187
solutions vs suspensions 86
Somerfeld, Arnold (1868–1951), atomic structure 69
soybeans, making stigmasterol from 333
spectrometry 303–05, 315–16
spectroscopy 302–05, 315–16
stars, nuclear processes in 139–40
state changes

Index

condensation 77, 78
evaporation 79–80
freezing 73–74, 128
identifying substances by 282–84
melting 40–41, 72–74
solidifying 40–41, 73
students' misconceptions 156
stearic acid, melting point 283–84
steroids 333
stigmasterol, discovery of 333
stoichiometric coefficients 45, 50
stromatolites 374
students' prior knowledge and misconceptions 16, 20–21
acids and alkalis 196–97
chemical analysis 271–72
chemical bonding 124–27
chemical change 27–28
energy changes 155–56
geology 351–52
organic chemistry 314–15
particle theory 60–62
periodic table 91
rates of reaction 169–70
redox and electrolysis 228–29
reversible reactions 187
stuff (matter)
identifying 306–07
terms used to describe 274–75
sub-micro scale 17–18, 98–99
chemical bonding 125, 129
(mis)understanding 19, 44, 197
substances, identifying 281–88, 306–09
suffixes 317, 318, 326
sugars, fermentation of 326–27, 328–29
sulfur dioxide 104, 179, 202, 289, 378

sulfur (S) 90
reaction with iron 35, 36, 55–56, 103–04
reaction producing 179
superheavy elements 100, 108
surface area, effect on rate of reaction 183–84
suspensions 86
symbol equations 44, 45, 46
symbols for ions 237
synthetic polymers 335
nylon 341–42
PE (poly(ethene)) 335–36, 337, 338
PET (poly(ethylene terephthalate)) 337, 338
polypropylene (PP) 337, 338
polystyrene (PS) 337, 338
polyvinyl chloride (PVC) 337, 338
synthetic superheavy elements (SHEs) 100, 108
temperature
effect on reaction rate 174–76
and energetics 154–55
and equilibrium position, Le Chatelier's principle 191–92
students misconceptions 170
theoretical energy level diagrams 161
thermite/thermit reaction 159–60, 242
thin layer chromatography (TLC) 279–80
Thomson, J.J (1856–1940), plum pudding model 19, 68
titration 217–20
'transition' metals, electron configurations for 116–17
Universal Indicator 203

carboxylic acids activity 330
rainbow reaction activity 205
reactivity of Group 1 metals activity 106
red cabbage activity 204
urea, synthesis of 313
van der Waals, Diderik (1837–1923) 144–45
volcanic eruptions 376–77, 379
volcanic gases 371, 376–77
volcanologists 383
water
decomposition reactions 166, 258, 263–64
dissolving solids in 61–62, 85, 158
effect of pH changes on nature 202
electrolysis of acidified 262–64
endothermic reaction with ammonium nitrate 158–59
hydroxide-acid reactions making 210–11
reaction with potassium 118–19
weathering 362–65
of gravestones 378–79
and the rock cycle 369–70
sedimentary rock formation 366–67
Wöhler, Friedrich, synthesis of organic compounds 313
word equations 43, 44, 45, 103
xenon (Xe) 111, 112, 118
Ramsay's isolation of 82, 95
yield 187
Haber process 191–92
zinc chloride, electrolysis 250–52, 260